Epigenetics and Systems Biology

Translational Epigenetics Series

Trygve O. Tollefsbol, Series Editor

Transgenerational Epigenetics

Edited by Trygve O. Tollefsbol, 2014

Personalized Epigenetics

Edited by Trygve O. Tollefsbol, 2015

Epigenetic Technological Applications

Edited by Y. George Zheng, 2015

Epigenetic Cancer Therapy

Edited by Steven G. Gray, 2015

DNA Methylation and Complex Human Disease

By Michel Neidhart, 2015

Epigenomics in Health and Disease

Edited by Mario F. Fraga and Agustin F. Fernández, 2015

Epigenetic Gene Expression and Regulation

Edited by Suming Huang, Michael Litt, and C. Ann Blakey, 2015

Epigenetic Biomarkers and Diagnostics

Edited by Jose Luis García-Giménez, 2015

Drug Discovery in Cancer Epigenetics

Edited by Gerda Egger and Paola Barbara Arimondo, 2015

Medical Epigenetics

Edited by Trygve O. Tollefsbol, 2016

Chromatin Signaling

Edited by Olivier Binda and Martin Fernandez-Zapico, 2016

Chromatin Regulation and Dynamics

Edited by Anita Göndör, 2016

Neuropsychiatric Disorders and Epigenetics

Edited by Dag H. Yasui, Jacob Peedicayil and Dennis R. Grayson, 2016

Epigenetics and Systems Biology

Edited by

Leonie Ringrose
Integrated Research Institute (IRI) for the Life Sciences and Department of Biology, Humboldt-Universität zu Berlin, Berlin, Germany

ELSEVIER

ACADEMIC PRESS
An imprint of Elsevier
elsevier.com

Academic Press is an imprint of Elsevier
125 London Wall, London EC2Y 5AS, United Kingdom
525 B Street, Suite 1800, San Diego, CA 92101-4495, United States
50 Hampshire Street, 5th Floor, Cambridge, MA 02139, United States
The Boulevard, Langford Lane, Kidlington, Oxford OX5 1GB, United Kingdom

British Library Cataloguing-in-Publication Data
A catalogue record for this book is available from the British Library

Library of Congress Cataloging-in-Publication Data
A catalog record for this book is available from the Library of Congress

ISBN: 978-0-12-803075-2

For Information on all Academic Press publications
visit our website at https://www.elsevier.com/books-and-journals

Publisher: Mica Haley
Acquisition Editor: Peter Linsley
Editorial Project Manager: Lisa Eppich
Production Project Manager: Julia Haynes
Designer: Mark Rogers

Typeset by MPS Limited, Chennai, India

Contents

List of Contributors

Rea Antoniou-Kourounioti Computational & Systems Biology, John Innes Centre, Norwich Research Park, Norwich, United Kingdom
Barbara M. Bakker University of Groningen, University Medical Center Groningen, Groningen, The Netherlands
Aurélien Bancaud LAAS, CNRS, University of Toulouse, Toulouse, France
Kerstin Bystricky LAAS, CNRS, University of Toulouse, Toulouse, France
Ho-Ryun Chung Max Planck Institute for Molecular Genetics, Epigenomics, Berlin, Germany
Marc Corrales Centre for Genomic Regulation (CRG), The Barcelona Institute of Science and Technology, Barcelona, Spain; Universitat Pompeu Fabra (UPF), Barcelona, Spain
Dirk De Vos University of Antwerp, Antwerp, Belgium
Ian B. Dodd University of Adelaide, Adelaide, SA, Australia
Guillaume J. Filion Centre for Genomic Regulation (CRG), The Barcelona Institute of Science and Technology, Barcelona, Spain; Universitat Pompeu Fabra (UPF), Barcelona, Spain
Olivier Gadal LAAS, CNRS, University of Toulouse, Toulouse, France
Luca Giorgetti Friedrich Miescher Institute for Biomedical Research, Basel, Switzerland
Edith Heard Institut Curie, PSL Research University, CNRS UMR3215, INSERM U934, Paris, France; Collège de France, Paris, France
Kayo Hibino National Institute of Genetics, Mishima, Shizuoka, Japan; Sokendai (Graduate University for Advanced Studies), Mishima, Shizuoka, Japan
Martin Howard Computational & Systems Biology, John Innes Centre, Norwich Research Park, Norwich, United Kingdom
Kazunari Kaizu RIKEN Quantitative Biology Center, Suita, Osaka, Japan
Kazuhiro Maeshima National Institute of Genetics, Mishima, Shizuoka, Japan; Sokendai (Graduate University for Advanced Studies), Mishima, Shizuoka, Japan
Benjamin L. Moore MRC Human Genetics Unit, University of Edinburgh, Edinburgh, United Kingdom
Mario Nicodemi Dipartimento di Fisica "E. Pancini", Universita' di Napoli "FedericoII", and INFN, Complesso Universitario di Monte S. Angelo, Napoli, Italy; Berlin Institute of Health, MDC-Berlin, Germany

Juliane Perner Cancer Research UK Cambridge Institute, University of Cambridge, Cambridge, United Kingdom
Ana Pombo Berlin Institute for Medical Systems Biology, Max Delbrueck Center for Molecular Medicine, Berlin-Buch, Germany
Marc Rehmsmeier Humboldt-Universität zu Berlin, Berlin, Germany
Leonie Ringrose Humboldt-Universität zu Berlin, Berlin, Germany
Colin A. Semple MRC Human Genetics Unit, University of Edinburgh, Edinburgh, United Kingdom
Kim Sneppen University of Copenhagen, Copenhagen, Denmark
Koichi Takahashi RIKEN Quantitative Biology Center, Suita, Osaka, Japan
Guido Tiana University of Milano and Istituto Nazionale di Fisica Nucleare, Milano, Italy
Fred van Leeuwen Netherlands Cancer Institute, Amsterdam, The Netherlands
Hanneke Vlaming Netherlands Cancer Institute, Amsterdam, The Netherlands

SECTION

Introduction

In writing and editing this book I hope to inspire epigeneticists and systems biologists each to engage with the other discipline, and to facilitate that process by giving examples that are comprehensible to both disciplines.

Epigenetics research aims to understand how a single cell, with a single genomic DNA sequence, can give rise to and maintain the extraordinary diversity of cell identities and functions that comprise the adult organism. Modifications on DNA and chromatin and the binding of other proteins and noncoding RNAs provide a regulatory layer that modulates genome function, so that one genome gives rise to several "epigenomes." Epigenetic mechanisms are profoundly implicated in human health and disease. Aberrant expression of epigenetic regulators has long been known to lead to cancer and developmental disorders [1]. More recently, key roles have been discovered for epigenetic processes in genome instability [2], metabolic health and disease [3], psychiatric disorders [4], degenerative disease [5], and host–pathogen interaction [6]. There is virtually no area of human health that is not affected by epigenetics, and pharmaceutical companies are responding rapidly to these discoveries. Indeed the world market for epigenetic technologies, diagnostics, and therapeutic drugs is expected to grow approximately threefold over the next decade (from $2.6 billion in 2013 to $7.8 billion in 2023) [7]. Several drugs that inhibit epigenetic enzymes have been approved for specific purposes and more have entered phase I and II clinical trials [8,9].

However, we are still far from a complete mechanistic understanding of epigenetic processes, and still further from understanding these mechanisms in quantitative terms. In recent years, vast amounts of epigenetic data have been generated by both high throughput and more reductionist approaches. We have reached a stage at which it is extremely difficult to formulate current knowledge in a manner that places rigorous constraints on whether proposed explanations of biological phenomena are feasible. A large gap in our understanding of many fundamental epigenetic processes is the lack of a coherent theoretical framework within which to study these systems. Systems biology uses mathematical and computational models to discover emergent properties of complex biological systems. Why does epigenetics need systems biology?

Epigenetic regulatory systems share several key features: They are *complex*, comprising multiple molecular components that regulate many genomic targets. They are *dynamic*, allowing flexibility in reaction to environmental, developmental, or disease signals. And they involve *stochastic processes*, such that the output of a given epigenetic regulatory event can vary from cell to cell, over time, and from individual to individual. It is clear that the descriptive models so common in biology, in which interactions are portrayed as static blobs sitting on other structures composed of other blobs (perhaps with a different shape or color), are not sufficient to capture and understand these complex and dynamic aspects. Instead we need mathematical models. Such models will enable us to understand and predict the behavior of these complex systems in terms of where in the genome, and when in time, epigenetic modifications will act, how they give rise to robust but flexible outputs in terms of gene expression, and what is the effect of diseases that perturb them.

If we wish to build mathematical descriptions, it is also clear that however detailed our understanding of a given regulatory system, we will make little progress if our understanding remains at a qualitative level [10]. Quantitative experiments open the door to comprehensive mathematical descriptions. Mathematical descriptions allow predictions. Testing predictions proves whether or not we have understood our system. Although modeling and simulation are standard tools of the trade in engineering and physics, they are surprisingly rare in the field of epigenetics and indeed in experimental biology in general.

Why is the successful combination of modeling and experiments so rare in biology? A major obstacle to uniting biology and mathematics is the barrier between disciplines. Wet-lab experimentalists are often averse to even the simplest calculation or equation, whereas those trained in the theoretical sciences often do not have any hands-on experience of the messy world of experimental biology. The primary research articles of one field can be virtually incomprehensible to the other, and there are very few review articles that bridge this gap. The iterative interplay between experiments and theoretical modeling that is required for successful application of systems biology means that collaborations are essential, but who are the experts in the other field and how can one find them? This book gives examples of successful combinations of systems biology and epigenetics, written by experts in each discipline. Each section covers one type of modeling, and one set of epigenetic questions to which such models have been successfully applied. We have aimed to present original research in a wider perspective and a simpler form than the primary publications, to explain the experiments in terms comprehensible to the modeler, and the models in terms comprehensible to the experimentalist. I hope that this book will give invaluable insights into the immensely powerful combination of epigenetics with systems biology.

Leonie Ringrose
Berlin, December 2016.

References

[1] Laugesen A, Helin K. Chromatin repressive complexes in stem cells, development, and cancer. Cell Stem Cell 2014;14(6):735–51.

[2] Zeller P, Padeken J, van Schendel R, Kalck V, Tijsterman M, Gasser SM. Histone H3K9 methylation is dispensable for Caenorhabditis elegans development but suppresses RNA:DNA hybrid-associated repeat instability. Nat Genet 2016;48(11):1385–95.

[3] van der Knaap JA, Verrijzer CP. Undercover: gene control by metabolites and metabolic enzymes. Genes Dev 2016 1;30(Nov (21)):2345–69.

[4] Nestler EJ, Peña CJ, Kundakovic M, Mitchell A, Akbarian S. Epigenetic basis of mental illness. Neuroscientist 2016;22(Oct (5)):447–63.

[5] Cabianca DS, Casa V, Bodega B, Xynos A, Ginelli E, Tanaka Y, et al. A long ncRNA links copy number variation to a polycomb/trithorax epigenetic switch in FSHD muscular dystrophy. Cell 2012;149(May (4)):819.

[6] Morandini AC, Santos CF, Yilmaz Ö. Role of epigenetics in modulation of immune response at the junction of host–pathogen interaction and danger molecule signaling. Pathog Dis 2016;74(Oct (7)). ftw082:1–8.

[7] https://www.visiongain.com/Report/961/Epigenetic-Therapies-and-Technologies-World-Market-Prospects-2013-2023.

[8] www.insightpharmareports.com/.

[9] http://www.epizyme.com/.

[10] Steffen PA, Fonseca JP, Ringrose L. Epigenetics meets mathematics: towards a quantitative understanding of chromatin biology. Bioessays 2012;34(10):901–13.

SECTION

Where Am I? Genomic Features and DNA Sequence Principles Defining Sites of Epigenetic Regulation: Machine Learning

In the cell nucleus, epigenetic regulators and transcription factors are faced with a genome-wide search for their specific targets. How do they achieve targeting with exquisite precision against the vast background of nonspecific potential binding sites? What information do these proteins have available to help their search? They might bind specifically to particular DNA sequences, or to other molecules that are already attached to specific sites such as RNAs, proteins, or modified histone tails. In addition, the effective search space might be reduced by compartmentalization or by the 3D arrangement of chromosomes. Each protein or complex will have a set of "rules" about what it likes to bind to, which will stabilize its binding when it gets to the right place.

This section focuses on machine-learning approaches. The approaches described here have in common that they extract a defined set of "rules" for targeting of epigenetic regulators from existing data, then use those rules to gain insight into biological function. Given the rules but not told where to look, can the computer make an accurate guess at where things will be or how they get there? Comparing this prediction to measured data tells us whether we have understood the rules, or whether we need to refine them. If we can successfully refine the rules and get better at the task, then the machine has learned something new.

Chapter 1 (Computational Identification of Polycomb/Trithorax Response Elements) explains how DNA sequence features can be used to identify sites in the genome at which specific regulators can act, and gives an excellent introduction to the concept of machine learning and the statistical, computational, and experimental challenges associated with it. Chapter 2 (Modeling Chromatin States) tackles the fundamental problem of classifying what is actually there in a given epigenetic landscape, how we can classify different "flavors" of chromatin and what is a useful functional unit. This chapter gives a superb introduction to Hidden Markov Models, and a critical evaluation of the advantages and limitations of the approaches described. Chapter 3 (Crossing Borders: Modeling Approaches to Understand Chromatin Domains and Their Boundaries) gives a fascinating insight into the question of three-dimensional genome structure. Given a set of "rules" representing only one-dimensional information (i.e., which histone modifications and proteins are bound along the genome), the model is able to accurately predict the way in which the genome folds in three dimensions, in turn giving insights into how these domain structures may arise. (See also Chapters 9–12 for alternative approaches to 3D genome architecture.) Chapter 4 (Inferring Chromatin Signaling From Genome-Wide ChIP-seq Data) tackles the long-standing question of cause and effect: Given a set of static observations of multiple epigenomic features, can we extract information about how they got there and who came first? The authors present an elegant approach to discern the order of events from large sets of genome-wide data, thus extracting information about the dynamics of the system by comparing different static data sets.

These four chapters also have in common that they critically tackle the problems inherent in interpreting genome-wide epigenomic data. The only (virtually) unambiguous genome-wide data set is DNA sequence. All other types of data relating to mapping of protein binding and histone modifications can give different results depending on experimental handling and data analysis methods. The following four chapters give invaluable expert insights into the challenges of dealing with this type of data and how to get meaningful information from it.

CHAPTER 1

Computational Identification of Polycomb/Trithorax Response Elements

Marc Rehmsmeier
Humboldt-Universität zu Berlin, Berlin, Germany

CONTENTS

How do you tell a mouse from an elephant? No, seriously. Assume you are facing the task of cataloging a collection of animal photographs. What would your criteria be in the above distinction? Surely everyone's first response would be: "That's easy: elephants are much, much bigger." True. Unfortunately, the photographs' sizes are all the same, and you might not be able to infer the true size of the animal in question from its picture. Other features come to mind no sooner than they are discarded: "Elephants have four legs!" I hear you shout with excitement. Indeed, they have. After some rumination, you solemnly declare: "Elephants have trunks." Yes! You have a winner here. To hedge yourself against photographs taken from poor angles, with only a side or the back of the animal showing, you come up with some less dramatic but useful features, such as "Mice sit closer to the ground, and, relatively speaking, they have much longer tails."

Now, what about cats and dogs? Tricky.

INTRODUCTION TO POLYCOMB/TRITHORAX RESPONSE ELEMENTS

Polycomb/Trithorax response elements (PRE/TREs for short, or PREs for very short) are an altogether different type of animal. They are several-hundred basepair long deoxyribonucleic acid (DNA) regions that mediate epigenetic memory of silent and active states by the Polycomb and Trithorax groups of proteins [1]. When Leonie Ringrose approached me on the subject many years ago, she had just heard of PREs in a presentation by Renato Paro. The message was as follows: PREs have similar functions but do not show sequence similarity; in other words: alignment programs cannot do much with them. At that time, only a handful of PREs had been identified in the fly genome, but from antibody studies on polytene chromosomes one strongly suspected that at least a hundred or so PRE loci in the fly genome

Epigenetics and Systems Biology. DOI: http://dx.doi.org/10.1016/B978-0-12-803075-2.00001-5

had to exist. The fly genome sequence had just become available [2], presenting an opportunity to computationally identify PREs genome-wide. A standard approach at the time would have been to identify unknown PREs by sequence similarity to known ones. However, due to the lack of sequence similarity mentioned above, there was no hope for success of such an approach.

2003 AD HOC APPROACH TO PRE PREDICTION, TOGETHER WITH ITS PARTICULAR MOTIVATIONS [3]

Coming back to the image of mice and elephants, how could we distinguish PREs (mice) from the rest of the genome (elephants)? Although the handful of known PREs did not show overall sequence similarity, they did have aspects in common: binding sites of factors that are involved in the recruitment of Polycomb repressive complexes (PRCs) and Trithorax group (TRXG) proteins. The binding sites were those of GAGA factor—GAF, with the consensus binding site described as GAGAG or GAGAGAGAGA [4]; the Pleiohomeotic protein—PHO, with the consensus binding site variably described as CNGCCATNDNND [5] or GCCATHWY [6], the consensus site being GCCA; and the Zeste protein—Z, consensus binding site YGAGYG [7]. An additional motif was EN 1 (GSNMACGCCCC) which had been shown to be essential for the silencing function of the *engrailed* PRE [8].[1]

All known PREs had a number of the motifs described above in their sequence. A possible approach would thus have been to scan the fly genome for occurrences of these motifs in unexpected densities, i.e., in clusters. However, there was one more complication: If the fly genome is full of elephants, are PREs mice? Or are they cats, and are there many dogs in there, too? As you will have guessed, the latter. GAF and Z do not only function as PRC/TRXG recruiters in PREs but also regulate other genes such as *hsp22* [10] and *white* [11] at those genes' promoters [3]. This appeared as an obstacle in the beginning, but bioinformatics came to the rescue with an established concept: that of a log-odds score. Outside statistics, odds are known in gambling, where they describe the relative amounts of stakes from the two betting parties, namely the bookmaker and the gambler. As bookmakers are clever people, or else they would go out of business, odds are not only incidentally closely related to the likelihoods of events. At "even money," the chance of winning the bet is (or should be) 50% for both parties. Long odds such as 20 to 1 have a much higher chance for the bookmaker to win, so he or she can confidently put 20 pounds in for every pound of yours. With short odds,

[1] Ambiguous bases in binding sites/motifs are given in IUPAC notation [9].

it is the other way around. I recommend Dick and Felix Francis' books on the topic for further study, e.g., [12]. In statistics, the odds are ratios of likelihoods. The logarithm of this ratio is then taken, because it has a very convenient effect: If the number in the numerator is larger than the one in the denominator (assuming that both these numbers are positive, which in the case of likelihoods they always are), i.e., if the ratio is larger than 1, the logarithm is positive. If the numerator is smaller than the denominator (with the same assumption), i.e., if the ratio is smaller than 1, the logarithm is negative. In the case of equal numbers and thus a ratio of 1, the logarithm is zero. The logarithm of the odds can then easily be interpreted by it being positive or negative (or, rather rarely, zero).

What do the binding sites of GAF, PHO, and Z and the EN 1 motif have to do with log-odds scores? To finally be able to answer that question, one also needs to know that the likelihood functions of such a score describe two models. For example, many readers will know score matrices that are used in aligning protein sequences. There, the likelihood function in the numerator (the "top part" of the ratio) is the probability of two amino acids in the same position under the assumption of homology (i.e., common ancestry), and the function in the denominator (the "bottom part" of the ratio) is the probability of observing two amino acids in the same position merely by chance, that is, when they are not evolutionarily related. In other words, we have two models, which in statistics are also called the alternative model (the one in the numerator) and the null-model (the one in the denominator). Calculating these two probabilities, taking the ratio and then the logarithm then quickly tells us whether our observation speaks more for the alternative (when the score is positive) or more for the null-model (when the score is negative).

Now, finally, we can formulate our idea: We have two types of animals—PREs and non-PRE promoters that look similar to PREs; see Fig. 1.1. There are our two models! The likelihoods are the probabilities of the various motifs (or motif combinations, as we'll see in a bit) under the assumption that the region under scrutiny is a PRE, or under the assumption that it is a non-PRE promoter that looks like a PRE, respectively. To score a whole region, the log-odds scores for each motif (or motif combination) are summed up, similar to summing up individual scores in sequence alignment, and the final sum is an indication of whether the region is a PRE or a non-PRE promoter that looks like a PRE. See Fig. 1.2. For more details on weights, see Fig. 1A and B in Ref. [3].

An important concern in modeling is whether the chosen features (occurrences of DNA motifs in our case) are good features, meaning that they allow a correct classification. In our animal example from the beginning, a trunk is

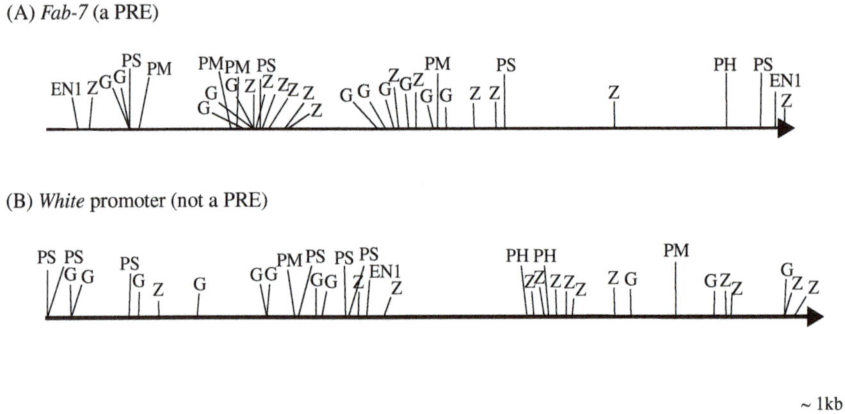

FIGURE 1.1

Motifs in PREs and in non-PREs: (A) *Fab-7* is a well-known PRE. (B) The *white* promoter contains clusters of motifs, but is not a PRE. (A and B) *PH, PM, PS*, PHO motifs. *G*, GAGA motif. *Z*, Zeste motif. *EN1*, engrailed PRE motif. Figure contributed by Leonie Ringrose.

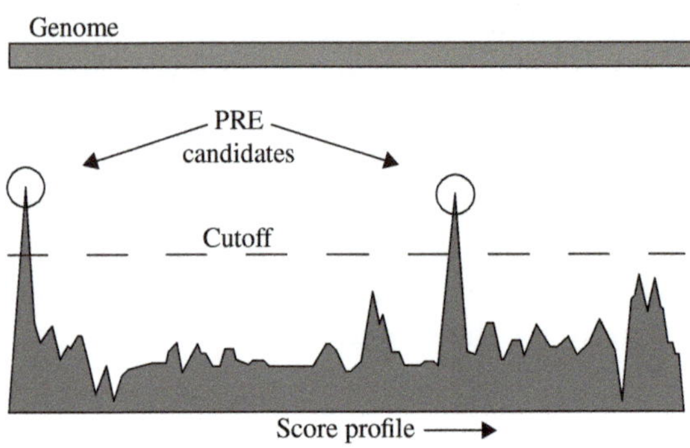

$c(m)$

$$\sum_m w(m)c(m)$$

$$w(m) := \log \frac{f(m|\,\text{PRE})}{f(m|\,\text{non-PRE})}$$

FIGURE 1.2

PRE prediction: The genome is scanned with a window of, for example, 500 bp (basepairs). Inside a sequence window, motifs (or motif pairs), *m*, are counted. Counts are then weighted with a weight, *w(m)*, and summed up. Each window produces one sum, all sums producing the score profile of the genome. The weights are calculated from motif (or motif pair) frequencies, *f*, in positive (PRE) and negative (non-PRE) training sets. Window scores that exceed a cutoff are PRE candidates.

a good feature for distinguishing elephants from mice. Are single DNA motifs good features for distinguishing PREs from non-PRE promoters that look like PREs? Not quite. And even more so, they are not good enough to pick out PREs from the vast genomic background, a whole zoo of bits of DNA. Luckily, the initiator of the project, Leonie Ringrose, had had an epiphany about what could be a good set of features: motif pairs, with the idea that proteins that bind them may cooperate or directly antagonize each other. So, instead of using single motif occurrences, we decided to try paired motif occurrences, where the constituents of a motif pair had to be in the range of 220 bp (basepairs) from each other—a distance that was inspired by nucleosome organization (147 bp making two turns of DNA around the histones, plus the lengths of the linker regions), but whose precise definition turned out to be of not too much importance.

In classification tasks one traditionally works in a balanced data scenario: Positives and negatives occur in more or less equal amounts. A balanced scenario is probably what you had in mind when you were reading my mice-and-elephants fable. Scanning a whole genome for a relatively small number of PREs, however, is very different: The data are highly imbalanced—the length of the whole fly genome (130 Mbp, megabasepairs) in comparison to 200 PREs of length 500 bp each is a ratio of 1300 to 1! This would not be a problem if for every mouse there were 1300 obvious elephants (except for the time needed to go through all of them, but that is what computers are for), but the genome is more like a zoo, with many kinds of animals and considerable chances that some or even many of them look like mice—think of other rodents for example. What are the odds then of being successful in our hunt for PREs?

EVALUATING CLASSIFICATION PERFORMANCE

Before we answer this particular question, we need to discuss how classification performance is evaluated in general. A frequent mistake in the interpretation of bioinformatics predictions is to mark any one result as cut in stone and to confuse different categories of misclassification. When we evaluate classification performance—asking how correct a particular classification is—we can only do this on known data. Using the animal example again, I could show you 100 images of mice and elephants or cats and dogs, and you would classify each image according to your criteria. If I know the truth about each picture (perhaps because they are photos that I took myself, seeing trunks where you cannot), I will be able to define four numbers: the number of correct mouse identifications, the number of correct elephant identifications, the number of elephants that were incorrectly identified as mice, and the number of mice that were incorrectly identified as elephants. In the field of classification, one of the two classes is usually called the positive class (let's say the mice, if they are the ones you are after), the other class being the negative class (the elephants), so the numbers would be numbers of true positives, true negatives, false positives, and false negatives, respectively. If the truth is dichotomous, that is, if each object belongs to one of the classes and to only one—so that no image can correctly be classified as mouse *and* elephant,

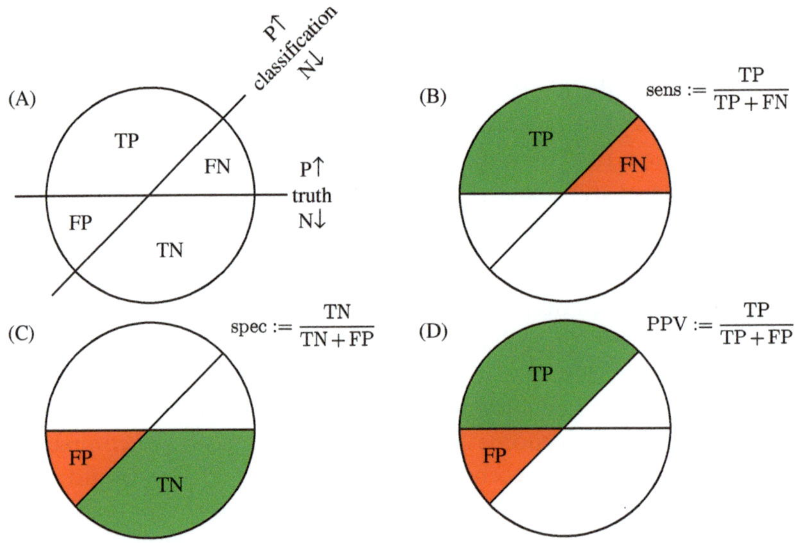

FIGURE 1.3

Classifier performance: (A) A classification gives rise to the "confusion matrix" values *TP, FP, TN,* and *FN. P*, positives, *N*, negatives, *TP*, true positives, *FP*, false positives, *TN*, true negatives, *FN*, false negatives. (B–D) Common performance measures. Relevant correct classifications are shaded green. Relevant incorrect classifications are shaded red. (B) *sens*, sensitivity (recall, true-positive rate). (C) *spec*, specificity. (D) *PPV*, positive predictive value.

nor as something else—these four numbers are unambiguously defined. Although they correctly and completely describe the performance of the classification, they are not very intuitive, and it is useful to derive other quantities from them which express more abstract concepts. Often used quantities are *sensitivity*, also known as *recall* or *true-positive rate*—the relative fraction of positives correctly identified, that is, the number of true positives divided by the number of all positives; *specificity*—the relative fraction of negatives correctly identified; and *positive predictive value (PPV)*—the relative fraction of correctly identified positives among all positive classifications. See Fig. 1.3. For an overview and for a discussion of classifier evaluation in a genome-wide context, see Ref. [13]. Many other summaries of the four original numbers exist, all expressing various aspects to various degrees (see, e.g., Table 1 in Ref. [13]).

An additional dimension in evaluating classification performance is that, frequently, there is not *the* classification, but a range of classifications along which one trades one performance aspect for another—normally sensitivity/recall for specificity or PPV, or the other way round. This can be achieved whenever a classifier operates on a scale of scores or similar values and the actual classification is achieved by a score cutoff that divides all elements under scrutiny into two sets: those elements that have been assigned a score equal to or above the cutoff, and those elements that have been assigned a score below the cutoff. For varying score cutoffs, we then have varying sensitivity/specificity or recall/PPV combinations. What cutoff should we choose then? That depends entirely on the given situation. When Leonie and I wanted to identify PREs genome-wide, only a handful of PREs were known, and we knew that even a small number of new PREs would be exciting. In addition, we wanted PRE candidates to be convincing, with excellent chances for them of being real. In classification terms, this translated into the requirement of keeping the number of false positives at a minimum, while counting on the prediction of a decent number of candidates.

RESULTS OF 2003 PRE PREDICTION

As the fly genome sequence was largely unknown in terms of its PRE potential (and what would have been the point of predicting PREs otherwise?), we modeled the non-PRE "background" of the genome statistically by counting nucleotide frequencies (i.e., answering how often, in relative terms, the four DNA bases A, C, G, and T occur). Based on these frequencies, we generated a large number of artificial genomes against which we ran our classifier. We then adjusted our score cutoff such that a classification on an average artificial genome would generate only 1 (one) false-positive prediction. Under the assumption that such an artificial genome models the real genome background reasonably well, we could expect 1 false-positive prediction at the same score cutoff. How many PRE candidates did we predict? 167! Far more than we could have hoped for and at the astonishing predicted-over-expected ratio of 167:1 (naturally so due to our 1 expected false positive), when one is normally satisfied with ratios of around 2:1 and very happy when they go above 3:1. The precise accuracy of our false-positive estimate did not matter very much in the end, as even a shift by an order of magnitude (10 false positives instead of 1) could not have tainted the result. We also hit many (though not all) of the known PREs in the Hox loci and at the *engrailed* gene, indicating good sensitivity/recall in our prediction.

The genome-wide predictions were so convincing to us that Leonie was willing to invest her time and (other people's) money in validating them experimentally. In a first experiment, she used chromatin immunoprecipitation (ChIP) to assess whether predicted PREs are bound by Polycomb (Pc) in vivo. The results showed that this is indeed the case; in comparison, a similarly sized set of negative controls showed only very little enrichment. Additional transgenic analysis of four candidate PREs showed Polycomb group (PcG) dependent repression of a reporter gene, pairing sensitivity of repression, and variegation of expression, which are all hallmarks of classical

PRE behavior. Since then, other candidates have been validated and extensively studied, such as the *vestigial (vg)* PRE [14–17].

NEW MOTIFS DISCOVERED

When one has identified such a large set of new elements—PREs in our case—it opens up new possibilities of analysis, also computationally. Asking what distinguishes one group of items from another group, or what the items of one group share, is common in machine learning and bioinformatics (in fact, in many areas of science). Indeed, this is what we asked in the beginning of this chapter: How do you tell a mouse from an elephant? One can also ask "What do mice have in common?," hoping that the features discovered can serve in a classification task. On the DNA sequence level, the question is most commonly approached by identifying short stretches of DNA, "motifs," that are "over-represented," or "enriched," meaning that they occur more often than "expected by chance." The chance expectation is calculated from a statistical model of nucleotide frequencies, or frequencies of dinucleotides or even longer building blocks, depending on what one wants to consider as a null-model, or "background." Additional constraints can be enforced in the discovery process, for example, that motifs need to occur in all of the sequences. Other aspects, such as how motifs are distributed over the sequences—evenly or not so evenly—can be taken into account as well.

Once we had predicted the 167 PREs, we ran a motif discovery tool, MEME [18], on the 30 top-scoring sequences to see what DNA motifs these sequences had in common (see also Fig. 5 in Ref. [3] for motif logos). Not surprisingly, but reassuringly, MEME found the GAF motif (GAGAGAGAGA) and the PHO motif (with the core GCCAT), demonstrating that discovered motifs have biological relevance. The Z motif (YGAGYG) was not identified, perhaps due to its degeneracy. In addition, MEME found three new motifs: a GTGT motif, a poly-T motif (stretches of the nucleotide T), and TGC triplets. Since then, the GTGT motif has been shown to be essential for silencing in the *vg* PRE [16] and to be bound by the sequence-specific DNA-binding protein Combgap (Cg) which contributes to the recruitment of PcG proteins [15]. It was also rediscovered as the CA repeat motif or CACA motif in a ChIP-on-chip study of genome-wide binding profiles of PcG and other proteins [19].

RECASTING PRE PREDICTION AS A MACHINE LEARNING PROBLEM

The algorithm for PRE prediction was an ad hoc solution, in the sense that it was geared specifically for this task, but our approach to PRE prediction can be interpreted in the wider field of machine learning. Very generally, the field of machine learning studies computer programs that improve their performance in a task by learning from experience [20]. More precisely (from Ref. [20]): "A computer program is said to learn from experience E with respect to some class of tasks T and performance measure P, if its performance at tasks in T, as measured by P, improves with experience E" [emphasis removed from quote]. Although this definition might look bureaucratic to some, it is a tool for gaining clarity about what we are trying to achieve, by providing us with a precise list of elements—T, P, and E. We can then recast our PRE prediction as a (machine) learning problem like this:

- Task T: Identifying PREs in the *Drosophila* genome
- Performance measure P: Recall and positive predictive value
- Training experience: A database of positive and negative examples.

The performance measure P here consists of two parts, recall and positive predictive value. Recall (or sensitivity), as described above, is the percentage of positives correctly identified as positives. Positive predictive value is the fraction of true positives among all predicted positives. Predicted PREs have been experimentally validated in a number of assays, such that, also in connection with our very stringent prediction cutoffs (*E*-value 1), we can expect positive predictive value to be close to 100%. As we do not know the precise number of PREs in the *Drosophila* genome, we can only roughly estimate recall: On the basis of genome-wide ChIP studies of PcG proteins, one can speculate that there are hundreds of PREs in the *Drosophila* genome ([21] and references therein). Of those, we have identified (nearly) 167, so that recall can be expected to be around 20% or higher. These two findings in other words: We believe that we have identified a sizeable portion of the PREs in the *Drosophila* genome and that the majority of the 167 candidate PREs are real.

Once one has realized that PRE prediction is a machine learning task, one can naturally ask whether other algorithms/approaches, such as support vector machines (SVMs), have better performance than our algorithm. Whenever we have been asked this question, then, I speculate, with the idea in mind that SVMs would surely be superior, probably because they have been successful in other classification tasks, and also because many people have heard of them. However, a superiority of SVMs over our approach is not evident a priori. Both approaches are classifiers, that is, they try to separate positive from negative instances (think of instances as points in space, annotated with plus and minus signs). SVMs do so by constructing "support vectors" that define an as-wide-as-possible margin of separation between the two classes, and they have the additional advantage that they can separate classes by a nonlinear (curvy) border, but they can also be employed as linear classifiers. See Fig. 1.4 for an illustration [22].

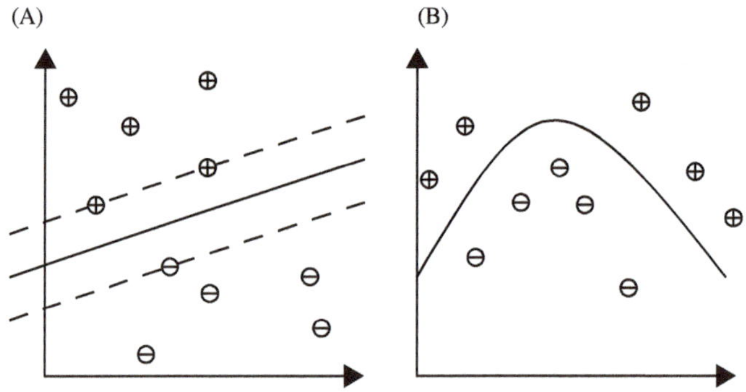

FIGURE 1.4

Linear and nonlinear separation: Circles are positive (plus sign) and negative (minus sign) instances in a two-dimensional feature space. (A) A linear separation of the two classes by the solid diagonal line. The dashed lines show the margin of separation. (B) A nonlinear separation by the solid curved line. *Adapted from Bredesen BA. Prediction of Polycomb/Trithorax response elements using support vector machines. 2013.*

Our ad hoc approach finds the centers of gravity in the two sets of training examples and tries to separate the two classes with a linear (i.e., straight) border. How any of the two approaches performs in a particular classification scenario, here the prediction of fly PREs, cannot be decided without putting it to the test, that is, applying it and then evaluating the results, either on the basis of previous knowledge or through further experimental procedures. Zeng et al. used an SVM as a filtering step in the prediction of PREs and demonstrated an increase in sensitivity to experimentally-defined target gene sets [23]. However, the evaluation was done on a gene-by-gene basis, and results on overlaps of PREs themselves were not reported. Our own experiments indicate improvements in PRE-level performance when comparing standard SVMs to the original approach, but the improvements are not dramatic (to be published elsewhere).

A vast number of alternative techniques have been studied in the field of machine learning, such as decision trees or various types of neural networks. All these are potential candidates for improving the performance of PRE prediction. No one, however, can say with confidence which, if any, of these methods would be better suited. In the end, the proof of the pudding is in the eating.

MISCLASSIFICATION COSTS AND THE TRADE-OFF DIMENSION

I mentioned earlier that, frequently, there is not *the* classification, but a range of classifications along which one trades one performance aspect for another. This also means that, for a particular method that produces such a range of classifications, one cannot improve two "antagonistic" measures at the same time—improvements in sensitivity/recall come with a cost in specificity or positive predictive value, simply because by moving a score cutoff or similar, one cannot go in two directions at once. A chance to break out of this trade-off problem is to use different methods, with which I not only mean a different algorithm, but also changes in other aspects, such as the training data used, or particular parameters. Although most method development in bioinformatics concentrates on new algorithms, these other aspects can be of equal or even greater importance. This brings me back to the pudding: What do we actually ask for in a classification? Obviously, it being perfect would close the matter, but that sounds unrealistic, so it must be about something else. Perhaps surprisingly, the answer depends on the situation—more precisely, it depends on the relative importance of correct and false classifications. For example, at the time of our original PRE prediction work, published in 2003, only few PREs were known. It was thus enough scientific progress to enlarge this set of known PREs to some extent, and it was not necessary to uncover the majority of PREs in the fly genome. At the same time, our resources for experimental validation of PRE candidates were limited, and we would not have had the time and money to validate hundreds of predictions. In more formal words, the cost associated with false-negative predictions (PREs we missed) was considered low, and the cost associated with false-positive predictions (candidates to be sorted out later by means of experimental falsification) was considered high. For this reason, we decided to employ a very strict prediction cutoff (remember our earlier *E*-value discussion), so that false positives would be avoided. Indeed, our experimental validations indicated that the majority of our candidate PREs were true positives, and—lucky as we were—we even had 167 of them, thus considerably enlarging the body of identified fly PREs. Your cost assessment, in your particular situation, might have entirely different motivations. Perhaps you have a high-throughput pipeline for experimental validation of predictions, so that you can validate or falsify predictions for little money in short time (false-positive predictions are not a problem). Or your entire career depends on finding this one master regulator that no one has found yet, but that everyone knows exists (false-negative predictions *are* a problem). Unfortunately, there are situations in which the trade-offs are simply too large, for example, when you depend on high recall/sensitivity and a low number of false-positive predictions. As indicated above, in such situations one needs to be creative and to find a way to step outside the trade-off dimension.

EVOLUTIONARY ANALYSIS AND SEARCH-SPACE REDUCTION

Technological progress often opens venues for new questions and answers (or new answers for old questions). This was also the case in the run-up to our 2008 paper [24], during which several *Drosophila* genome sequences became available, among them, and as the first, that of *Drosophila pseudoobscura* [25], a distant relative of *Drosophila melanogaster*. With the genome sequence in the computer and our PRE prediction program at the ready, what easier a task could there have been to have a go at the new sequence? We did, and the result was surprising: The number of candidate PREs more than doubled in comparison to *D. melanogaster*. Before you raise your hand: This is not due to a possibly much larger *D. pseudoobscura* genome size. In fact, the genomes are of comparable length, and *E*-values are calibrated for a particular genome size anyway. Also, we were able to confirm the larger number of PREs with immunostainings of polytene chromosomes (see Ref. [24] for details). Individual candidates were also validated in transgenic assays.

In addition to this surprising first result, we were also able to document examples of functional regulatory elements (PREs here) that were not conserved between *Drosophila* species in terms of sequence or even in terms of chromosomal position (again, see Ref. [24] for details). However, even though homology did not play as great a role as it usually does—for example, in the evolution of protein domains—we could put it to good use. Remember that *E*-values in our case are calculated based on genome size. We can also use the more abstract term of "search space" which describes the abstract statistical space that we search for something interesting, PREs in our case. This space can give rise to false-positive predictions, and the larger the search space, the more so. Luckily, this also works the other way around: A smaller search space means a smaller number of false positives.

In our analysis across *Drosophila* genomes, we reduced search space in the following way: Starting with a candidate PRE in one species, we identified the homologous location in the other species. We then looked in a region around this location for a PRE (we actually used a scale of regions, but the principle remains; see Ref. [24] for details). Generally speaking, we have a PRE candidate whenever a piece of genome scores higher than the threshold. The threshold, in turn, is calibrated with the size of the search space—larger search spaces require larger thresholds to keep the expected number of false positives (*E*-value) the same. Conversely, a small search space, like 10 kbp (kilobasepairs) regions around c. 200 loci, allows for a relaxed threshold in comparison to the whole genome. With this strategy, we were able to identify PREs that we had missed in the genome-wide search, thus increasing the number of true-positive predictions while keeping the number of false positives on the same level—a creative solution to the trade-off problem.

TODAY: GENOME-WIDE PROFILING DATA

The situation leading up to our 2003 publication was special: A handful of fly PREs had been identified; it was strongly suspected that there should be several hundred PREs in the genome; binding sites for PRC-recruiting proteins were known; useful bioinformatics and statistics concepts were available. These factors together gave us the motivation and the means to define 167 PRE candidates, most of which are likely bonafide PREs, judging from experimental validation. A few years later saw the advent of genome-wide profiling studies, first through chromatin immunoprecipitation (ChIP) followed by microarray chip analysis (ChIP-chip) and then through ChIP followed by next-generation sequencing (ChIP-seq), also in the area of Polycomb regulation.

HOW GOOD IS OUR METHOD WHEN EVALUATED UNDER THESE DATA?

The various profiling studies defined between several hundred and several thousand sites as PcG binding sites [26–28]. Our single-genome and multiple-genome predictions overlap with 20% and 34%, respectively, of the regions defined in the Schwartz et al. dataset [26]. Considering the union of the above three ChIP datasets, 25% of our predictions overlap with an experimentally defined region (data not shown). If one restricts the overlap analysis to regions of accessible chromatin, as defined by DNase I footprinting [29], 46% of our predictions overlap with experimentally defined Pc binding sites (data not shown). Between 90% and 97% of experimentally defined sites are accessible (data not shown). The ChIP-chip studies [26,27] identified between 170 and 201 sites; the ChIP-seq study [28] identified 2265 sites. The experimental datasets discussed here were among the earliest ChIP-chip and ChIP-seq datasets and serve as representatives, but it will be worthwhile to include studies that have been published since (future work).

SENSITIVITY AND SPECIFICITY OF GENOME-WIDE PROFILING

What do these numbers tell us? Before we can answer this question, we have to think about the meaning of a ChIP-chip or ChIP-seq result. First of all, it is important to note on what material an experiment has been performed. For example, in Refs. [26,28], cultured cell lines were used (Sg4 in Ref. [26] and S2 in Ref. [28], both derivatives of Schneider's line 2 [30]), and in Ref. [27], embryos, and larval imaginal discs. The origin of cells in the Sg4 and S2 lines are unknown [30], but in any case represent a single cell or tissue type only and this not in an in vivo environment. Material from whole embryos represents a mixture of cell types. Imaginal discs correspond to a specific developmental time and are developmentally committed to specific structures. Thus, the results of any ChIP-chip or ChIP-seq study are restricted in their generality: Signals from underrepresented cell-types (in embryos) will be lost in the results; the restriction to a specific developmental time or cell/tissue type likewise strongly suggests that relevant binding events, which might occur at other developmental times or in other cells/tissues, will be missed. In addition to these "biological" problems, there are technical limitations such as ChIP-seq signals biased by chromatin accessibility [31]. Indeed, as mentioned above, between 90% and 96% of experimentally defined sites are accessible (data not shown). It has also recently been shown that active promoters give rise to false-positive "phantom peaks" [32], which generally questions the validity of ChIP-seq datasets. Also, since ChIP-chip and ChIP-seq studies are genome-wide, they have their own statistical problems in terms of sensitivity and specificity due to the necessity of defining signal cutoffs and the inevitability of misclassification. Perhaps even more importantly, ChIP studies do not provide means of functional validation. Compare this with the operational definition of an enhancer: Enhancers are stretches of DNA that, when put in front of a reporter gene, recapitulate the expression pattern of their original target gene. An enhancer has to do something before it merits being called an enhancer. ChIP, in contrast, addresses only one hallmark of regulation, the binding of a protein. It also probably cannot distinguish between stable and unstable bindings [33]. Pc enrichment could also be due to mechanisms other than direct recruitment such as chromatin looping [34], similar to enhancer-promoter looping [33]. In another line of study, the Ringrose lab have shown that in the case of the downstream *vg* PRE, the sequence that has the highest PRE prediction is what recruits Pc; after recruitment, the protein accumulates just adjacent to this entry site. This suggests that at least in this case, we have predicted the element that initially recruits Pc, but not the place where it eventually accumulates (Leonie Ringrose, personal communication). As a consequence, a ChIP study of Pc enrichment can miss the recruitment site and thus the actual PRE.

Taking all these observations together leads to the realization that the experimental definition of Pc binding sites through genome-wide profiling efforts is by no means unequivocal. It is likely that not all Pc sites thus defined are bonafide PREs, and it is very likely that many bonafide PREs have not been identified in profiling studies. It is clear then that computationally identified PRE candidates do not have to have profiling study counterparts to be considered real PREs, and not all ChIP-defined PRE candidates are necessarily real. Computational and experimental approaches have to be judged with respect to their specific chances and potential pitfalls. Profiling studies are limited validators of PRE functionality due to their restriction to Pc enrichment and either individual cell lines or mixtures of cell types but have the merit of their genome-wide scope. Computational PRE identification does not have the burden of cell type restriction, but its success depends on a well-defined classifier. Good training and test datasets, statistical modeling, cross-validation, and experimental validation—ideally in transgenic assays that appraise the hallmarks of Polycomb regulation—will go far in rendering computational PRE identification reliable and useful. In this context, genome-wide profiling datasets will be most useful as improved sources for classifier training and cross-validation, by providing more training/validation instances, and by covering PRE classes other than those targeting homeotic genes, as was the case in the first round of genome-wide identification of PREs [3].

CONCLUSION

This chapter has introduced you to important aspects of the computational and experimental identification of PREs/TREs. This will enable you to judge the complex relationships between statistical, computational, and machine learning issues on the one side and experimental issues on the other. Ideally, you will be able to apply similar ideas in other areas of epigenetics and beyond. Incidentally (or not), the work presented here is an example of a systems-biology life-cycle, in which phases of theoretical work and phases of experimental work follow each other, each of them learning from their predecessor, forever improving our understanding of an important part of epigenetic regulation.

References

[1] Steffen PA, Ringrose L. What are memories made of? How Polycomb and Trithorax proteins mediate epigenetic memory. Nat Rev Mol Cell Biol 2014;15(5):340–56.

[2] Adams MD. The genome sequence of *Drosophila melanogaster*. Science 2000;287(5461): 2185–95. (80-).

[3] Ringrose L, Rehmsmeier M, Dura JM, Paro R. Genome-wide prediction of polycomb/trithorax response elements in *Drosophila melanogaster*. Dev Cell 2003;5(5):759–71.

[4] Strutt H, Cavalli G, Paro R. Co-localization of Polycomb protein and GAGA factor on regulatory elements responsible for the maintenance of homeotic gene expression. EMBO J 1997;16(12):3621–32.

[5] Mihaly J, Mishra RK, Karch F. A conserved sequence motif in Polycomb-response elements. Mol Cell 1998;1(7):1065–6.

[6] Fritsch C, Brown JL, Kassis JA, Müller J. The DNA-binding polycomb group protein pleiohomeotic mediates silencing of a *Drosophila* homeotic gene. Development 1999;126(17):3905–13.

[7] Hur MW, Laney JD, Jeon SH, Ali J, Biggin MD. Zeste maintains repression of Ubx transgenes: support for a new model of Polycomb repression. Development 2002;129(6):1339–43.

[8] Americo J, Whiteley M, Brown JL, Fujioka M, Jaynes JB, Kassis JA. A complex array of DNA-binding proteins required for pairing-sensitive silencing by a polycomb group response element from the Drosophila engrailed gene. Genetics 2002;160(4):1561–71. PubMed PMID: 11973310; PubMed Central PMCID: PMC1462036.

[9] Dixon HBF, Bielka H, Cantor CR. Nomenclature for incompletely specified bases in nucleic acid sequences. Recommendations 1984. J Biol Chem 1986;261(1):13–17.

[10] van Steensel B, Delrow J, Bussemaker HJ. Genomewide analysis of *Drosophila* GAGA factor target genes reveals context-dependent DNA binding. Proc Natl Acad Sci USA 2003;100(5):2580–5.

[11] Kostyuchenko M, Savitskaya E, Koryagina E, Melnikova L, Karakozova M, Georgiev P. Zeste can facilitate long-range enhancer-promoter communication and insulator bypass in *Drosophila melanogaster*. Chromosoma 2009;118(5):665–74.

[12] Francis D, Francis F. Even money. Penguin Publishing Group; London, England; 2009.

[13] Saito T, Rehmsmeier M. The precision-recall plot is more informative than the ROC plot when evaluating binary classifiers on imbalanced datasets. PLoS One 2015;10(3):e0118432. http://dx.doi.org/10.1371/journal.pone.0118432.

[14] Perez L, Barrio L, Cano D, Fiuza UM, Muzzopappa M, Milan M. Enhancer-PRE communication contributes to the expansion of gene expression domains in proliferating primordia. Development 2011;138(15):3125–34.

[15] Ray P, De S, Mitra A, Bezstarosti K, Demmers JAA, Pfeifer K, et al. Combgap contributes to recruitment of Polycomb group proteins in *Drosophila*. Proc Natl Acad Sci USA 2016;113(Apr (14)):3826–31.

[16] Okulski H, Druck B, Bhalerao S, Ringrose L. Quantitative analysis of polycomb response elements (PREs) at identical genomic locations distinguishes contributions of PRE sequence and genomic environment. Epigenetics Chromatin 2011;4:4.

[17] Herzog VA, Lempradl A, Trupke J, Okulski H, Altmutter C, Ruge F, et al. A strand-specific switch in noncoding transcription switches the function of a Polycomb/Trithorax response element. Nat Genet 2014;46(9):973–81.

[18] Bailey TL, Elkan C. Fitting a mixture model by expectation maximization to discover motifs in biopolymers. Proc Int Conf Intell Syst Mol Biol 1994;2:28–36.

[19] Schuettengruber B, Ganapathi M, Leblanc B, Portoso M, Jaschek R, Tolhuis B, et al. Functional anatomy of polycomb and trithorax chromatin landscapes in *Drosophila* embryos. PLoS Biol 2009;7(1):e13.

[20] Mitchell TM. Machine learning. New York: McGraw-Hill; 1997. 414 pp.

[21] Kassis JA, Brown JL. Polycomb group response elements in *Drosophila* and vertebrates. Adv Genet 2013;81:83–118.

[22] Bredesen BA. Prediction of Polycomb/Trithorax response elements using support vector machines. 2013.

[23] Zeng J, Kirk BD, Gou Y, Wang Q, Ma J. Genome-wide polycomb target gene prediction in *Drosophila melanogaster*. Nucleic Acids Res 2012;40(July (13)):5848–63.

[24] Hauenschild A, Ringrose L, Altmutter C, Paro R, Rehmsmeier M. Evolutionary plasticity of polycomb/trithorax response elements in *Drosophila* species. PLoS Biol 2008;6(10):2130–43.

[25] Richards S, Liu Y, Bettencourt BR, Hradecky P, Letovsky S, Nielsen R, et al. Comparative genome sequencing of *Drosophila pseudoobscura*: chromosomal, gene, and cis-element evolution. Genome Res 2005;15(1):1–18.

[26] Schwartz YB, Kahn TG, Nix DA, Li XY, Bourgon R, Biggin M, et al. Genome-wide analysis of Polycomb targets in *Drosophila melanogaster*. Nat Genet 2006;38(6):700–5.

[27] Oktaba K, Gutiérrez L, Gagneur J, Girardot C, Sengupta AK, Furlong EEM, et al. Dynamic regulation by polycomb group protein complexes controls pattern formation and the cell cycle in *Drosophila*. Dev Cell 2008 Dec;15(6):877–89.

[28] Enderle D, Beisel C, Stadler MB, Gerstung M, Athri P, Paro R. Polycomb preferentially targets stalled promoters of coding and noncoding transcripts. Genome Res 2011;21(2):216–26.

[29] Thomas S, Li X-Y, Sabo PJ, Sandstrom R, Thurman RE, Canfield TK, et al. Dynamic reprogramming of chromatin accessibility during *Drosophila* embryo development. Genome Biol 2011;12(5):R43.

[30] Schneider I. Cell lines derived from late embryonic stages of *Drosophila melanogaster*. J Embryol Exp Morphol 1972;27(2):353–65.

[31] Ramachandran P, Palidwor GA, Perkins TJ. BIDCHIPS: bias decomposition and removal from ChIP-seq data clarifies true binding signal and its functional correlates. Epigenetics Chromatin 2015;8:33.

[32] Jain D, Baldi S, Zabel A, Straub T, Becker PB. Active promoters give rise to false positive "Phantom Peaks" in ChIP-seq experiments. Nucleic Acids Res 2015;43(14):6959–68.

[33] Levine M. Transcriptional enhancers in animal development and evolution. Curr Biol 2010;20(17):R754–63.

[34] Kadauke S, Blobel GA. Chromatin loops in gene regulation. Biochim Biophys Acta 2009;1789(1):17–25.

GLOSSARY

Polycomb response elements (PREs) several-hundred basepair long DNA regions that mediate epigenetic memory of active and silent states by the Polycomb and Trithorax groups of proteins.

Recruitment a process during which proteins or protein complexes are brought into the vicinity of, for example, DNA, facilitated by recruiter proteins.

Motifs recurring, fixed-length, and usually short patterns of nucleotides.

Bioinformatics the development, study, and application of computational and statistical methods to questions from molecular biology and related fields.

Log-odds score a number that expresses relative likelihoods by calculating the logarithm of ratios (odds).

Alternative model the explanation that is the alternative to a null model, modeling the presence of something interesting.

Null model an explanation that models randomness, the absence of anything interesting.

Feature any kind of property that is used in a classification task to distinguish between sets of instances.

Epiphany an experience of sudden and striking insight (source: Wikipedia).

Balanced classification scenario a classification scenario in which the classes are of comparable size.

Imbalanced classification scenario a classification scenario in which the classes differ substantially in size.

True positive a positive instance that has been correctly identified.

True negative a negative instance that has been correctly identified.

False positive a negative instance that has been wrongly classified as positive.

False negative a positive instance that has been wrongly classified as negative.

Sensitivity/recall the relative fraction of positives correctly identified.

Specificity the relative fraction of negatives correctly identified.

Positive predictive value (PPV) the relative fraction of true positives among all positive classifications.

Chromatin immunoprecipitation (ChIP) a molecular-biology procedure that assesses whether and to what degree a piece of DNA is bound by a protein of interest.

Motif discovery a bioinformatics procedure that identifies distinguishing motifs among a set of DNA, RNA, or protein sequences.

Machine learning a field in computer science which studies computer programs that improve their performance in a task by learning from experience.

E-value expected number of false-positive predictions.

Support vector machine a machine learning method that constructs a hyper-plane for the separation of instances from different classes.

Homology the state of evolutionary relatedness through common ancestry.

Search space the abstract statistical space that is searched for something interesting.

Genome-wide profiling the genome-wide assessment of DNA-protein binding events.

Chip-chip chromatin immunoprecipitation followed by microarray chip analysis.

Chip-seq chromatin immunoprecipitation followed by next-generation sequencing.

Phantom peaks spurious ChIP-seq signals at active promoters.

LIST OF ACRONYMS AND ABBREVIATIONS

bp basepairs (unit)

Cg Combgap (protein)

ChIP chromatin immunoprecipitation

ChIP-chip, ChIP-on-chip chromatin immunoprecipitation followed by microarray chip analysis

ChIP-seq chromatin immunoprecipitation followed by next-generation sequencing

D. melanogaster *Drosophila melanogaster*

D. pseudoobscura *Drosophila pseudoobscura*

DNA deoxyribonucleic acid

EN Engrailed (protein)

E-value expected value

GAF GAGA factor (protein)

Hox genes homeotic, homeobox-containing genes

hsp22 *heat shock protein 22* (gene)

kbp kilobasepairs (unit)

Mbp megabasepairs (unit)

MEME Multiple EM for Motif Elicitation (software)

Pc Polycomb (protein)

PcG Polycomb group (proteins)

PHO Pleiohomeotic (protein)

PPV positive predictive value

PRC Polycomb repressive complex

PRE, PRE/TRE Polycomb/Trithorax response element

SVM support vector machine

TRXG Trithorax group (proteins)

vg *vestigial* (gene)

Z Zeste (protein)

CHAPTER 2

Modeling Chromatin States

Marc Corrales[1,2] and Guillaume J. Filion[1,2]
[1]Centre for Genomic Regulation (CRG), The Barcelona Institute of Science and Technology, Barcelona, Spain [2]Universitat Pompeu Fabra (UPF), Barcelona, Spain

CONTENTS

THE PURPOSE OF MODELING CHROMATIN STATES

Chromatin is the meld of proteins and DNA found in the nucleus of cells. Its roles in gene regulation, chromosome compaction, DNA maintenance, and replication have emerged with our increasing ability to recognize its complexity. It is therefore not surprising that a myriad of models have been proposed to understand its organizational and functional principles.

Early observations at the microscope revealed that chromatin has at least two components. Using different dyes, it was possible to highlight different parts of the chromosome, which were called *euchromatin* and *heterochromatin* [1]. This nomenclature is still in use today, but the definition has kept changing to suit the latest trend. With the advent of biochemistry, heterochromatin became DNAse-resistant chromatin. With the advent of sequencing, heterochromatin became repeat-rich chromatin. With the advent of the histone code [2], heterochromatin became chromatin coated by tri-methylated histone H3 lysine 9 (H3K9me3).

Even though heterochromatin is in general DNAse-resistant, repeat-rich, silent, and coated by H3K9me3, the counterexamples started to accumulate. When it was discovered that H3K9me3 domains are transcribed [3], it became clear that the same word "heterochromatin" represents different objects. But how many? And with what function? Those were the questions in the mind of many researchers at the time bioinformatics was gaining momentum. As the amount of genome-wide datasets increased, it became possible to peek under the hood and have a clearer picture of what chromatin is made of.

Epigenetics and Systems Biology. DOI: http://dx.doi.org/10.1016/B978-0-12-803075-2.00002-7

THE COMMON APPROACH

The human mind does not directly make sense of genome-wide chromatin data. In order to reveal regularities and relationships in the data, biologists have started to apply machine learning algorithms. Machine learning can be loosely defined as the field of computer science that develops algorithms in order to learn and make predictions from the data. The role of those analyses and models is to transform the data into messages that are easier to process and to convey. In modeling chromatin states, the methods that are required belong to a class called "unsupervised learning" [4]. In unsupervised learning, the researchers do not really know what they are looking for, they use algorithms to help them identify hidden patterns in the data. There are many unsupervised learning methods, each answering a different question, each making different assumptions, and each with its own limitations.

The Holy Grail would be an unsupervised method telling the true number of groups in a dataset. On the contrary, such a method does not exist. As a "group" in the mundane sense is not a mathematical concept, one always needs to specify an explicit or implicit criterion of homogeneity. If this criterion is very strict, the classification will contain many groups, and if this criterion is more relaxed, the classification will contain fewer groups. For instance, the visible spectrum is a perfect continuum of frequencies, so there is no meaningful classification from the mathematical point of view. However, there are many classifications in use, with 7 colors (traditional), 1867 colors (the PANTONE catalog), or over 16 million colors (computer graphics). None of these classifications is wrong, each of them is tailored to a different need. The upshot is that the number of colors is not a property of light, but an abstraction designed for a particular purpose. It is important to keep this in mind when using computational approaches to characterize chromatin.

Depending on the study, the number of reported chromatin states varies from 5 [5] to 51 [6]. Even though each study involves a different species, different features and different methods, the most significant difference between them is the intended purpose of the classification: The first was to give a bird's eye view of the genome, the second was to identify functional elements.

Reviewing all the classification methods used to characterize chromatin states would be outside the scope of this chapter. Instead, we will use hidden Markov models (HMMs) to illustrate how these methods work. HMMs are part of the standard machine learning toolbox, and they constitute a natural entry point to understand chromatin classifications. Before going into more detail, it is worth pondering on the observation that the most widely used nonsupervised learning technique is clustering [7], yet few studies have used clustering to identify chromatin states. This is no accident. Clustering methods are inherently "spaceless" so they never use the information that chromatin forms contiguous domains on the chromosome. Chromatin types are not just homogenous groups; they are homogenous groups arranged on a line (the chromosome), and this is exactly what HMMs are made to recognize.

According to the formalism of HMMs, every piece of the chromosome is in one of a finite number of states. These states are not observable (this is the "hidden" in HMM), but they have visible characteristics, which are the proteins that bind the given piece of chromosome. Using HMMs, one can infer the state of the chromatin from the local protein binding. The most desirable property of HMMs is that they take into account the transitions between states. In this sense, HMMs are "context-aware." The information on neighboring loci is implicitly incorporated in the final call of the chromatin state. Fig. 2.1 shows an example of how HMMs can be used to model chromatin states. In this toy example, there are only two states in which proteins are either both frequently present, or both frequently absent. Conflicts occur in the cases that only one protein is present. The model uses the whole ChIP-seq dataset to resolve the conflicts and yields the sequence of states most consistent with the observed data given the assumptions (i.e., there are only two possible states).

HMMs are deeply rooted in the theory of probabilities. This means that the user needs to focus on one single problem: finding the right distribution of the signal. Clustering approaches are an example of the opposite. Their implementation is more straightforward, but the results depend heavily on the chosen distance, and many options are equally meaningful (or meaningless). As a result, users have several possibilities of arranging their data into groups, but no easy way to know which is the best. This situation does not arise when using HMMs because every segmentation has an associated probability, and parsimony requires to choose the most likely. Technicalities aside, HMMs return the best segmentation of the data given the assumptions.

This can sometimes backfire on the analyst because "best given the assumptions" does not mean "best overall." An HMM based on wrong assumptions will conflict with existing data and make inaccurate predictions. Certainly the assumptions must be changed, and one could keep increasing the number of states until the model fits the data. Even though the final number of states would be optimal, it would still depend on the other assumptions of the model, so it would have no legitimate claim for being the "true" number of states.

Let us consider a simple example with two chromatin proteins and two states. In state 1, proteins are both present in high amount, and in state 2, they are both depleted. Importantly, let us further assume that the proteins form a

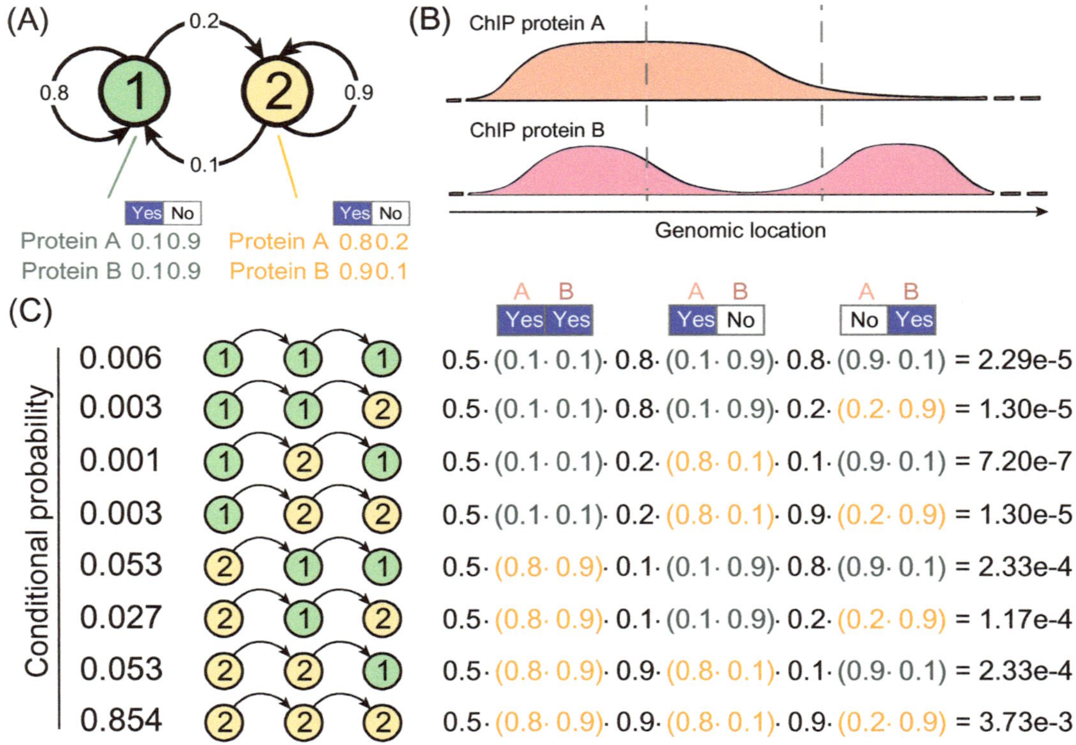

FIGURE 2.1 Detailed example of an HMM used to model chromatin states.

(A) Model specifications. In this simplified example, we assume that there are only two possible chromatin states referred to as 1 and 2 (in green and yellow, respectively), and two chromatin proteins referred to as A and B. In state 1, both proteins have a frequency of occurrence equal to 0.1, and in state 2 their frequency of occurrence is 0.8 and 0.9, respectively. The transitions between states 1 and 2 are as shown by the corresponding arrows. The initial state is 1 or 2 with equal probability. (B) Idealized ChIP-seq profile with three windows. Protein A is present in the first two windows and protein B is present on the first and the third. The last two windows conflict with the model assumption because only one protein is present. (C) Detailed state inference. As there are only three genomic windows, we can exhaustively enumerate the eight possible sequences of states and compute their probability. The possible state sequences are shown on the left. For each genomic window, the probabilities of transition are written in black, and the probability of the observations is written in green or yellow (according to the state). The probabilities of the states given the occurrences of proteins A and B are computed by rescaling so that the sum equals 1. They are shown in the column at the left of the state sequences. There is an 85.4% chance that the sequence of states is [2,2,2]. This example shows how HMMs solve conflicts between the model and the observations by using all the available information. The inferred sequence of states is the one that minimizes the inconsistencies between the model and the data.

complex, so that their binding is dependent. The higher the amount of one of the proteins, the higher the amount of the other, even in each state considered separately. Fig. 2.2 shows the fate of two unrelated investigators using an HMM to model this situation. The first assumes rightly that the proteins may be correlated, whereas the second wrongly assumes that they are independent. The investigators otherwise make identical assumptions. Following a standard approach, they use Akaike's Information Criterion (AIC) to select the best model. The AIC measures the quality of the fit, while penalizing complex models. The first investigator concludes that a model with two states and eight parameters gives a good description of the system (which is how the data was generated). The second investigator concludes that a model with 31 states and 560 parameters is the most appropriate. Importantly, the second investigator will be confident that such a large amount of states is required, because overly complex models were penalized by the AIC. On the contrary, the penalty on complexity is of little help here: The search space of the second investigator does not contain the true solution, and the only way to approach it is to increase the number of states. Measures such as the AIC

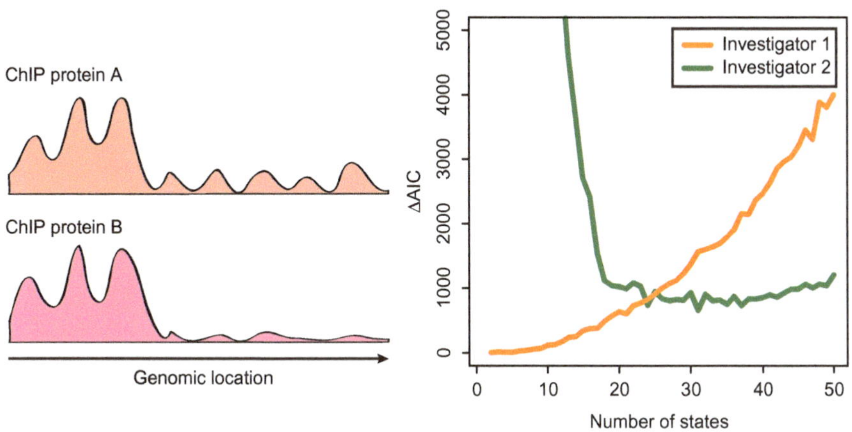

FIGURE 2.2 Wrong assumptions inflate the number of chromatin states.
Left: Proteins A and B form a complex. As a result, their variations are correlated, even in regions where they are both enriched or both depleted. Right: The system was simulated as a 2-state HMM with correlated Gaussian emissions (i.e., bivariate Gaussian variables with correlation 0.9). The first investigator (orange) fitted the data using HMMs with correlated Gaussian emissions. The second investigator (green) fitted the data using HMMs with uncorrelated Gaussian emissions. Both computed Akaike's Information Criterion (AIC), which measures the quality of the fit and gives a penalty for every parameter in the model. For normalization purposes, the AIC of the best model is subtracted from all the values (a lower AIC means a better model). The first investigator concludes that a model with 2 states is the best. The second investigator concludes that a model with 31 states is the best. Assuming that emissions are not correlated led to choose an overly complex model. Note that penalizing the complexity does not make the answer right.

and the optimality property of HMMs give a false impression of safety: The selected model can still be too complex.

This example summarizes the main difference between the study of Filion et al. [5] and that of Ernst and Kellis [6], and it explains why the number of states can be so different between two studies. Although the first study explicitly assumes that chromatin proteins form complexes and tend to covary on the genome, the second study implicitly assumes independence. That does not mean that the first study holds the right answer and that there are exactly five chromatin states. When it comes to modeling chromatin states, everybody is in the case of the second investigator of the example above. Given the simplicity of the assumptions, there are all the reasons to believe that the "truth" does not lie anywhere in the search space. We are thus condemned to add more states to our models, unaware that slightly more complex assumptions may actually simplify the model tremendously. As a consequence, discussions about the "true" number of chromatin states based on model selection criteria are irrelevant. Our personal opinion is that investigators should simply choose a number of chromatin states that is useful for their scientific question.

The example above also illustrates the importance of understanding the relationships between chromatin features when modeling chromatin states. Accordingly, the latest trend has been to use network approaches. The advantage is that complex relationships can be encoded in a compact and human-readable way. As interesting as it is, the detail of these developments is somewhat technical and outside the scope of this chapter, so we refer the interested reader to the original literature [8–13].

WHAT HAVE WE LEARNED FROM THOSE MODELS?

The principal result of genome-wide chromatin studies was to expand the narrow classification in either euchromatin or heterochromatin. The current view conveys a more detailed and flexible model that acknowledges the increasing number of exceptions. Despite the diversity of methodological approaches and model organisms, several principles of chromatin organization are generalizable.

Chromatin is organized in continuous domains. Consistent with the idea that coregulated genes cluster along the chromosome [14,15], combinations of chromatin proteins cover large portions of DNA, frequently covering several neighboring genes. The size of the domains depends largely on the species and on the chromatin type. To some extent, it also depends on the coarseness of the classification. For instance, an active domain with several expressed genes can be subdivided into additional states corresponding to the transcription start sites, gene bodies, and enhancers.

One of the most surprising observations is that hundreds of histone marks and chromatin proteins constitute relatively few chromatin states. The *Drosophila* genome was segmented in five [5], or nine [16] types of chromatin. The human genome was first segmented in 51 types [6], which were later fused to form only seven states [17]. Chromatin components are extremely redundant, and each chromatin feature carries a lot of information about the others.

As all the models have more than two chromatin states, it is natural to ask what are the modern equivalents of heterochromatin and euchromatin. Heterochromatin is traditionally subdivided into facultative and constitutive heterochromatin [18]. Constitutive heterochromatin is located near the centromeres and the telomeres. Early cytological evidence suggested that it is tightly compact [1,19]. In terms of composition, studies before the genomics era showed that it is coated by heterochromatin protein 1 (HP1), by the histone H3 lysine 9 methyltransferase Su(Var)3-9, and by H3K9me3—or their respective orthologs [17,20]. Constitutive heterochromatin is generally rich in repeated sequences called satellites, the composition and abundance of which varies between species [18].

A chromatin type with these properties is present in all the proposed segmentations [5,6,9,16,21]. Yet, it is doubtful that this chromatin type is a general repressor of transcription. For instance, in *Drosophila melanogaster*, 45% of pericentric heterochromatin genes were found to be active [22]. Moreover, the heterochromatin mark histone H3 lysine 9 dimethylation (H3K9me2) covers large portions of the genome that are considered euchromatic [22]. Most likely, pericentric heterochromatin represses only a subset of the genes, but how the rest escapes repression still remains to be elucidated.

The case of facultative heterochromatin is more puzzling. Among the major components are Polycomb proteins, initially identified as silencers of homeotic genes [23]. They are responsible for the deposition of histone H3 lysine 27 methylation (H3K27me), a modification highly correlated with gene silencing [23,24]. Polycomb repressive complexes (PRC1 and PRC2) coat continuous genomic regions referred to as Polycomb domains that are also present in all the chromatin models mentioned above. Significantly, there are more silent genes than can possibly fit in those domains, so at first sight, Polycomb proteins alone cannot account for most of the silencing. In fact, the vast majority of silent genes sit in a chromatin state that is mysteriously devoid of proteins and histone marks. This chromatin type was given several names, such as "black chromatin" [5], "silent domains" [22], "low-activity regions" [17], or "quiescent" chromatin state [25], but the characteristics are the same. It is the principal chromatin type covering eukaryotic chromosomes in both *Drosophila* and humans, and accordingly it contains the majority of the silent genes. It is refractory to the binding of transcription factors, even though their binding motifs are present. This suggests that nonhistone proteins have low accessibility to DNA in this chromatin type. There is, however, one piece of evidence against this view: When expressed ectopically, the yeast GAL4 transcription activator binds its motifs equally efficiently in black chromatin as in the rest of the genome [5]. This casts shadows over the accessibility hypothesis and suggests instead that the binding of transcription factors in the nucleus is not limited by accessibility to DNA.

At the time of this review, there are no studies explaining the biology of this chromatin type. No mechanism explains the silent state of the genes nor the reason why DNA binding proteins expressed and bound to their motifs in other loci fail to do so in this chromatin context. As most silent genes are covered neither by Polycomb-H3K27me3 nor by HP1-H3K9me3, the principal silencing mechanism still remains poorly understood.

Euchromatin is also present in all the models [5,9,17,22]. Significantly, it always consists of at least two distinct types associated with either housekeeping or regulated genes. By integrating different regulatory inputs, promoters ensure the correct expression in quantity, time, and space of the genes they control [26,27]. When they are active, most of them carry the histone H3 lysine 4 dimethylation or trimethylation (H3K4me2/3), and histone H3 lysine 9 acetylation (H3K9ac), irrespective of their type. Techniques to map transcription start sites showed that *Drosophila* housekeeping promoters exhibit a disperse pattern of initiation along the promoter. The initiation of developmental and tissue-restricted genes is more focused (all the transcripts typically start at the same nucleotide), which suggests that different mechanisms of initiation are taking place [27]. In mammals, equivalent initiation patterns were previously observed [28], and the prevalence of broad initiating promoters was reported, but the functional associations with housekeeping or restricted expression was established later [29,30]. The current consensus is that housekeeping genes have broad or dispersed initiation and ordered nucleosomes in all the species. Tightly regulated genes feature a sharp or focused initiation, disordered nucleosomes, and many contain a TATA box.

The differences between housekeeping and regulated genes are not limited to the mechanisms of initiation. All the studies in *Drosophila* report differences in structure, in histone marks on the gene body and in local protein occupancy [5,16,22]. Housekeeping genes tend to be short (typically under 2 kb), whereas regulated genes are bigger (typically above 3 kb) and generally contain a large first intron [5,22]. In the classification in five chromatin types [5], the distinction between those classes of genes corresponds closely to the "yellow" and "red" chromatin types. This means that the protein composition of the chromatin at the promoter and the gene body are not the same. Housekeeping genes are covered in "yellow" chromatin and are bound by MRG15, a chromodomain protein that binds histone H3 tri-methylated at lysine 36 (H3K36me3) present in the body of active genes. Regulated genes are covered in "red" chromatin and are bound by the vast majority of the assayed proteins. The modENCODE classification has an equivalent of "red" chromatin called "state 3" [16,22]. The "state 3" chromatin covers intronic regions from active genes, it is enriched in chromatin remodelers and in the histone variant H3.3, it also has high nucleosome turnover and high DNAse hypersensitivity. The chromatin of "state 3" carries histone modifications usually found on enhancers, such as histone H3 lysine 27 acetylation (H3K27ac), histone H3 lysine 4 monomethylation (H3K4me1), histone H3 lysine 18 acetylation H3K18ac, and histone H3 lysine 36 monomethylation (H3K36me1).

The surprising finding that some loci are bound by most of the chromatin proteins and transcription factors led to the concept of HOT, or highly occupied target [5,16,21,22]. The heterogeneous occupancy of the genome was first observed in *Drosophila* [31]. DamID, a technology to map chromatin protein occupancy, was used to map seven transcription factors, five cofactors, and two heterochromatin proteins, revealing a strong overlap among the factors on a 2.9-Mbp DNA fragment.

Another study in *Drosophila* confirmed the existence of HOTs by analyzing 21 transcription factor binding patterns during development [32]. The authors observed that functionally diverse transcription factors bind short DNA regions in the absence of the binding motifs. Only after the large-scale initiatives that analyzed the genome-wide composition of chromatin in model organisms and humans (modENCODE and ENCODE consortia, respectively) did it become clear that HOTs are a prevalent phenomenon occurring in worms, flies, and humans. HOTs were observed in all the studies. They always appear as short regions bound by most of the assayed transcription factors, even if they are functionally unrelated and even in the absence of binding motifs [17,21,22,33].

A question that occurs naturally is whether HOTs are an artifact. The main argument for their existence is that they were observed with different mapping methods. ChIP and DamID both have their limitations, but those are very different, so they are not expected to produce the same aberrant results. The observation that ChIP-seq signal can be found in the absence of the target protein [34,35] is most likely unrelated, as this "ghost" ChIP signal is found exclusively on the body of active genes, whereas HOTs are also found in intergenic regions. Even if HOTs are not an artifact, it is hard to imagine how so many proteins can fit in relatively small regions of the genome. It may be that the targets are occupied by different proteins in different cells, or that the folding of HOTs is such that the binding of a protein at a locus creates visible signal at the loci in close proximity. Here the static and linear view of chromatin imposed by genome browsers is probably at odds with a more complex biological reality.

There is no consensus on the functional role of HOTs, and it is not even clear whether the HOTs identified by different teams are the same biological entity. Kvon et al. [36] showed that in *Drosophila*, HOTs can work as cell type specific enhancers during embryo development. Another study analyzed a dataset containing the binding profiles of 65 transcription factors and 19 cofactors during *Drosophila* development [37]. The authors concluded that HOTs overlap with housekeeping genes and with enhancers of developmentally regulated genes. Finally, a study focusing on human and *Caenorhabditis elegans* and analyzing extreme HOTs (top 1% instead of 5% used in *Drosophila* studies), reported that HOTs correspond to CpG-rich promoters of housekeeping and broadly expressed genes [38]. In this one, the authors did not find enhancer signatures at HOTs. In short, it is not clear whether HOTs overlap with enhancers, regulated genes, or housekeeping genes. With different criteria, different proteins and different model organisms, the disagreement is perhaps not surprising.

The existence of HOTs suggests that the majority of the signal detected in mapping studies may not be driven by protein–DNA interactions but by protein–protein interactions [21]. It has also been proposed that HOTs result from an increased accessibility of DNA [32]. However, there is some confusion about the effect of histones on the binding of transcription factors. The key point is that the specific binding motifs of the transcription factors are not present in HOTs. It is clear that nucleosomes can prevent specific interactions between a transcription factor and its target site. But to our knowledge, there is no data showing that nucleosomes prevent nonspecific interactions, which are due to electrostatic forces rather than contacts in the major groove of the DNA [39]. How nucleosomes modify those interactions remains to be addressed. Until these data are available, the view that HOTs emerge from increased accessibility of DNA is only speculation.

To summarize, we have learned that chromatin is richer than the previous classification in heterochromatin and euchromatin. The concept imposing itself is the chromatin state: a domain of the genome covered by recurrent combinations of chromatin proteins and histone posttranslational modifications. Chromatin components are more redundant than expected, so that a handful of states are sufficient to give a global picture. The genome-wide variations of occupancy are wider than expected: Most of the transcription factor binding events take place in a small portion of the genome.

These conclusions raise more questions than they bring answers. On the one hand, the majority of the genome is in a "desert" state, devoid of protein binding and histone marks. How is this state generated and maintained? More importantly, how is it made permissive in order to allow transcription? On the other hand, several regions of the genome show an unexpected enrichment in the binding of transcription factors. The accumulation of transcription factors at HOTs is shockingly high. In addition, the majority of transcription factors show no apparent functional relationships, and they bind in the absence of their cognate motif. How are the proteins targeted there? To advance our understanding of those issues, a step forward is needed in the way we model chromatin. Now that we have a better overview of the pieces that shape the system, it is necessary to develop models that will allow us to understand how a chromatin state is set up, how it is maintained and how it is modified.

STATIC VERSUS DYNAMICAL MODELS

All the endeavors to model chromatin mentioned above are inherently static. They attempt to give an accurate snapshot of chromatin, like a map that would coincide closely with a territory. It is not surprising that this approach fails to explain how chromatin changes. The genome, as a chemical object, responds to stimuli and transforms itself during development. A map is not the appropriate model to describe a territory that keeps changing. As a consequence, the models have revealed very little about how chromatin changes upon response to stimuli or differentiation.

Static pictures are useful to define a nomenclature and a common language for scientists, but they are fundamentally unadapted to discuss mechanisms. For instance, all the studies above recognize that silent genes lie in a chromatin state containing very few proteins. But is this chromatin type a cause or a consequence of transcriptional silence? This kind of question has accumulated massively since the rise of genome-wide technologies. One may view the static pictures of the chromatin models as the images of a film and hope that the plot will be revealed when the film is played. Following the actors, we would understand their actions and their purposes. We argue that this may not be the case: The film may be a loop with no start and no end. The study of inducible genes and signaling cascades has created a biased view where events unfold from a start point to an end point, from causes down to consequences. This is fundamentally incompatible with the view that chromatin is the support of epigenetic memory [2]. A memory system must self-reinforce, it must contain coherent feedback loops where the actors are both cause and consequence of each other.

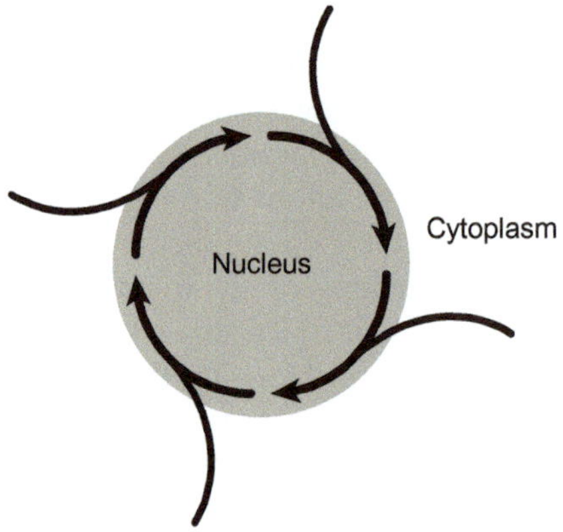

FIGURE 2.3 Chromatin is a circle.
Many chromatin components are self-reinforcing. Upon binding to a locus, they recruit other components that promote their own binding. As such, binding events in the chromatin are both cause and consequence. In contrast, cytoplasmic signaling cascades are linear, so causes and consequences are clearly distinct. These cascades can induce stable changes to the chromatin network, even if they are activated transiently (see Fig. 2.4).

Self-reinforcement is very common in chromatin. For instance, Su(var)3-9, the enzyme that methylates H3K9 in *Drosophila* [40] is targeted via HP1 to regions where H3K9 is methylated [41]. Is the binding of HP1 a cause or a consequence of the methylation of H3K9? The answer is "both." More interesting is to understand how the HP1 loop is broken or formed at a particular locus. Similarly, mixed-lineage leukemia protein 1 MLL (also known as KMT2A, for lysine methyltransferase 2A), methylates H3K4 and directly binds to H3K4me3 with high specificity [42]. H3K4me3 also recruits the transcription factor 2D complex (TFIID) [43], which itself interacts with MLL [44] and facilitates the methylation of H3K4. Once again, causes and consequences are indistinguishable because every event reinforces the others.

A signaling pathway is a straight line, but chromatin is a circle (Fig. 2.3). Hierarchic or static models are nonstarters to understand how a chromatin type is set, maintained or changed. In order to understand chromatin dynamics, the models must incorporate time. Taking time into account can cast a different light on the processes at work and pinpoint conceptual flaws in our understanding.

CHROMATIN STATES ARE ATTRACTORS

The natural framework to incorporate time is the theory of dynamical systems. This approach has proved essential to tackle countless problems in natural sciences. In addition, bifurcation theory [45] gives a unified framework to understand the emergence of phenomena at a macroscopic scale. Finally, the formalism of dynamical systems allows us to rigorously model genotype–phenotype relationships [46,47].

A dynamical model consists of an array of equations describing the forces at work on a system and the evolution of some physical variables of interest. Equally important is the initial state of the system that usually has a decisive influence on the long-term dynamics. In the context of chromatin, the variables of interest are the distributions of chromatin components, and the equations describe their influence on each other. These equations derive from the laws of physics, from the genotype of the organism and from its environment. A particular mutation may slow down the production of a chromatin component, reduce the affinity between two proteins, etc., so different genomes correspond to systems with slightly different equations. Similarly, many parameters of the equations are environmental variables susceptible to change in ways that affect the dynamics of the system. But the fate of the organism is not necessarily sealed by its genome and its environment: Initial conditions and random noise also contribute to its trajectory.

Let us illustrate with an example how this approach differs from the static models of chromatin. Assume that two chromatin proteins A and B bind in synergy to some locus. The more A binds, the more it recruits B and vice versa. For instance, one may think of A and B as HP1 and Su(var)3-9. Let us also assume that the locus has a maximum number of binding sites for both A and B, so that concentrations are interpreted as fraction of occupied binding sites. We can write the following equations:

$$\begin{cases}[\dot{A}] = (1-[A])[B]^2 - [A]/10 \\ [\dot{B}] = (1-[B])[A]^2 - [B]/10\end{cases} \tag{2.1}$$

The first equation says that A is recruited to free binding sites by dimers of B, and that it spontaneously dissociates from the locus at a constant rate with scaling coefficient 1/10. The second equation describes a symmetric situation for the recruitment of B. The equations were chosen only for illustration purposes, we do not claim that this model accurately describes a real-world locus. What matters are the emerging properties of the system, which are shared among many dynamical systems of this type [47], and which we will describe now.

By solving the equation $[\dot{A}] = 0$, we obtain a set of points where the total binding of A does not change (Fig. 2.4). For these concentrations of A and B at the locus, the amount of A that enters the locus is equal to the amount that leaves. Symmetrically, by solving the equation $[\dot{B}] = 0$, we obtain another curve, which intersects the first at three points. Those points represent

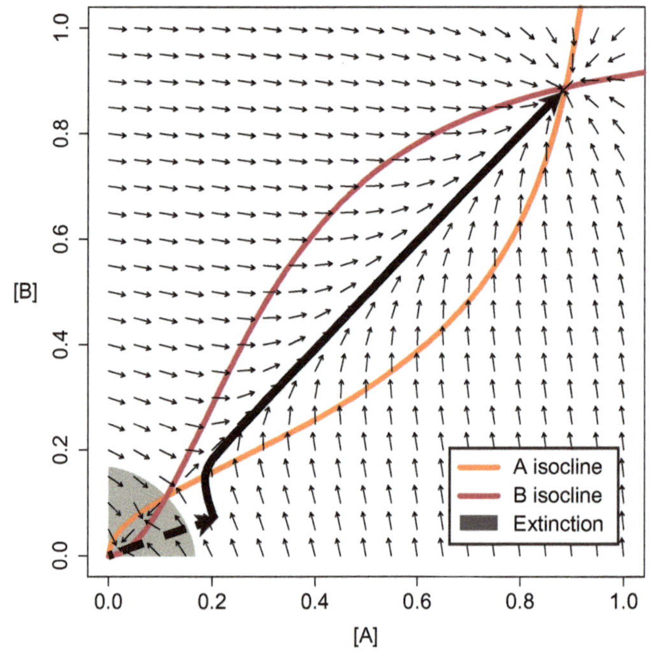

FIGURE 2.4 Phase portrait.

For every occupancy of A and B, Eq. (2.1) is used to compute the derivative. The small arrows represent this value as an infinitesimal change in this plane. The salmon and pink curves represent points of the space where the arrows are either vertical or horizontal, respectively, i.e., the occupancy of either A or B does not change. The curves intersect on three points, where occupancies of A and B are stable. From the flow of the arrows, it is clear that only two of the three points are attractors: one at the bottom-left, the other one at the top-right. The gray area is the basin of attraction of the bottom-left attractor. If the initial occupancies are in this region, both proteins will disappear and the occupancy at equilibrium will approach 0%. If a perturbation, for instance coming from signaling cascades, moves the occupancy of the proteins outside this basin (dotted large arrow), the occupancy of both proteins will reach over 80% (large arrow).

occupancies such that the total binding of A and B does not change and are called steady states.

Two of the three steady states are attractors. This means that if we wait long enough, the locus will be in one of two possible states. In the first, at the bottom-left corner, A and B are both absent. In the second, close to the top-right corner, A and B both bind over 80% of their binding sites. Attractors possess mutually exclusive basins of attraction. If the initial condition is in the basin of attraction of the first attractor, the occupancy will inevitably evolve toward the first attractor, and vice versa. This also holds if an external force modifies the locus so that the state switches basin of attraction: The occupancies will evolve toward the other attractor. For instance, if the locus is initially at 0% occupancy, it is sufficient that A and B transiently occupy slightly over 10% of their binding sites to switch regime and eventually bind more than 80% of them. What causes the transient binding of A and B does not matter. Also note that there are many possibilities to leave the basin of attraction of the first attractor, but all of them have the same final outcome.

This example illustrates the following points: (1) Only a few combinations of occupancies are possible, those are the attractors. Each attractor is a potential chromatin state. (2) Stable changes of chromatin states can be induced by transient events. (3) A transition is a change of fate in the sense that the outcome is the same irrespective of the sequence of molecular events. The last point is the most important for this discussion. Dissecting the sequence of events during a change of chromatin state may give less "mechanistic insight" than one might expect with the hierarchical, linear model in mind. To understand the epigenome, it would be more useful to fathom how the relationships between chromatin components predetermine the outcome of future perturbations.

The idea that chromatin is a dynamic equilibrium is gaining momentum. For instance it has recently been shown that silencing is caused by pervasive and dynamic methylation of H3K27 [48]. Only models taking time into account can capture the properties of such systems [49,50]. A lot more could be said about the benefits of dynamical systems to model chromatin. On the contrary, dynamical systems scale poorly with the amount of variables. A single parameter can add or remove an attractor, with huge consequences. For instance, the factor 1/10 in Eq. (2.1) shapes the phase portrait shown in Fig. 2.4. If the value were 1/3, the curves would not cross and the only possible long-term outcome would be 0% occupancy. Small errors on the estimation of the parameters can make the model useless, and with an increasing number of parameters, the risk of serious flaws increases too. In practice, such models are restricted to a handful of chromatin components.

We will continue to learn from "big data," but our opinion is that modeling chromatin at a much smaller scale is also necessary. The initial models based on unsupervised learning were very useful to update a euchromatin–heterochromatin classification that was no longer accurate. They have also been useful to map functional elements, such as enhancers, but they have not told us how chromatin works because they are fundamentally descriptive. In order to really understand the workings of chromatin, we need functional models, translating hypotheses into testable predictions. The success of data-driven methods should not downsize the importance of hypothesis-driven approaches for modeling chromatin states.

References

[1] Heitz E. Das Heterochromatin der Moose. 1928.

[2] Jenuwein T, Allis CD. Translating the histone code. Science 2001;293(5532):1074–80.

[3] Volpe TA, Kidner C, Hall IM, Teng G, Grewal SIS, Martienssen RA. Regulation of heterochromatic silencing and histone H3 lysine-9 methylation by RNAi. Science 2002;297(5588):1833–7.

[4] Hastie T, Tibshirani R, Friedman J. The elements of statistical learning: data mining, inference, and prediction. New York: Springer Science & Business Media; 2013. 536 p.

[5] Filion GJ, van Bemmel JG, Ulrich B, Wendy T, Jop K, Ward LD, et al. Systematic protein location mapping reveals five principal chromatin types in *Drosophila* cells. Cell 2010;143(2):212–24.

[6] Ernst J, Kellis M. Discovery and characterization of chromatin states for systematic annotation of the human genome. Nat Biotechnol 2010;28(8):817–25.

[7] Jain AK, Murty MN, Flynn PJ. Data clustering: a review. ACM Comput Surv 1999;31(3):264–323.

[8] van Bemmel JG, Filion GJ, Rosado A, Talhout W, de Haas M, et al. A network model of the molecular organization of chromatin in *Drosophila*. Mol Cell 2013;49(4):759–71.

[9] Hoffman MM, Buske OJ, Wang J, Weng Z, Bilmes JA, Noble WS. Unsupervised pattern discovery in human chromatin structure through genomic segmentation. Nat Methods 2012;9(5):473–6.

[10] Juan D, Perner J, Carillo de Santa Pau E, Marsili S, Ochoa D, et al. Epigenomic co-localization and co-evolution reveal a key role for 5hmC as a communication hub in the chromatin network of ESCs [Internet]. 2014. Available from: http://dx.doi.org/10.1101/008821.

[11] Lundberg SM, Tu WB, Raught B, Penn LZ, Hoffman MM, Lee S-I. ChromNet: learning the human chromatin network from all ENCODE ChIP-seq data. Genome Biol 2016;17(1):82.

[12] Zhou J, Jian Z, Troyanskaya OG. Global quantitative modeling of chromatin factor interactions. PLoS Comput Biol 2014;10(3):e1003525.

[13] Lasserre J, Chung H-R, Vingron M. Finding associations among histone modifications using sparse partial correlation networks. PLoS Comput Biol 2013;9(9):e1003168.

[14] Thévenin A, Ein-Dor L, Ozery-Flato M, Shamir R. Functional gene groups are concentrated within chromosomes, among chromosomes and in the nuclear space of the human genome. Nucleic Acids Res 2014;42(15):9854–61.

[15] Hurst LD, Pál C, Lercher MJ. The evolutionary dynamics of eukaryotic gene order. Nat Rev Genet 2004;5(4):299–310.

[16] Kharchenko PV, Alekseyenko AA, Schwartz YB, Minoda A, Riddle NC, Ernst J, et al. Comprehensive analysis of the chromatin landscape in *Drosophila melanogaster*. Nature 2011;471(7339):480–5.

[17] ENCODE Project Consortium. An integrated encyclopedia of DNA elements in the human genome. Nature 2012;489(7414):57–74.

[18] Saksouk N, Simboek E, Dejardin J. Constitutive heterochromatin formation and transcription in mammals. Epigenetics Chromatin 2015;8(1):3.

[19] Creighton HB, McClintock B. A correlation of cytological and genetical crossing-over in *Zea mays*. Proc Natl Acad Sci U S A 1931;17(8):492–7.

[20] Elgin SCR, Reuter G. Position-effect variegation, heterochromatin formation, and gene silencing in *Drosophila*. Cold Spring Harb Perspect Biol 2013;5(8):a017780.

[21] Gerstein MB, Lu ZJ, Van Nostrand EL, Cheng C, Arshinoff BI, Liu T, et al. Integrative analysis of the *Caenorhabditis elegans* genome by the modENCODE project. Science 2010;330(6012):1775–87.

[22] modENCODE Consortium, Roy S, Ernst J, Kharchenko PV, Kheradpour P, Negre N, et al. Identification of functional elements and regulatory circuits by *Drosophila* modENCODE. Science 2010;330(6012):1787–97.

[23] Lewis EB. A gene complex controlling segmentation in *Drosophila*. Nature 1978;276(5688): 565–70.

[24] Schwartz YB, Pirrotta V. Polycomb silencing mechanisms and the management of genomic programmes. Nat Rev Genet 2007;8(1):9–22.

[25] Hoffman MM, Ernst J, Wilder SP, Kundaje A, Harris RS, Libbrecht M, et al. Integrative annotation of chromatin elements from ENCODE data. Nucleic Acids Res 2013;41(2):827–41.

[26] Lenhard B, Sandelin A, Carninci P. Metazoan promoters: emerging characteristics and insights into transcriptional regulation. Nat Rev Genet 2012;13(4):233–45.

[27] Hoskins RA, Landolin JM, Brown JB, Sandler JE, Takahashi H, Lassmann T, et al. Genome-wide analysis of promoter architecture in *Drosophila melanogaster*. Genome Res 2011;21(2):182–92.

[28] Carninci P, Sandelin A, Lenhard B, Katayama S, Shimokawa K, Ponjavic J, et al. Genome-wide analysis of mammalian promoter architecture and evolution. Nat Genet 2006;38(6): 626–35.

[29] Akalin A, Altuna A, David F, Erik A, Xianjun D, Jan B, et al. Transcriptional features of genomic regulatory blocks. Genome Biol 2009;10(4):R38.

[30] Rach EA, Winter DR, Benjamin AM, Corcoran DL, Ting N, Jun Z, et al. Transcription initiation patterns indicate divergent strategies for gene regulation at the chromatin level. PLoS Genet 2011;7(1):e1001274.

[31] Moorman C, Sun LV, Wang J, de Wit E, Talhout W, Ward LD, et al. Hotspots of transcription factor colocalization in the genome of *Drosophila melanogaster*. Proc Natl Acad Sci U S A 2006;103(32):12027–32.

[32] MacArthur S, Stewart M, Xiao-Yong L, Jingyi L, Brown JB, Chu HC, et al. Developmental roles of 21 *Drosophila* transcription factors are determined by quantitative differences in binding to an overlapping set of thousands of genomic regions. Genome Biol 2009;10(7):R80.

[33] Foley JW, Sidow A. Transcription-factor occupancy at HOT regions quantitatively predicts RNA polymerase recruitment in five human cell lines. BMC Genomics 2013;14:720.

[34] Park D, Lee Y, Bhupindersingh G, Iyer VR. Widespread misinterpretable ChIP-seq bias in yeast. PLoS One 2013;8(12):e83506.

[35] Teytelman L, Thurtle DM, Rine J, van Oudenaarden A. Highly expressed loci are vulnerable to misleading ChIP localization of multiple unrelated proteins. Proc Natl Acad Sci U S A 2013;110(46):18602–7.

[36] Kvon EZ, Stampfel G, Yáñez-Cuna JO, Dickson BJ, Stark A. HOT regions function as patterned developmental enhancers and have a distinct cis-regulatory signature. Genes Dev 2012;26(9):908–13.

[37] Slattery M, Ma L, Spokony RF, Arthur RK, Kheradpour P, Kundaje A, et al. Diverse patterns of genomic targeting by transcriptional regulators in *Drosophila melanogaster*. Genome Res 2014;24(7):1224–35.

[38] Chen RA- J, Stempor P, Down TA, Zeiser E, Feuer SK, Ahringer J. Extreme HOT regions are CpG-dense promoters in *C. elegans* and humans. Genome Res 2014;24(7):1138–46.

[39] Revzin A. The biology of nonspecific DNA protein interactions. Boca Raton, FL: CRC Press; 1990. 272 p.

[40] Rea S, Eisenhaber F, O'Carroll D, Strahl BD, Sun ZW, Schmid M, et al. Regulation of chromatin structure by site-specific histone H3 methyltransferases. Nature 2000;406(6796):593–9.

[41] Audergon PN, Catania S, Kagansky A, Tong P, Shukla M, Pidoux AL, et al. Epigenetics. Restricted epigenetic inheritance of H3K9 methylation. Science 2015;348(6230):132–5.

[42] Chang P-Y, Hom RA, Musselman CA, Zhu L, Kuo A, Gozani O, et al. Binding of the MLL PHD3 finger to histone H3K4me3 is required for MLL-dependent gene transcription. J Mol Biol 2010;400(2):137–44.

[43] Vermeulen M, Mulder KW, Denissov S, Pijnappel WW, van Schaik FM, Varier RA, et al. Selective anchoring of TFIID to nucleosomes by trimethylation of histone H3 lysine 4. Cell 2007;131(1):58–69.

[44] Choi Y, Asada S, Uesugi M. Divergent hTAFII31-binding motifs hidden in activation domains. J Biol Chem 2000;275(21):15912–16.

[45] Crawford JD. Introduction to bifurcation theory. Rev Mod Phys 1991;63(4):991–1037.

[46] Huang S. Reprogramming cell fates: reconciling rarity with robustness. Bioessays 2009;31(5):546–60.

[47] Cortini R, Barbi M, Caré BR, Lavelle C, Lesne A, et al. The physics of epigenetics. Rev Mod Phys [Internet] 2016;88(2). Available from: http://dx.doi.org/10.1103/revmodphys.88.025002.

[48] Lee H-G, Kahn TG, Simcox A, Schwartz YB, Pirrotta V. Genome-wide activities of Polycomb complexes control pervasive transcription. Genome Res 2015;25(8):1170–81.

[49] Dodd IB, Micheelsen MA, Sneppen K, Thon G. Theoretical analysis of epigenetic cell memory by nucleosome modification. Cell 2007;129(4):813–22.

[50] Angel A, Song J, Dean C, Howard M. A Polycomb-based switch underlying quantitative epigenetic memory. Nature 2011;476(7358):105–8.

GLOSSARY

Akaike's Information Criterion criterion of model-selection based on information theory and aiming to obtain an optimal balance between predictive power and complexity.

Attractor state of a dynamical system toward which trajectories tend to evolve. Attractors are by definition points of equilibrium.

Basin of attraction ensemble of states of a dynamical system that evolve toward a given attractor. By definition a basin of attraction can contain only one attractor.

Chromatin the combination of the DNA fiber and associated proteins with special emphasis on histone proteins and their modifications.

Chromatin state recurrent combination of proteins and histone modifications covering contiguous DNA segments.

ChIP-seq technique using antibodies against a protein or histone mark of interest to map the genomic loci where it is enriched.

Dynamical model mathematical model in which time is a measurable variable. Such models are often specified through differential equations derived from the assumptions of the analyst.

DamID technique using protein fusion with Dam (DNA Adenine methyltransferase) to map the genomic loci where it is enriched. DamID and ChIP have the same purpose, but different experimental principles.

Hidden Markov model mathematical model in which some observables depend on a hidden state. Transitions between the states are dictated by the Markov property, i.e., they depend only on the current state.

Housekeeping gene here we use the term housekeeping gene to refer to the genes that are transcribed in the majority of tissues and time points.

Regulated gene here we use regulated to refer to those genes that exhibit a restrained expression in time and/or tissue.

Search space here search space means the ensemble of models that are compatible with the assumptions of the analyst. Fitted models are always in the search space by definition.

Unsupervised learning branch of machine learning that intends to reveal similarities and differences in the data in order to highlight some underlying organization and structure.

LIST OF ACRONYMS AND ABBREVIATIONS

AIC Akaike's Information Criterion
ChIP-seq chromatin immunoprecipitation and sequencing
H3K4me1/2/3 mono/di/tri-methylated histone H3 lysine 4
H3K9me1/2/3 mono/di/tri-methylated histone H3 lysine 9
H3K27me1/2/3 mono/di/tri-methylated histone H3 lysine 27
H3K36me1/2/3 mono/di/tri-methylated histone H3 lysine 36
HMM hidden Markov model
HOT highly occupied target
HP1 heterochromatin protein 1
MLL mixed-lineage leukemia protein 1
PRC1 polycomb repressive complex 1
PRC2 polycomb repressive complex 2
TFIID transcription factor 2D complex

CHAPTER 3

Crossing Borders: Modeling Approaches to Understand Chromatin Domains and Their Boundaries

Benjamin L. Moore and Colin A. Semple
MRC Human Genetics Unit, University of Edinburgh, Edinburgh, United Kingdom

CONTENTS

INTRODUCTION

In spite of their central importance to critical cellular processes such as transcription and recombination, there is still much to learn about the substance and functions of higher order chromatin structure and nuclear organization. Here we explore some of the best understood strata in the structural hierarchy, to address a series of questions. What are the relationships between different levels of chromatin structure? How do chromatin domains at different scales relate to one another? And what governs the formation of domain boundaries? Overall, many agree that the aim is a "coherent picture that integrates linear and three-dimensional views of chromosome organization" [1], but how do we get there? At the time of his death the physicist Richard Feynman's blackboard reportedly contained the phrase: "What I cannot create, I do not understand." It is becoming clear that modeling approaches have much to offer in the area of chromosome structure, and that by accurately recreating known characteristics of chromatin within theoretical models we can learn guiding structural principles and probe underlying systems.

Chromosome conformation capture (3C), introduced by Dekker et al. [2], was the first sequencing-based method capable of assaying nuclear architecture, using formaldehyde cross-linking of nuclear proteins to trap genomic regions that were physically colocated for sequencing. The end result is a relative measure of interaction frequency between any two regions of interest, in theory directly proportional to their distance in three-dimensional space. The evolution of the 3C method eventually resulted in an assay capable of assaying pairwise interaction frequencies between thousands of sites across an entire genome. Such an assay was published by Lieberman-Aiden et al. and named Hi-C [3]. With the falling costs of sequencing subsequent Hi-C papers have increased their sequencing depth, culminating to the point at

Epigenetics and Systems Biology. DOI: http://dx.doi.org/10.1016/B978-0-12-803075-2.00003-9

which analyses are starting to be performed at the level of individual restriction fragments, genome-wide [4,5]. The interaction maps produced by Hi-C were found to contain several inherent biases. Restriction fragment properties, such as their length, GC content and mappability, were confounding interaction frequency estimates and therefore required normalization before subsequent analysis [6,7]. Statistical techniques were developed to correct for these latent variables [8,9], whereas experimentalists have looked to improve on the experimental procedure itself, developing new variants of the assay.

Lieberman-Aiden et al. [3] described low-resolution structures they named "A" and "B" nuclear compartments. These are genomic regions with a median size of around 5 megabases that showed properties typical of euchromatin and heterochromatin, respectively. Regions within the A compartment were observed via 3D-FISH (microscopy aided by fluorescent probes) to be centrally positioned in the nucleus and ChIP-seq data showed several hallmarks of transcriptional activity. B compartments conversely contained heterochromatic and often lamina-associated regions, with little transcription and repressive histone modifications such as H3K9me3 (tri-methylation of lysine residue 9 on histone H3) [3]. These A and B compartments were identified through a continuous principle component (PC) eigenvector profile, derived from normalized Hi-C contact matrices [3]. This approach can be intuitively understood as formulated by Lajoie et al. [10]:

1. A tartan pattern on normalized Hi-C matrices indicates two preferentially contacting and exclusive compartments (Fig. 3.1).
2. Assume a function (c) that maps a given genomic bin to its compartment, using a positive number for compartment A and a negative for compartment B.
3. The interaction frequency between bins i and j is thus c(i)c(j). (Note that this rule alone is sufficient to generate a tartan pattern: if i and j are in the same compartment, the product will be positive, and for compartments of opposing type, negative.)
4. Our symmetric Hi-C matrix thus contains c(i)c(j) and in this formulation, PCs analysis finds the basis for minimizing the mean-squared error between c(i)c(j) and c(i).

Importantly, this measure holds more information than a simple two-state classification, rather the continuous values can be interpreted as relative levels of compartment identity, and hence degrees of compaction or activity [9].

The falling cost of high-throughput sequencing has enabled increasingly deep sequencing of Hi-C experiments. Sequencing depth is the main resolution-limiting resource for this assay. To increase the analysis resolution while maintaining the same level of coverage requires an exponential increase in the total amount of sequencing [3]. Nevertheless such deep sequencing was achieved in a handful of landmark studies. In experiments totaling around two billion total sequencing reads, Dixon et al. [4] produced Hi-C contact maps in human and mouse cell lines at 40 kb resolution. At around the same time, Nora et al. [11] published an even higher-resolution 5C dataset covering a 4.5 Mb region of the mouse X chromosome. In both of these studies, the authors uncovered what are now known as "topologically associating domains" or TADs, observable as off-diagonal blocks in a contact map which exhibit higher than expected self-interaction frequency. With a mean size of around 1 Mb, TADs were recognized as a novel layer of higher order chromatin organization at a level below the larger A/B compartment domains. TADs have since been reported in a variety of metazoan organisms from dog [12] to *Drosophila* [13].

Dixon et al. [4] defined a TAD calling algorithm based on the directional bias of contacts within a genomic region, and used a hidden Markov model (HMM) to infer strongly up- or downstream-biased regions, reasoning that domain boundaries are present when a strongly upstream biased region is adjacent to a region of opposite bias (Fig. 3.1). These boundaries themselves were investigated and found to show suggestive functional enrichments for DNA binding proteins including CTCF (CCCTC-binding factor) a versatile transcription regulator, long thought to act as an insulator of chromatin state. Deletion of a CTCF site has since been found to disrupt the corresponding TAD border, while removal of some other enriched factors had little effect [11,14]. Dixon et al. [4] also performed comparative analysis, reporting large and significant overlap of domain boundary positions both between cell types and between human and mouse.

Since then, several studies have investigated the functional implications of TADs. A proposed biological explanation is that TADs delimit functional contacts, such as those between enhancers and promoters, and so could inhibit spurious contacts with other nearby genetic elements [15], and perhaps even constitute regulons (groups of genes coregulated together). However, the size of TADs means they often span multiple genes, commonly with unrelated functions, so it seems unlikely they function as regulons in general [16]. The functional significance of TADs is the subject of ongoing debate within the chromatin biology field.

THE EXPANDING UNIVERSE OF STRUCTURAL DOMAINS

Since the discovery of TADs, multiple publications have proposed either complementary or related classes of chromatin domains. For example, Filippova et al. [17] developed a tuneable algorithm which identifies "alternative topological domains." The authors use dynamic programming to search

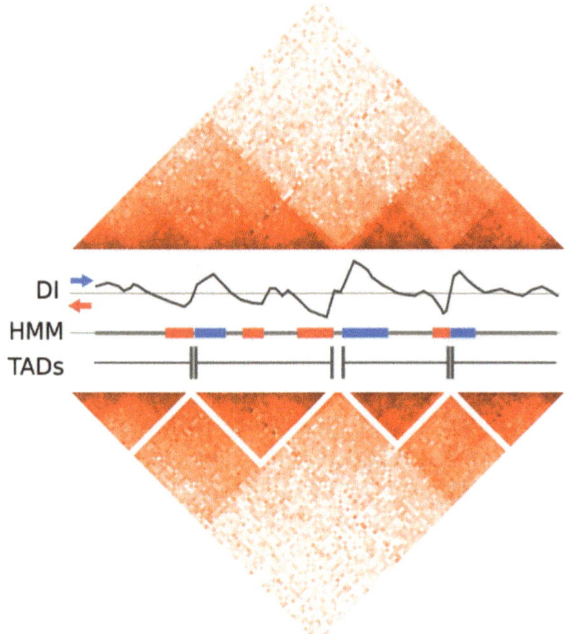

FIGURE 3.1
The original analysis pipeline for calling topologically associating domains. The directionality index (DI) is calculated for each bin based on the ratio of upstream/downstream contacts [4]. A hidden Markov model (HMM) is then used to infer the most likely state sequence that emitted the DI variable. Finally a simple rule is applied whereby a run of high-confidence upstream-biased state calls marks the end of a domain. New domains begin with any subsequent downstream-biased state. Gaps between TAD calls can be observed and these are regarded as border regions up to a size threshold of 400 kb, with larger regions regarded as unclassified.

for an optimal set of nonoverlapping boundary pairs that maximize intradomain contacts in mouse ESC and human fibroblast Hi-C data. The algorithm includes a length scaling factor that is used to penalize domain size; by varying this, the authors define a subset of domains showing heightened persistence across resolutions [17]. These multiscale domains were found to be smaller, on average, than those previously reported by Dixon et al. [4], even when applied to the same Hi-C experimental data (with a mean size of 200 kb as opposed to 1 Mb). However the domains of Filippova et al. [17] show increased intradomain contacts and stronger boundary enrichments relative to previously described TADs, indicating this algorithm may generate a more accurate representation of topological domains in mammalian genome organization. Intriguingly, this study also reports quantitative evidence for hierarchical genome organization, finding that smaller domains combine into larger meta-domains as the scaling factor penalty decreases [17]. A study of *Drosophila* embryonic cell chromosomes found a similarly hierarchical organization of physical domains, and also was able to relate these to epigenomic features (histone modifications and fine-scale chromatin accessibility), uncovering specific sets of enrichment signatures representing active, null, polycomb-associated, and telomeric domains [13]. This study provided evidence for the first time that contact domains are linked with average epigenomic enrichments. An observation since seen for a variety of hierarchical domain structures [18].

Recent high-resolution studies have been able to resolve ever finer levels of substructure. Rao et al. [5] refined the concept of chromosome compartments to "subcompartments," dividing simple A/B compartment divisions into a total of five subtypes. The authors were also able to identify "contact domains" of median size 185 kb, many of which were associated with identifiable individual looping events [5]. This domain size is close to those of Filippova et al. [17] and the authors suggest that previously observed large TADs may be the result of insufficient sequencing; that is, not all boundaries could be detected using 40 kb binned contact maps thus multiple contact domains were unintentionally combined into large domains. Additionally, these subcompartment types exhibit average epigenomic enrichments along the lines of those reported in *Drosophila* by Sexton et al. [13] and so potentially provide an overarching concept that connects the above-mentioned TADs, alternative domains and epigenomic features. In addition, evidence has also been presented for "meta-TADs," larger structural arrangements encompassing a number of TADs that seem to bridge the gap between TADs and the much larger compartment domains [18].

Examples have long been known of specific enhancer elements that are brought into close proximity with the promoter(s) they are regulating; under this model, these contacts form a "loop" structure between two potentially distal loci [19] (Fig. 3.2). It is now thought that the majority of active promoters are engaged with multiple, often cell type specific, regulatory looping events [20].

The CTCF protein [21] is a notable component of many long-range looping events, already mentioned as a component of TAD boundaries. CTCF is strongly conserved in higher eukaryotes, ubiquitously expressed and embryonic lethal when deleted with roles in regulating transcription, imprinting, dosage-compensation and acting as an insulator [21].

Rao et al. [5] identified around 10,000 individual looping events in the human genome, almost all linking loci over distances of less than 2 Mb, and around 30% connected predicted enhancers and promoters. Furthermore, 86% of these loops involved CTCF bound regions, with roughly the same overlapping proportion involving cohesin subunits RAD21 and SMC3. (Cohesin is a multiprotein complex originally studied for its role in sister

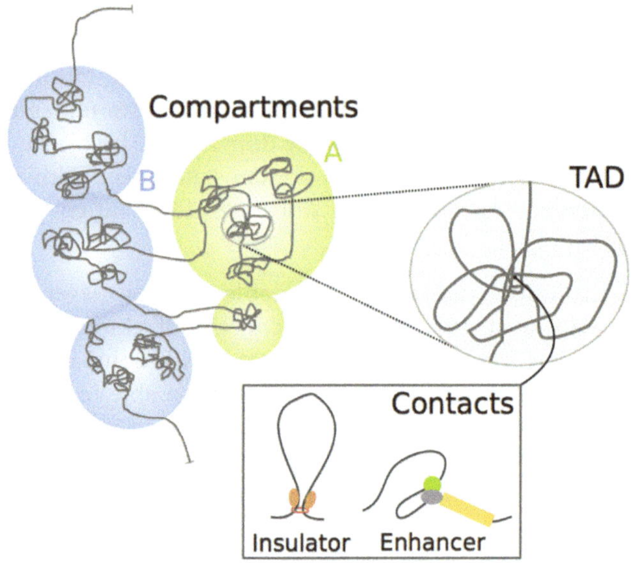

FIGURE 3.2

Hierarchical levels of higher order chromatin organization. Functional contacts, such as loops between bound CTCF insulators, occur within TADs which in turn are found within A or B compartments.

chromatid cohesion, but since discovered to affect chromatin structure in transcriptional regulation and DNA repair.) The authors thus propose that a CTCF-binding motif can act as an "anchor" that can then be bound by a transitive complex of other architectural proteins [5]. A majority of these loops (65%) also demarcated a topological domain. Another striking finding of this research was that loops almost always occur between bound CTCF motifs with a convergent orientation [5]. Although the evidence linking CTCF and genome architecture is substantial, it should be noted that from a global perspective as few as 15% of all CTCF ChIP-seq peaks were found to occur at TAD boundaries in human and mouse cells and around 25% of TAD borders had no observable CTCF binding [4]. These facts indicate that CTCF alone is neither necessary nor sufficient for the formation of higher order chromatin structures such as TADs. Indeed, the degree of insulation at a given genomic site has been reported to correlate with the degree of cobinding of a range of architectural proteins including not only CTCF but cohesin, condensing, and the transcription complex TFIIIC, among others [22,23].

EXPERIMENTAL TECHNIQUES

As we have seen, interphase chromosomes adopt a hierarchical structure, and recent studies have examined their chromatin organization at different scales, from subgenic regions associated with histone modifications and DNA-binding proteins, up to multimegabase-scale regulatory domains such as TADs and nuclear compartments. However, there is still a lack of detailed, quantitative approaches to integrate these different strata into unified models, discover prominent interactions between strata, and better understand the basis of genomic function. In this chapter we combine large volumes of matched subgenic level chromatin features and Hi-C datasets across three human cell types to construct unified, predictive models of chromatin structure. We are able to assess whether the available lower level features alone are sufficient to explain higher order chromatin structure and to identify the most influential features underlying higher order structures.

Raw Hi-C reads were downloaded for three widely used human cell lines through the Gene Expression Omnibus (GEO) [24] with identifiers: GSE35156 (H1 hESC; an ESC line), GSE18199 (K562; an erythroleukemia cell line), and SRX030113 (GM12878; a lymphoblastoid cell line derived from transformed white blood cells). These paired reads were mapped independently to the human reference genome (build hg19/GRCh37). Mapping was performed using the hiclib python library [8] and bowtie2 [25] with the very-sensitive flag. An iterative mapping approach was used to maximize the number of aligning fragments [8]. Each fragment end was aligned first using short terminal subsequences. Those unmapped or with ambiguous mapping were then taken forward into the next iteration and extended until the entire fragment end had been aligned. Those remaining pairs with one or more unmapped ends were discarded. This approach is designed to maximize uniquely alignable fragment ends, while avoiding mismappings from reads that cross a restriction fragment junction [10]. After mapping, interactions are first aggregated into restriction fragments then by regular binning at various resolutions (particularly 40 kb, 100 kb, and 1 Mb). Several filters were applied at this stage, with the following cases removed [10]: reads directly adjacent to a restriction enzyme site (within 5 bp), identical read pairs (presumed PCR duplicates), and very large restriction fragments (>100 kb) which are likely from a repetitive or poorly assembled region.

Iterative correction and eigenvector expansion (ICE) is an approach to normalization and processing Hi-C data and is implemented as part of the hiclib library written in python [8]. The iterative correction algorithm performs matrix balancing with the aim of generating a doubly stochastic matrix from raw interaction counts [10], that is, such that symmetric matrix A has both row and columns of equal sum. In practice, this effectively enforces "equal visibility" of each fragment, correcting for previously described biases in interaction recovery such as GC-content and fragment length [6], but without explicitly modeling these latent variables.

Additional functionality provided by ICE is the eigenvector expansion of normalized contact maps. Eigenvectors from observed/expected matrices were chosen for consistency with Lieberman-Aiden et al. [3] as opposed to the related eigenvectors calculated in Imakaev et al. [8] from the corrected maps alone. Briefly, observed contacts (O) are divided by an expected matrix (E), which is generated by averaging the super- and subdiagonals of the O matrix. That is, the E matrix gives the expected value of interactions at a given distance, hence the O/E matrix is a normalized contact map without the distance decay seen in raw Hi-C contact matrices. Examples of these maps are shown in Fig. 3.1.

Importantly, for the eigenvector expansion step the first two PCs were calculated, and that with the highest absolute Spearman correlation with GC content is taken to reflect A/B compartmentalization. PC eigenvectors were then orientated to positively correlate with GC, ensuring positive values reflected A compartments and negative values B compartments.

After uniformly reprocessing each Hi-C dataset and calling compartment eigenvector profiles we can compare these between three human cell lines. We find compartment profiles have a striking concordance (Fig. 3.3), despite the variable sources of both sample material and experimental data. This strong correlation of higher order chromatin structure (at the level of compartments) between three very different human cell types shows that the vast majority of genomic regions appear to be constitutively present in either A or B compartments regardless of cell lineage. This close correspondence also validates our uniform processing of these different datasets, and suggests our pipeline is successfully accounting for differences in sequencing depth and other batch effects. The pairwise genome-wide Pearson correlation coefficients between these independent measures are all between 0.75 and 0.80 and are highly statistically significant.

TADs were called using the software provided in Dixon et al. [4] and their recommended parameters. The TAD calling algorithm is a multistage process. Firstly, a statistic called the "directionality index" (DI) is calculated for each genomic region or bin [4]. The equation for calculating the DI of a given bin involves the variable U, which represents the sum of reads mapped up to 2 Mb upstream of a given 40 kb bin, and D likewise for downstream contacts. Here E is the expected number of downstream or upstream contacts (equal under the null hypothesis), hence $E = U + D/2$ [4].

Genome-wide ChIP-seq datasets for 22 DNA binding proteins and 10 histone modifications were generated by the ENCODE Consortium [26] along with DNase I hypersensitivity and H2A.Z occupancy, for each of the three best studied ENCODE cell lines: H1 hESC, K562, and GM12878 (described above). These data were preprocessed using MACSv2 [27] to produce signal

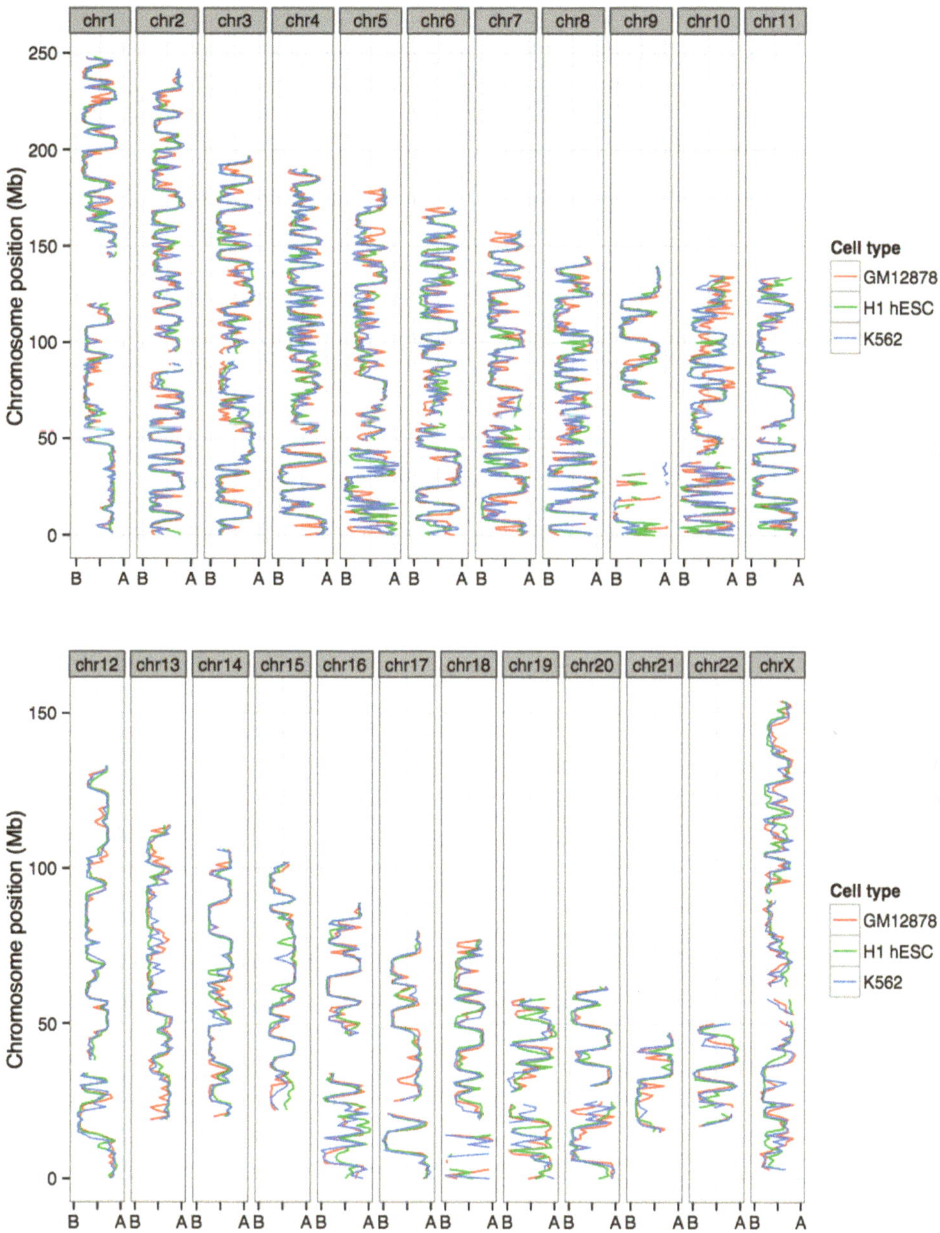

FIGURE 3.3

Uniform reanalysis of Hi-C datasets. Compartment profiles are observably well correlated between human cell types and across all chromosomes. Compartment eigenvectors are plotted along the lengths of each human chromosome (chrY and chrM are omitted). In each case the overlaid profiles show strong concordance between the three different human cell types under study.

fold-change relative to input chromatin for each bin. In most cases a paired input control was generated by the same laboratory for each ENCODE ChIP-seq experiment. GC content was also calculated over the same genomic intervals and used in the feature set to give 35 total inputs matched across the three cell types [23].

Data generated by ENCODE consortium members also has proven utility in modeling techniques based on machine learning. Notably ENCODE associated groups have released models which classify the human genome into discrete "chromatin states," such as actively transcribed regions or gene promoters [28] using HMMs. This work inspired our own machine learning framework applied in the context of higher order chromatin structure [23].

MODELING HIGHER ORDER CHROMATIN STRUCTURE

"Decision trees" are a long established approach in machine learning, where an object is classified according to whether it does or does not pass a test at each node of a tree like series of tests. Each object progresses across the tree to be finally classified as the sum of all test outcomes. The Random Forest (RF) approach extends this concept to an ensemble of decision trees, where each of a large number of trees processes a random subsample of the total dataset [29]. At each node in each tree a subset of predictor variables is randomly sampled, and the variable found to optimize the split at that node is used to make a binary split in the data [29]. At successive nodes this process is repeated to "grow" each tree, until we have a large number of trees making somewhat different predictions based upon somewhat different collections of variables. We can combine all predictions across trees via averaging or voting, to produce an ensemble prediction for the forest. Attractive features of the RF approach include the ability to capture nonlinear relationships between variables, and the capacity to estimate prediction error rates using the ("out-of-bag") data not used in training the model. It is also possible to straightforwardly assess the most influential variables in RF regression models. As shown below, the importance of a given variable can be measured using the mean decrease in accuracy seen between models based upon a randomly permuted version and the original versions of that variable.

RF regression [29] was used as implemented in the R package randomForest [30]. The package makes use of an ensemble of trees (of size ntrees), each built from a subsample of the training set. In growing each tree, a small number of variables (mtry) are tested at each bifurcation node, and that which minimizes the variance in child node subsets is selected as optimal. Having trained a group of trees, these can then be used as predictive tools by inputting a vector of features to each tree and averaging the output values across

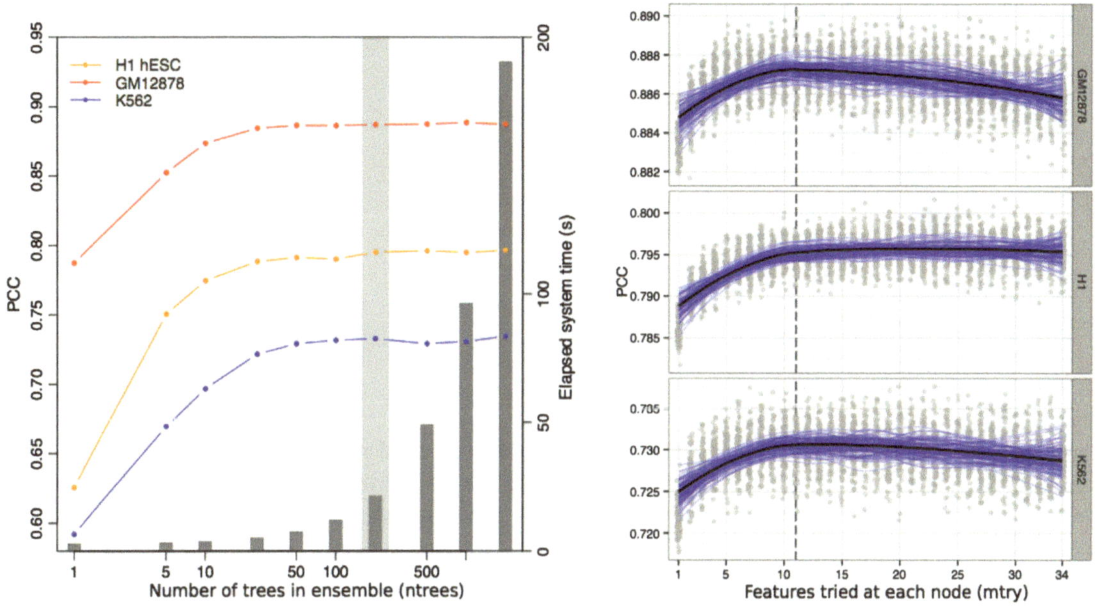

FIGURE 3.4

Random Forest parameters are largely insensitive. Modeling success, measured as the Pearson correlation coefficient (PCC) between observed and predicted eigenvector values, is insensitive to variation in two user- chosen Random Forest parameters over a broad range. Optimizations for ntrees (the number of trees in the forests) and mtry (the number of features tested at each node) are shown for three different models, with typical default values of 200 trees and a third of input variables highlighted. The left panel shows the model accuracy (PCC; red, yellow, and blue lines) for the three models built for the three cell types under study, over a variety of forest sizes (ntrees; gray histogram bars) and requiring varying time to compute (elapsed time; right *y*-axis). The right panel shows model accuracy (PCC) plotted over the numbers of predictor variables (features) used by the trees at each node. The vertical band of shading in the left panel and the vertical dashed line in the right panel represent the values of ntrees and mtry used in this study, respectively.

the forest. RF regression was used as it is known to be one of the most powerful regression methods developed to date, typically providing low levels of bias and low variance predictions without the need for variable selection [31]. Additionally the RF method is an example of "algorithmic modeling" [32] in that it makes no assumptions about the underlying data model. Of the few user chosen parameters, the number of features to test at each node (mtry) was set to n (where $3n$ is the number of input features) and the number of trees in the forest (ntrees) was chosen as 200. These parameters are thought to be insensitive over a broad range of values, and do indeed appear to be here, as shown in Fig. 3.4.

The success of the RF modeling approach used to predict compartment eigenvectors was measured by four different metrics. Prediction accuracy was assessed by the Pearson correlation coefficient between the out of bag

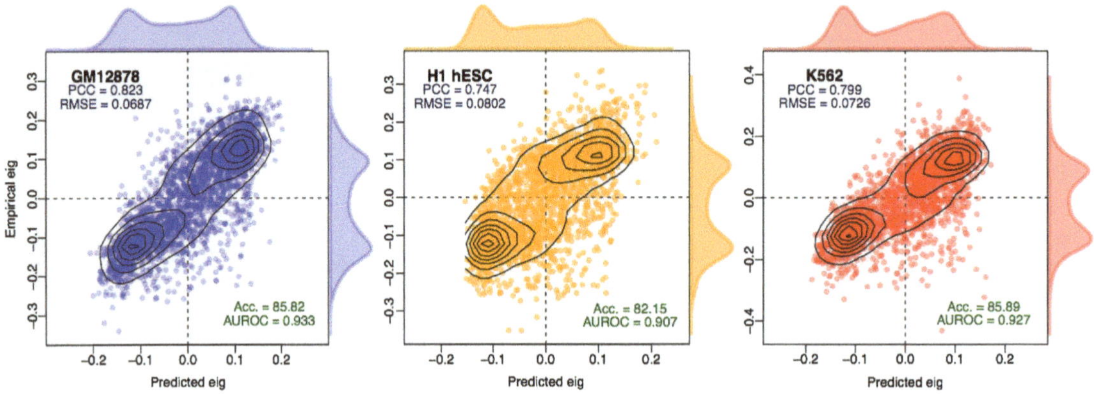

FIGURE 3.5

Compartment eigenvector model predictions are strongly correlated with observed values. Models constructed for all three cell types (GM12878, H1 hESC, K562) show strong and significant correlations between observed (*y*-axes) and predicted (*x*-axes) eigenvector values for all 1 Mb segments of the genome. The scatterplot density contours and marginal density plots (depicted outside the scatterplot frames) all show the characteristic clusters of positive (compartment A) and negative (compartment B) eigenvector values within observed and predicted datasets. In each plot the Pearson correlation coefficient (PCC) and root mean-squared error (RMSE) report the degree of success of the regression model, whereas accuracy (Acc.) and area under the receiver operating characteristic (AUROC) give the classification accuracy of binary outcomes.

predictions and observed eigenvectors, and the root mean-squared error of the same data. Classification error, when predictions where thresholded into nuclear compartments ($A > 0$; $B < 0$) was also calculated using accuracy (% correct classifications or True Positives) and area under the receiver operating characteristic curve. Together these give a comprehensive overview of the model performance, both in terms of regression accuracy of the continuous eigenvector, and in how that same model could be used to label discrete chromatin compartments (Fig. 3.5). Models can be further explored via cross-application of cell type specific models, a single RF regression model was learned from all 1 Mb bins for a given cell type. This was then used to predict compartment eigenvectors for all bins from each of the other two cell types [23].

Having built accurate predictive models, we next dissect the relative variable contributions made from our range of input features and compare these across cell types. An overview of the top 10 most highly ranked features in cell type specific models shows some agreement but also notable differences between cell types (Fig. 3.6). Only one input feature, the heterochromatin associated histone modification H3K9me3, is present in the top 10 most important variables of each model. H3K9me3 is one of the few features to be negatively correlated with compartment eigenvectors, hence offers orthogonal information to the majority of other, positively correlated input variables [23]. Of those important variables shared between two cell type

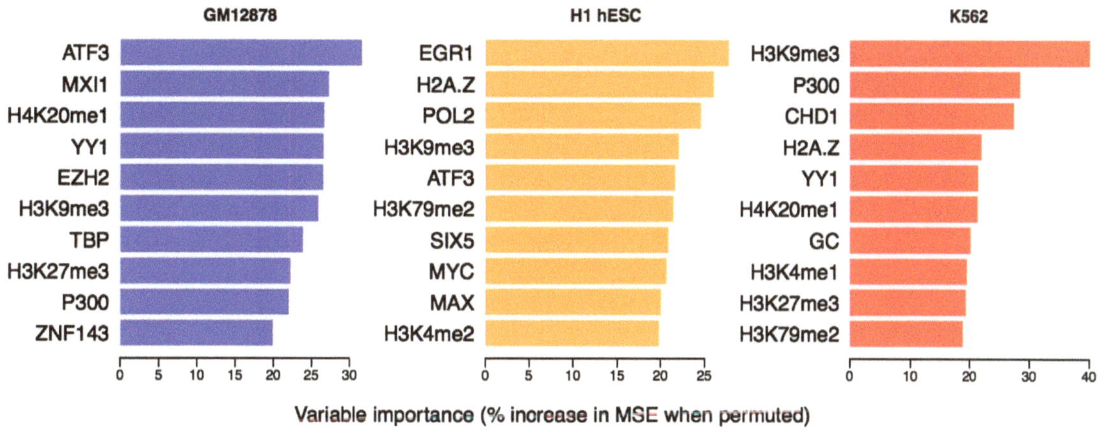

FIGURE 3.6

The most influential 10 variables for each cell type specific model. Variable importance was measured as the mean decline in the accuracy of the model [reflected in the percentage increase in mean squared error (MSE)] when based upon a randomly permuted version, relative to the original version of a given variable.

models, H3K27me3 is also a repressive histone modification deposited by polycomb repressive complex 2 (PRC2), whereas H2A.Z is a histone variant again linked to polycomb-regulated genes and essential for embryonic development. Furthermore EZH2, the catalytic subunit of PRC2, is also included in the feature set and is highly ranked in the GM12878 (lymphoblastoid) cell type model (Fig. 3.6). Other interrelated and important variables include the transcription factors MYC and MAX, which are found in the top 10 influential variables in H1 hESC, and the transcriptional repressor MXI1, found to be an informative variable in GM12878. Recent results suggest MYC binds open chromatin as a transcriptional amplifier in embryonic stem cells [33], with MAX and MXI1 acting as antagonistic coregulators. These biological relationships between variables may help explain the observed differences between models: different representatives of the same correlated clusters of input variables appear to have been selected in each model (this is explored in Moore et al. [23]).

FORMALIZING COMPARTMENT CALLING AND BOUNDARY PREDICTION

The continuous compartment eigenvector is most commonly used to classify A/B compartments by thresholding based on sign: Typically the eigenvector is orientated such that positive values reflect A compartments and negative values B compartments [3]. However, given that compartments are understood to be generally broad and alternating domains along a chromosome, often aligning with other large domains such as LADs (nuclear lamina-associated domains), an improved classification method might penalize the calls of very short compartment calls, which may be the result of noise. For this reason, instead of using raw eigenvector values we consider observed values as emissions from unobserved underlying states. We built a HMM to represent these states through a well-described probabilistic framework [34]. HMMs are widely used in bioinformatics to predict the probability that a particular unobserved feature (in our case a nuclear compartment domain) is present at some point in the genome, based on a variety of other observed features (in our case the varying eigenvector values as we move along a chromosome). We designed a HMM to have two unobserved states with univariate Gaussian distributed emissions (representing our A and B compartments; Fig. 3.7). To parameterize these states, we used the iterative Baum–Welch algorithm, an expectation maximization procedure (a method to find the maximum likelihood estimates of parameters in statistical models) designed for HMMs. Having parameterized the HMM for each cell type, we then use the Viterbi algorithm (an algorithm used to find the most likely sequence of hidden states resulting in a sequence of observed events) to infer the most likely

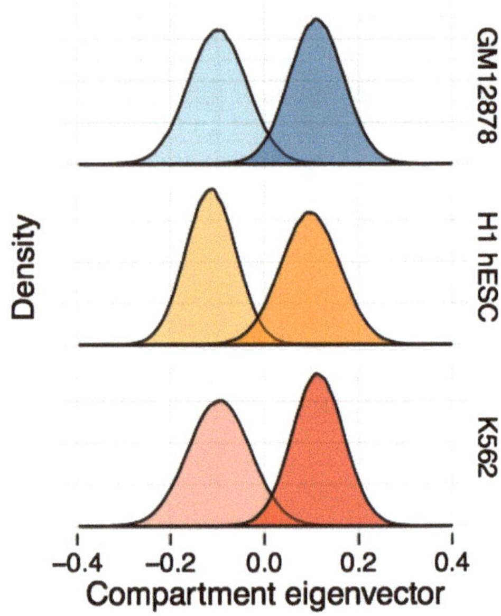

FIGURE 3.7

Modeling nuclear compartment domains. Probability distributions (*y*-axes) for compartment eigenvector values (*x*-axes) for the two HMM states modeled (i.e., compartments A and B) are shown for each cell type. Distributions centered below zero (lighter colors) show the distribution for the B compartment state, distributions centered above zero (darker) represent the A compartment state.

state sequence to have generated our observed data. This two-state sequence is then used to assign compartment identities to genomic regions.

A schematic of this procedure is shown in Fig. 3.8. In practice, this approach acts to de-noise our nuclear compartment calls, that is, whether we assign a section of a chromosome to the A or B nuclear compartment. Whereas a

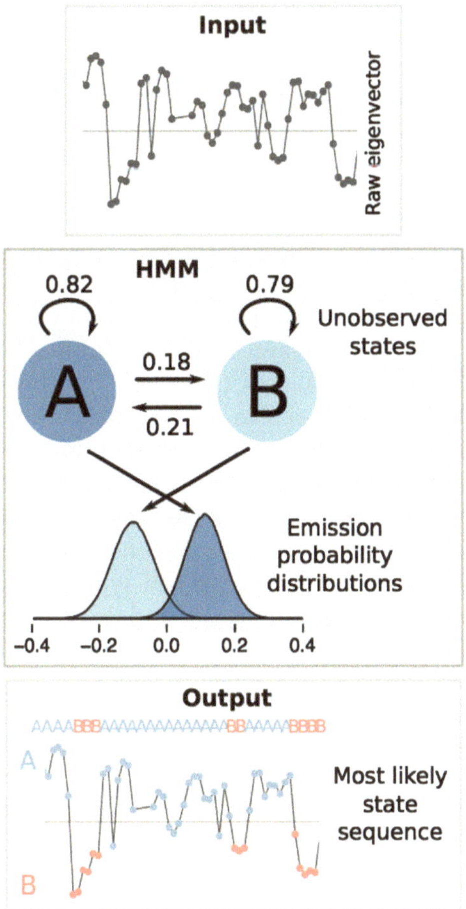

FIGURE 3.8

Overview of the HMM method to define nuclear compartment domains. Schematic of the HMM method for compartment calls, showing transition probabilities and emission distributions learned in the GM12878 cell type. The top panel shows the input data: the varying eigenvector values observed along a chromosome. The middle panel describes the two state HMM; each state has transition probabilities (the probabilities of changing state or maintaining the same state) shown as arrows, and each state has its own emission probability distribution which models the variability in eigenvector values within and between compartments. The lower panel shows the output A and B compartment calls made for successive regions of the input data by the HMM.

change from positive to negative (or vice versa) eigenvector values at a single point along the chromosome would (under a simple thresholding procedure) have resulted in a single-region compartment, these may now be modeled as noisy emissions from a single unobserved state. As an additional benefit of this approach, compartment domain boundaries were also identified: as the points at which transitions occurred between compartment states. Given the large multimegabase scale of compartment domains, and therefore the transitional regions between them, the transitional points identified were extended ±1.5 Mb to give a 3 Mb window corresponding to a boundary region. Smaller regions of the genome can also be assigned to TADs using an analogous approach: using a HMM to de-noise domains called from Hi-C interaction patterns [4]. Reimplementing this approach also provided a set of boundaries at the start and end of each TAD, providing a basis to compare boundary compositions between TADs and compartment domains.

Prior to our studies TAD boundary regions were reported to correspond with pronounced peaks and troughs of particular features within 500 kb of the predicted boundary [4]. Exploration of this phenomenon using a set of 24 mouse ESC chromatin features (and a smaller number of human ESC features) revealed enrichment peaks of CTCF, and the histone modifications H3K4me3 and H3K36me3, as well as a pronounced dip in H3K9me3, suggesting that high levels of transcription may contribute to boundary formation [4]. However, the peaks and dips of these features lacked any estimates of statistical significance. It was also unclear whether other features might show unusual patterns in TAD boundary regions, and how the constellation of features involved might vary between cell types. Moreover, the features associated with boundaries separating A and B compartments calculated from Hi-C eigenvectors had not been studied to our knowledge. The datasets assembled here, consisting of 35 matched chromatin features across three cell types, allowed us to conduct the first comparative study of the constituents of human TAD and compartment boundary regions.

We generated average TAD boundary enrichment or depletion profiles by averaging input features into 50 kb bins spanning 450 kb in either direction from the central TAD boundary bin (Fig. 3.9). Given their greater scale, compartment boundaries were examined in 100 kb bins over 3 Mb. Fig. 3.9 shows definitively that many different features appear to show an enriched peak at TAD boundaries genome-wide, but does not indicate the statistical significance of these patterns. To test the significance of the enrichment or depletion of a chromatin feature over a given boundary class, a two tailed Mann–Whitney test was used to compare the boundary bin with the 10 outermost bins of the window (5 from either side). Therefore we tested whether there was any significant difference in rankings of input feature signal between boundary bins and peripheral bins over all boundary instances

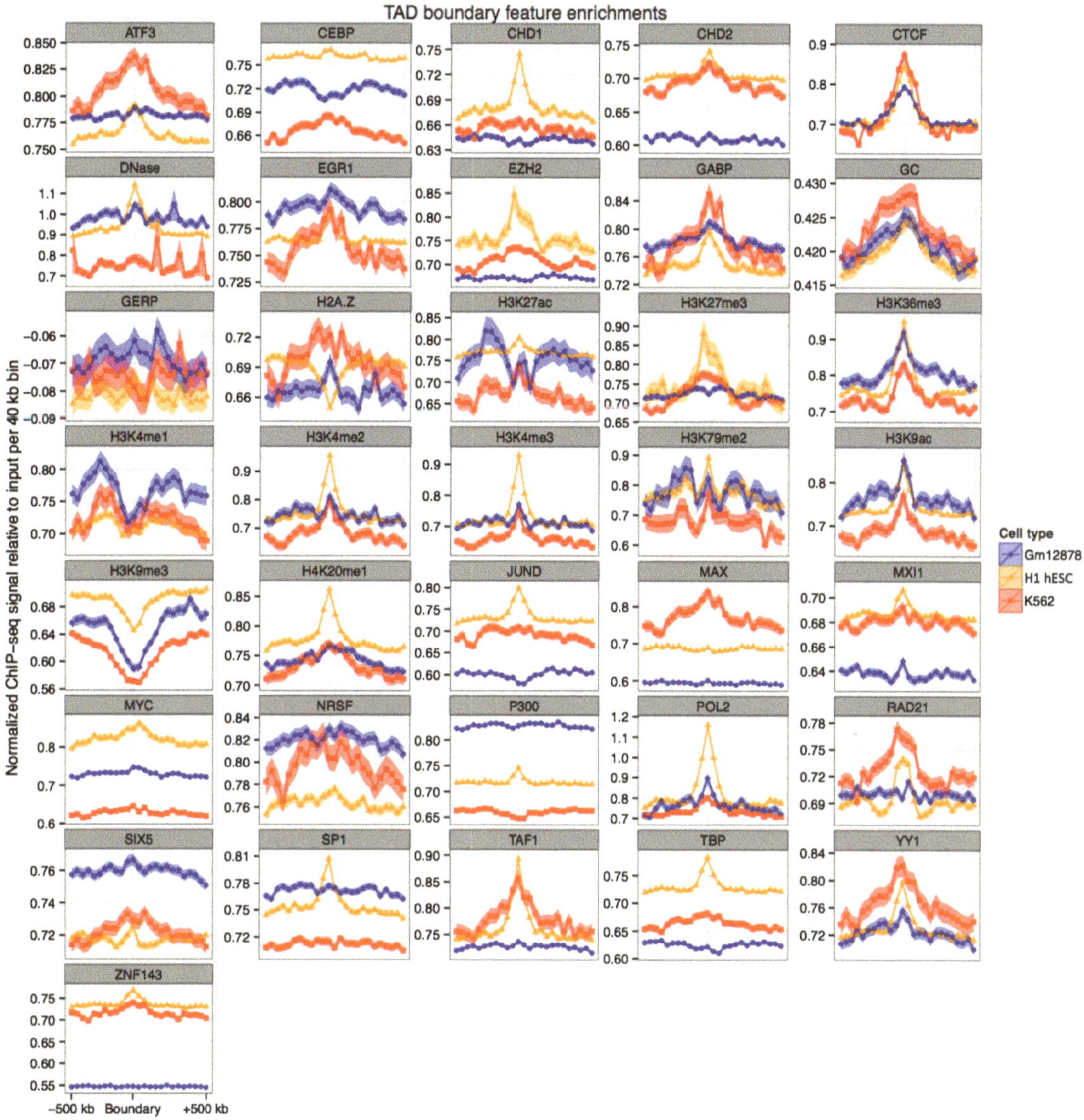

FIGURE 3.9

TAD boundary enrichments and depletions. A collection of 36 matched chromatin features were averaged over 1 Mb windows centered on TAD boundaries genome-wide (40 kb bins) for each cell type. Ribbons represent the 95% confidence intervals of the mean at each position.

per class. The significance level at $a = 0.01$ was then Bonferroni-adjusted for multiple testing correction, and results with P-values exceeding this threshold were deemed significantly enriched or depleted at a given boundary, as indicated in Fig. 3.10B.

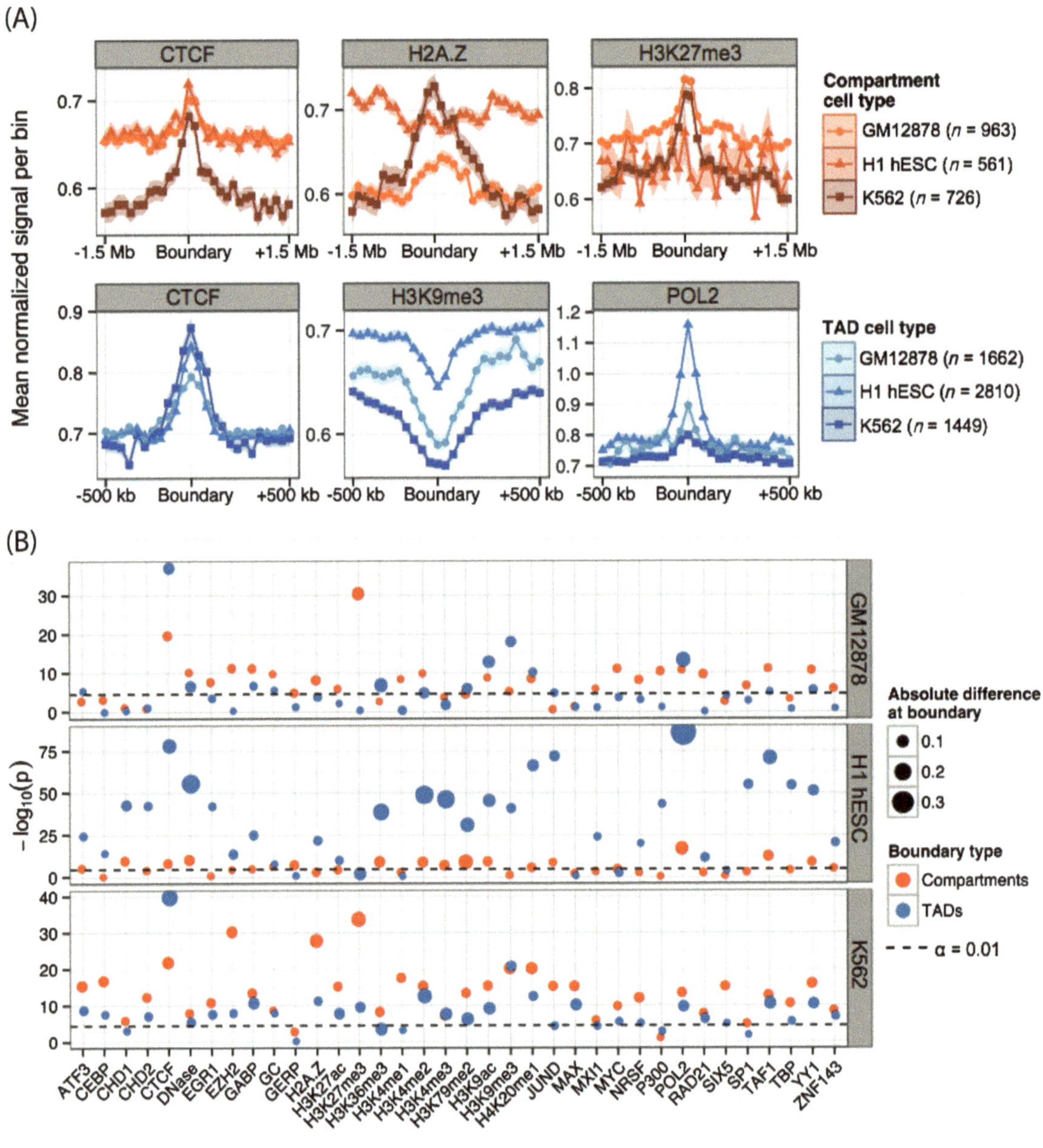

FIGURE 3.10

Compartment domain and TAD boundary composition in three human cell types. (A) Selected profiles for locus-level feature enrichments and depletions are shown for compartment boundaries (CTCF, H2A.Z, and H3K27me3) and TAD boundaries (CTCF, H3K9me3, and POL2), as a mean normalized ChIP-seq signal relative to input chromatin per bin (±1 standard error). TAD boundaries were examined over 40 kb bins over the 1 Mb flanking each boundary and compartment boundaries were examined in 100 kb bins over 3 Mb. (B) The significance of enrichment or depletion (*y*-axes; $\log_{10}(p)$ for two-tailed Mann–Whitney tests) of each feature was calculated for the boundary bin relative to the five flanking bins on either side for TAD (blue points) and compartment (red points) boundaries. Points are scaled by the absolute mean difference in signal over the boundary relative to the mean of peripheral bins: the larger the point the bigger the difference generating the enrichment/depletion. In each case the dashed line represents the multiple testing corrected threshold for statistical significance ($P < 0.01$).

Significant boundary enrichments for both CTCF and YY1 (Yin Yang 1; a DNA-binding protein which can act as a transcriptional repressor) are evident in all cell types at both compartment and TAD boundaries (Fig. 3.10), which is intriguing given the evidence that YY1 and CTCF cooperate to affect long distance interactions [35]. Cobinding of CTCF with YY1 has also been shown to identify a subset of highly conserved CTCF sites [36]. This colocation may also therefore be a contributing factor in the establishment of TAD boundaries, which appear to be broadly conserved across mammals [4]. Thus TAD and compartment domain boundaries are frequently and significantly marked by a variety of features, and compartment boundaries can be successfully predicted as a side effect of modeling the continuous compartment profile eigenvector.

OUTLOOK

In this work we use locus-level chromatin information to model higher order characteristics of nuclear architecture, such as chromosomal compartments and topological domains. Our data show that accurate predictions of Hi-C derived chromosome compartment eigenvectors using locus-level chromatin features alone are entirely achievable. Generalization across cell types further suggests that chromosome compartments could be inferred for those cell types without any available Hi-C data but with available ChIP-seq for a handful of chromatin features. In the same vein, chromosome compartments are known to be related to other aspects of higher order chromatin organization, including replication timing domains, nuclear LADs and nucleolus association domains (NADs) [37]. We therefore suggest a similar modeling approach could prove successful for these domains of interest. An exciting idea is that an integrative model capable of identifying these LADs and NADs could forward this information to a subsequent three-dimensional reconstruction algorithm, which could then use this information to generate a comprehensive, structurally rich perspective on nuclear architecture.

Another consideration is that we resolved TAD domains to 40 kb bins, far removed from the 19 base pair long CTCF motifs which can generate physical domains. Indeed, given the recent release of some very deeply sequenced Hi-C datasets [5] an improved method of predicting domains might start from individual ChIP-seq peaks and consider pairs of appropriately orientated CTCF motifs. In addition, any predictive model of such domains would do well to consider the hierarchical nature of chromatin organization (exemplified by metaTADs [18]) rather than seeking simple linear discretization of the chromatin fiber into nonoverlapping domains.

Across all domains, many questions remain about how they are constructed and maintained. Two competing ideas are that boundary elements, akin to the classic chromatin insulators, block intradomain contacts and the spread of heterochromatin and hence create chromatin domains; however, another suggestion is that boundary regions are rather less important and in fact an unavoidable consequence of adjacent self-interacting domains, which are perhaps instead held together through internal enhancer–promoter interactions and other contacts. In favor of functional boundary elements, both knockdown of CTCF [38] and deletion of a specific boundary element [11] have been shown to increase inter-TAD contacts, suggesting boundaries do indeed contribute to domain delineation.

We have reported an array of boundary enrichments and depletions (Fig. 3.10), which at minimum suggests some directed biological process is in effect at boundaries [23]. Nonetheless not all observed boundary enrichments and depletions are expected to have a detectable function; it has been shown for example that removal of the H3K27me3 mark has no effect on domain boundaries [11]. One potential functional consequence of boundaries could be that genes positioned adjacent to or over a domain boundary might be most amenable to dynamic regulation, for example by associating or disassociating from the nuclear lamina. Enrichments of active gene promoters have been noted at domain boundaries in this work (e.g., Fig. 3.10) and in previous studies [4]. Alternatively, this boundary enrichment could be due to promoter–promoter looping inducing domain boundaries [39].

Deletion of specific boundary elements, while increasing intra-TAD interactions, is insufficient to cause adjacent domains to completely merge [11], suggesting the presence of other factors mediating domain stability. In addition, the majority of CTCF binding sites, currently thought to be the principal architects of domain boundaries, fall within TADs rather than at their boundaries (approximately 85% of human CTCF sites are nonboundary sites [4]). This strongly suggests CTCF binding alone is insufficient to bring about a domain boundary. Further it has been shown that the majority of enhancer-promoter contacts are tissue invariant [40], hence if functioning as anchors of structural domains, these constitutive contacts could account for the high levels of domain conservation reported previously [4] and in this work (Fig. 3.3). As is the case with many biological phenomena, the question of whether boundary regions or internal contacts are responsible for chromatin domains is reductive, and it seems likely that both boundary insulation and intradomain contacts work together to maintain chromatin domains. Higher-resolution chromatin conformation capture data and improved domain calling algorithms will undoubtedly enable more powerful boundary-predictive models in the near future, which in turn could allow broad comparisons of inferred higher order chromatin structure without the application of costly and time-consuming genome-wide C-methods.

References

[1] Bickmore WA, van Steensel B. Genome architecture: domain organization of interphase chromosomes. Cell 2013;152(6):1270–84.

[2] Dekker J, Rippe K, Dekker M, Kleckner N. Capturing chromosome conformation. Science 2002;295(5558):1306–11.

[3] Lieberman-Aiden E, van Berkum NL, Williams L, Imakaev M, Ragoczy T, Telling A, et al. Comprehensive mapping of long-range interactions reveals folding principles of the human genome. Science 2009;326(5950):289–93.

[4] Dixon JR, Selvaraj S, Yue F, Kim A, Li Y, Shen Y, et al. Topological domains in mammalian genomes identified by analysis of chromatin interactions. Nature 2012;485(7398):376–80.

[5] Rao S, Huntley M, Durand N, Stamenova E, Bochkov I, Robinson J, et al. A 3D map of the human genome at kilobase resolution reveals principles of chromatin looping. Cell 2014;159(7):1665–80.

[6] Yaffe E, Tanay A. Probabilistic modeling of Hi-C contact maps eliminates systematic biases to characterize global chromosomal architecture. Nat Genet 2011;43(11):1059–65.

[7] Hu M, Deng K, Qin Z, Liu JS. Understanding spatial organizations of chromosomes via statistical analysis of Hi-C data. Quant Biol 2013;1(2):156–74.

[8] Imakaev M, Fudenberg G, McCord RP, Naumova N, Goloborodko A, Lajoie BR, et al. Iterative correction of Hi-C data reveals hallmarks of chromosome organization. Nat Methods 2012;9(10):999–1003.

[9] Dekker J, Marti-Renom MA, Mirny LA. Exploring the three-dimensional organization of genomes: interpreting chromatin interaction data. Nat Rev Genet 2013;14(6):390–403.

[10] Lajoie BR, Dekker J, Kaplan N. The Hitchhiker's guide to Hi-C analysis: practical guidelines. Methods 2015;72:65–75.

[11] Nora EP, Lajoie BR, Schulz EG, Giorgetti L, Okamoto I, Servant N, et al. Spatial partitioning of the regulatory landscape of the X-inactivation centre. Nature 2012;485(7398):381–5.

[12] Vietri Rudan M, Barrington C, Henderson S, Ernst C, Odom DT, Tanay A, et al. Comparative Hi-C reveals that CTCF underlies evolution of chromosomal domain architecture. Cell Rep 2015;10(8):1297–309.

[13] Sexton T, Yaffe E, Kenigsberg E, Bantignies F, Leblanc B, Hoichman M, et al. Three-dimensional folding and functional organization principles of the *Drosophila* genome. Cell 2012;148(3):458–72.

[14] Narendra V, Rocha PP, An D, Raviram R, Skok JA, Mazzoni EO, et al. CTCF establishes discrete functional chromatin domains at the Hox clusters during differentiation. Science 2015;347(6225):1017–21.

[15] Fraser J, Williamson I, Bickmore WA, Dostie J. An overview of genome organization and how we got there: from FISH to Hi-C. Microbiol Mol Biol Rev 2015;79(3):347–72.

[16] Pombo A, Dillon N. Three-dimensional genome architecture: players and mechanisms. Nat Rev Mol Cell Biol 2015;16(4):245–57.

[17] Filippova D, Patro R, Duggal G, Kingsford C. Identification of alternative topological domains in chromatin. Algorithms Mol Biol 2014;9:14.

[18] Fraser J, Ferrai C, Chiariello AM, Schueler M, Rito T, Laudanno G, et al. Hierarchical folding and reorganization of chromosomes are linked to transcriptional changes in cellular differentiation. Mol Syst Biol 2015;11(12):852.

[19] Kadauke S, Blobel GA. Chromatin loops in gene regulation. Biochim Biophys Acta—Gene Regul Mech 2009;1789(1):17–25.

[20] Sanyal A, Lajoie BR, Jain G, Dekker J. The long-range interaction landscape of gene promoters. Nature 2012;489(7414):109–13.

[21] Ong CT, Corces VG. CTCF: an architectural protein bridging genome topology and function. Nat Rev Genet 2014;15(4):234–46.

[22] Van Bortle K, Nichols MH, Li L, Ong CT, Takenaka N, Qin ZS, et al. Insulator function and topological domain border strength scale with architectural protein occupancy. Genome Biol 2014;15(5):R82.

[23] Moore BL, Aitken S, Semple CA. Integrative modeling reveals the principles of multi-scale chromatin boundary formation in human nuclear organization. Genome Biol 2015;16:110.

[24] Barrett T, Wilhite SE, Ledoux P, Evangelista C, Kim IF, Tomashevsky M, et al. NCBI GEO: archive for functional genomics data sets—update. Nucleic Acids Res 2013;41:D991–5.

[25] Langmead B, Salzberg SL. Fast gapped-read alignment with Bowtie 2. Nat Methods 2012;9(4):357–9.

[26] ENCODE Consortium. An integrated encyclopedia of DNA elements in the human genome. Nature 2012;489(7414):57–74.

[27] Zhang Y, Liu T, Meyer CA, Eeckhoute J, Johnson DS, Bernstein BE, et al. Model-based analysis of ChIP-Seq (MACS). Genome Biol 2008;9(9):R137.

[28] Ernst J, Kellis M. ChromHMM: automating chromatin-state discovery and characterization. Nat Methods 2012;9(3):215–16.

[29] Breiman L. Random forests. Mach Learn 2001;45(1):5–32.

[30] Liaw A, Wiener M. Classification and regression by randomForest. R News 2002;2:18–22.

[31] Dasgupta A, Sun YV, König IR, Bailey-Wilson JE, Malley JD. Brief review of regression-based and machine learning methods in genetic epidemiology: the Genetic Analysis Workshop 17 experience. Genet Epidemiol 2011;35(S1):S5–S11.

[32] Breiman L. Statistical modeling: the two cultures. Stat Sci 2001;16(3):199–231.

[33] Nie Z, Hu G, Wei G, Cui K, Yamane A, Resch W, et al. c-Myc is a universal amplifier of expressed genes in lymphocytes and embryonic stem cells. Cell 2012;151(1):68–79.

[34] Eddy SR. What is a hidden Markov model? Nat Biotechnol 2004;22(10):1315–16.

[35] Atchison ML. Function of YY1 in long-distance DNA interactions. Front Immunol 2014;5:45.

[36] Schwalie PC, Ward MC, Cain CE, Faure AJ, Gilad Y, Odom DT, et al. Co-binding by YY1 identifies the transcriptionally active, highly conserved set of CTCF-bound regions in primate genomes. Genome Biol 2013;14(12):R148.

[37] Bickmore WA. The spatial organization of the human genome. Annu Rev Genomics Hum Genet 2013;14:67–84.

[38] Zuin J, Dixon JR, van der Reijden MI, Ye Z, Kolovos P, Brouwer RW, et al. Cohesin and CTCF differentially affect chromatin architecture and gene expression in human cells. Proc Natl Acad Sci U S A 2014;111(3):996–1001.

[39] Li G, Ruan X, Auerbach RK, Sandhu KS, Zheng M, Wang P, et al. Extensive promoter-centered chromatin interactions provide a topological basis for transcription regulation. Cell 2012;148(1–2):84–98.

[40] Bouwman BA, de Laat W. Getting the genome in shape: the formation of loops, domains and compartments. Genome Biol 2015;16(1):154.

LIST OF ACRONYMS AND ABBREVIATIONS

3C chromosome conformation capture
3D-FISH 3D fluorescence in situ hybridization
ChIP-seq the combination of chromatin immunoprecipitation (ChIP) assays with sequencing
ESC embryonic stem cells
Hi-C chromosome conformation capture allied to high throughput sequencing
LADs nuclear lamina-associated domains

CHAPTER 4

Inferring Chromatin Signaling From Genome-Wide ChIP-seq Data

Juliane Perner[1] and Ho-Ryun Chung[2]

[1]Cancer Research UK Cambridge Institute, University of Cambridge, Cambridge, United Kingdom [2]Max Planck Institute for Molecular Genetics, Epigenomics, Berlin, Germany

CONTENTS

INTRODUCTION

Chromatin is a complex made of proteins, DNA, and other molecules in the cell nucleus. The main protein constituents are the core histone proteins H2A, H2B, H3, and H4, which together with DNA form nucleosomes [1]. The histones are subject to posttranslational modifications [2]. These histone modifications are set, read, and erased by proteins and complexes called chromatin modifiers [3].

Histone modifications form recurring patterns, termed chromatin states, along the genome [4,5]. These chromatin states coincide with genomic elements, such as promoters and enhancers and correlate with function, such as repressed or active genomic regions. The existence of these patterns and the recurrence of chromatin states suggest *systematic* processes that generate them.

Histone modifications are very local and usually passive signals. It is the chromatin modifiers that interpret the histone modification signals. The outcome can be very diverse: A chromatin modifier might write or erase a histone modification or lead to the recruitment of other factors. These interactions between histone modifications and chromatin modifiers resemble chromatin signaling pathways [6,7]. One important downstream target of chromatin signaling is the transcriptional machinery around RNA Polymerase II. Consequently, the levels of histone modifications at the promoter predict gene expression [8] and in particular the transcription rate [9]. This observation can be extended to chromatin modifiers [10]. Hence, histone modifications and chromatin modifiers are close to the transcriptional process itself. In fact, they can be viewed as an integral part of transcription.

The intricate wiring within chromatin-signaling pathways confers a robust transfer of information from regulatory mechanisms to the transcriptional

Epigenetics and Systems Biology. DOI: http://dx.doi.org/10.1016/B978-0-12-803075-2.00004-0

machinery. These pathways use histone modifications as localized information carriers, in which chromatin modifiers dynamically alter the chromatin state dependent on the current histone modification state and other signals. The chromatin states found by integrating genome-wide localization data for several histone modifications can, therefore, be considered as the ensemble and/or time average. To understand the mechanisms that underlie the information transfer and to unravel the entry points for the regulatory layer, it is important to identify the individual *direct* interactions within these chromatin signaling networks.

The identification of the individual chromatin-signaling interactions is far from trivial. First, important histone modifications and chromatin modifiers may yet be unknown. Second, the interplay between the different components is complex and is stabilized by feedback loops to confer robustness [6]. Third, the interactions between histone modifications and chromatin modifiers are transient and thus need experimental methods that measure changes with high temporal resolution. Such experimental methods exist, but lack the spatial resolution required to discern the very localized activity characteristic for chromatin signaling pathways. To reduce complexity, interactions have been studied at a single-gene locus or in vitro. Although insights from these studies advance our understanding of individual interactions, in most cases, evidence is lacking that these interactions are generalizable to other gene loci or to the in vivo situation. Moreover, many interactions are studied in the context of pairwise association studies using high-throughput genome-wide location data. Despite overcoming some of the caveats of the single-gene studies, a drawback of this approach is that the derived interaction is not guaranteed to be direct but could also be mediated by unaccounted intermediate interaction partners. Controlling for the effect of these intermediate interaction partners is very important for the inference of *direct* interactions. However, such a control is mostly not applied because the intermediate interaction partners are unknown and/or it is unclear how to implement and apply such controls experimentally.

All the considerations discussed above call for a systems biology approach to discern the interactions in chromatin signaling pathways. In other words, a systematic approach is needed to uncover *direct* interactions accounting for *indirect* interactions using spatially resolved genome-wide localization data for as many histone modifications and chromatin modifiers as possible. A widely used experimental technique to study DNA–protein binding in vivo and on a genome-wide scale is chromatin immunoprecipitation followed by sequencing (ChIP-seq). Large collections of ChIP-seq data on histone modification and chromatin modifier locations are now available [11–13] and can be used as a starting point to undercover chromatin-signaling interactions.

The envisioned systems biology approach corresponds to a network reconstruction method that infers interactions between components of an interconnected system from measurements of its components. These methods include network inference based on correlation [14] and mutual information [15], which do not address the effects of indirect, i.e., transitive, interactions. However, extensions to these methods that take into account the indirect effects are possible [16]. Other methods are based on maximum entropy [17], Bayesian network inference [18], or partial correlation [10,19–21], which explicitly account for indirect effects.

The methods based on maximum entropy and Bayesian network inference require discrete number of classes. This, in turn, requires a choice of the number of classes and the numerical thresholds that partition the data into these classes. Moreover, Bayesian networks model only directed acyclic graphs, i.e., graphs with nodes and edges, where each edge is directed from one node to another, and there is no path starting at a particular node that loops back to this particular node-there are no loops in the graph. The explicit exclusion of loops restricts their applicability to biological networks, which are characterized by the presence of stabilizing feedback loops. The partial correlation/ Gaussian Graphical Model framework [19] works directly with the ChIP-seq counts and models an undirected graph, i.e., graphs, in which each edge connects one node to another without any specified direction, which also allows for loops.

Here, we describe how to learn interactions between histone modification data and chromatin modifiers from large sets of ChIP-seq data using the partial correlation/Gaussian Graphical Model framework.

EXPERIMENTAL TECHNIQUES

ChIP-seq [22] encompasses two major steps: ChIP and the subsequent sequencing of the precipitated DNA fragments (Fig. 4.1). ChIP is a well-established experimental method to identify regions where the protein of interest interacts with or binds to the DNA. It is applied to cells ensuring the presence of all necessary interaction partners within the naturally occurring environment.

The ChIP protocol starts with conserving the interactions between all proteins and the DNA by formaldehyde cross-linking. In the case of histone modifications, this may not be necessary, but it is critical in case of chromatin modifiers, because they are often only indirectly associated with the DNA via their interactions with histones, their modifications, and other DNA binding proteins.

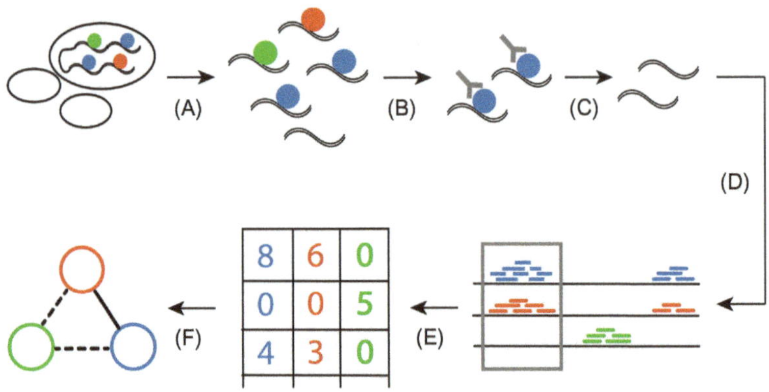

FIGURE 4.1 From ChIP-seq to chromatin signaling.
Upper panel shows the different steps of the ChIP-seq protocol: Proteins (colored spheres) are crosslinked to DNA in a population of cells (black spheres). (A) Chromatin is then extracted from the cells and fragmented. (B) Fragments carrying the antigen of interest (blue spheres) are enriched by immunoprecipitation using an antibody (gray Y). (C) The cross-linking is reversed and the DNA fragments (small black lines) are washed and purified. (D) The resulting DNA fragments are sequenced and the tags (blue small lines) are computationally mapped to a reference genome. Lower panel shows the processing of different ChIP-seq experiments to recover chromatin signaling interactions: After mapping to the reference genome, the accumulation of tags indicates binding of the antigens to specific genomic locations. (E) A matrix of binding profiles is generated by, for each antigen, counting the number of tags in a set of genomic regions. For example, the counts for the three different antigens in the genomic region that is indicated by the gray box result in a row of the matrix of binding profiles. (F) The resulting matrix of binding profiles, in which each column represents an antigen and each row a genomic location, is passed to a network reconstruction method. The resulting network shows cobinding (solid line) and mutual exclusivity (dashed line) between the different antigens.

The chromatin is then fragmented using, e.g., sonication resulting in small chromatin pieces of DNA crosslinked to proteins (Fig. 4.1A). The efficiency of the fragmentation differs depending on the compaction of the chromatin, i.e., open euchromatin or closed heterochromatin. As a consequence, fragmentation usually does not result in a homogeneous representation of the genome. The effect of this intrinsic bias and how to correct for it is discussed in more detail in the "Modeling" section.

The chromatin fragments are then precipitated (or pulled out of solution) with an antibody that specifically recognizes its antigen, i.e., the protein and/or covalent histone modification (Fig. 4.1B). This step enriches rather than selects antigen-containing chromatin fragments. Next, the chromatin–antibody complexes are washed, which decreases unspecific chromatin fragments. Then cross-linking is reversed and the DNA fragments are purified (Fig. 4.1C). The resulting DNA fragments are ligated to DNA adapters that are required for sequencing and amplified using the polymerase chain reaction

(PCR) method. High-throughput DNA sequencing determines the identity of these DNA fragments. In this approach, millions of short tags are sequenced in parallel. These tags correspond to the sequence at either one (single-end) or both ends (paired-end) of the DNA fragment. Efficient computational methods search for matches of the sequence of each fragment along a reference genome, effectively recovering its original position (Fig. 4.1D). The location of each single fragment in itself holds only limited information about the binding sites of an antigen. This is due to the enrichment rather than selection nature of the ChIP step. It is the accumulation of ideally distinct fragments at a genomic locus that is a measure for the presence of the antigen. The number of distinct fragments should be roughly proportional to the frequency of cells that harbored the antigen at this genomic location.

There are some biases that need to be considered when deriving interactions from ChIP-seq data. The number of tags mapping to each region may vary systematically due to technical or biological biases. Technical biases include differential PCR amplification and the mappability of the genomic regions, whereas biological biases arise e.g., from differential fragmentation in euchromatic versus heterochromatic regions and copy number variations. To correct for these biases, most ChIP-seq experiments have an input control experiment. Here, the chromatin fragments before the ChIP are recovered, the crosslinks are reversed, and the DNA fragments are sequenced. Thus, only unspecific fragments will be sequenced and can be compared to the main experiment to identify false signals.

Other biases are much harder to account for, e.g., the preferential precipitation of certain genomic regions in the absence of the antigen or by unspecific antibodies, such as IgG [23]. Also, some antibodies show so-called cross-reactivity, which means they cannot distinguish between very similar proteins. This is especially known for antibodies for detecting histone modifications [24,25], e.g., the trimethylation of the lysine residue at position four of the histone H3 (H3K4me3) and the corresponding dimethylation (H3K4me2). Thus, when comparing the binding locations of the two proteins their colocation might be artificially increased simply because of the antibody's cross-reactivity. Although the antibodies have been widely improved, cross-reactivity is a bias that needs to be kept in mind when interpreting associations derived from ChIP-seq data.

With respect to inferring chromatin signaling interactions, there are more general shortcomings of the ChIP-seq data that restrict our analysis and interpretation. Naturally, the number of proteins that we can analyze is limited by the availability of specific good quality antibodies. Further, traditional ChIP-seq interrogates each protein separately. Thus, the observation that two antigens bind to the same location in independent ChIP-seq experiments is not sufficient to infer that the two antigens are actually there at the same time. This is a consequence from the fact that ChIP-seq measures DNA binding in a population of many different cells. Nevertheless, the advantages of ChIP-seq, i.e., the possibility to identify antigen binding sites in vivo and genome wide at high spatial resolution, make it suitable to derive biological hypotheses on chromatin signaling.

MODELING

ChIP-seq experiments reveal the genomic loci that were bound by an antigen of interest. The binding profile of an antigen of interest can then be derived by simply counting the number of tags falling into the regions of interest. In the following, we will look at a ±2000-base pair region around promoters (Fig. 4.1E). To correct for some of the technical biases discussed above, the counts are normalized by the corresponding input experiment. Here, the ratio of counts obtained in the ChIP and the counts obtained in the input is defined as

$$\hat{x}_{ij} = \frac{x_{ij} + \alpha_j}{z_i + \beta}, \tag{4.1}$$

where x_{ij} corresponds to the counts for the ChIP against antigen j in region i, z_i to the input counts, $\alpha_j > 0$ to a ChIP-specific, and $\beta > 0$ to an input-specific pseudocount. Here, we set $\alpha_j=\beta=1$, which renders $\hat{x}_{ij} > 0$ and prevents division by zero in case $z_i=0$. Further, the input-normalized values $\hat{x}_{ij}$ are log-transformed. This serves two purposes: First, the distribution of log-transformed input-normalized values $\hat{x}_{ij}$ is closer to a normal (Gaussian) distribution and second, the impact of very large $\hat{x}_{ij}$ on the Pearson correlation coefficient is greatly reduced. For the use in the model below, we also center and scale the data to have zero mean and unit variance.

Given a collection of ChIP-seq experiments on histone modifications and chromatin modifiers, the task of network reconstruction methods is to infer the interactions by evaluating the colocation of each pair of proteins (Fig. 4.1F). Thus, as a first step the similarity between binding profiles of the proteins has to be measured. In a typical pairwise association study, this is done by computing the Pearson correlation coefficient between the numbers of observed sequence tags across different genomic locations. The Pearson correlation coefficient evaluates the linear dependence between two random features. It can be estimated from the observed data by dividing the sample covariance of the two features by the product of their sample standard deviations. The correlation coefficient ranges from −1 to 1, with −1 representing a negative association, 1 representing a positive association, and 0 representing no association between the two random features.

The pairwise analysis using the Pearson correlation between the binding profiles results in groups of proteins where the proteins in each group are highly correlated to each other but less correlated with proteins in other groups (Fig. 4.2A). However, the internal structure does not become apparent within the highly correlated groups. The high mutual correlation of these proteins may be due to indirect interactions. For example, a high correlation between protein A and C could be the result of an interaction of protein A and C with another protein B rather than a direct interaction between A and C. Thus, in order to obtain the chromatin signaling interaction, we need to correct for indirect interactions.

To account for indirect interactions given a large collection of ChIP-seq experiments, we compute the so-called partial correlation coefficient. Similar to the Pearson correlation coefficient, it evaluates the dependencies between two observed binding profiles. However, in contrast to the Pearson correlation coefficient, it quantifies the dependencies after accounting for all other possible explanations in the set of all remaining ChIP-seq experiments (specific examples are given below). The resulting partial correlation matrix is much sparser and reveals the internal structure within the highly correlated groups (Fig. 4.2B).

The partial correlation coefficient is computed as the Pearson correlation coefficient $p_{x,y|z}$ between the residuals $r_{x|z}$ and $r_{y|z}$ of each pair x and y of

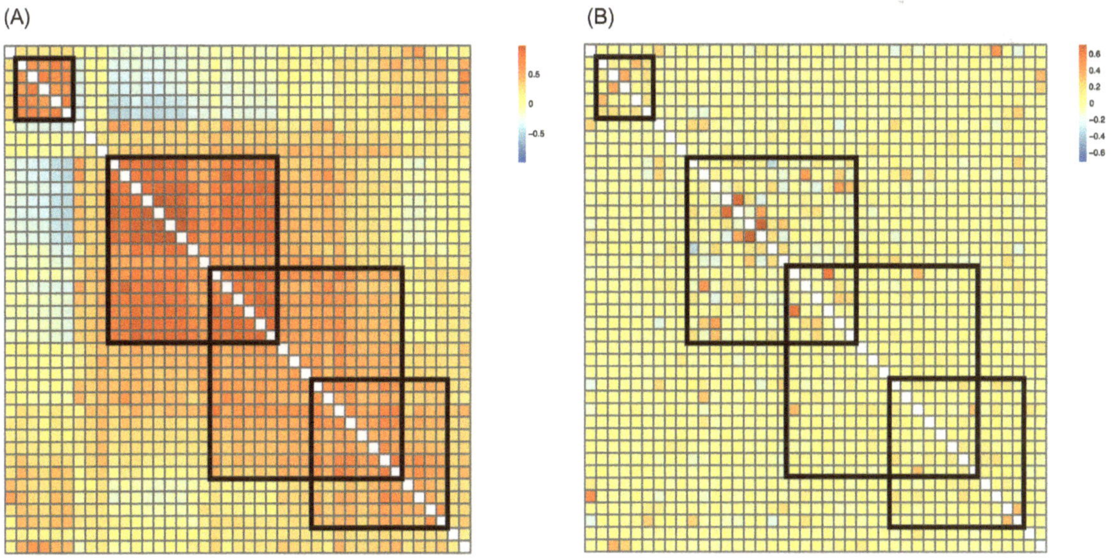

FIGURE 4.2 Correlation versus partial correlation.
(A) Clustered correlation matrix. The color indicates the correlation coefficient as shown in the color key to the right. The boxes highlight clusters of proteins/histone modifications with high mutual correlation. Clustering was performed on the correlation matrix using the R functions *dist* and *hclust*. (B) Clustered partial correlation matrix. The ordering and the boxes are the same as in (A).

binding profiles after regressing each of them on the remaining binding profiles z:

$$r_{x|z} = x - f_x(z) \tag{4.2}$$

$$r_{y|z} = y - f_y(z) \tag{4.3}$$

$$p_{x,y|z} = \frac{\text{cov}(r_{x|z}, r_{y|z})}{\text{sd}(r_{x|z})\text{sd}(r_{y|z})}, \tag{4.4}$$

where $f_x(z)$ and $f_z(z)$ correspond to the predicted values of x and y obtained by regressing against z.

For example, the polycomb repressive complex 2 (PRC2) consists of a group of silencing proteins that include enzymes that modify chromatin and other proteins that bind to these modifications. Three proteins, namely Enhancer of Zeste Homolog 2 (EZH2), Suppressor of Zeste 12 Homolog (SUZ12), and Embryonic Ectoderm Development (EED) form the core catalytic complex of PRC2 [26]. EZH2 catalyzes the trimelthylation of lysine 27 of the histone protein H3 (H3K27me3) [27–30] and EED binds to H3K27me3 [31], but SUZ12 does not (Fig. 4.3A). Thus, H3K27me3 ChIP-seq levels should correlate with those of EZH2, EED, and SUZ12. However, only in the case of EZH2 and EED this is due to a direct interaction with H3K27me3. The partial correlation framework answers the question of how much more information SUZ12 provides on H3K27me3 and vice versa than the information provided by EZH2 and EED on H3K27me3 and SUZ12, respectively (Fig. 4.3). This is achieved by modeling SUZ12 ($f_x(z)$; Fig. 4.3B) and H3K27me3 ($f_y(z)$; Fig. 4.3C) levels as weighted linear combinations of EZH2 and EED. The residual information of SUZ12 ($r_{x|z}$) and H3K27me3 ($r_{y|z}$) are correlated only if SUZ12 harbors additional information about H3K27me3 and vice versa (Fig. 4.3D) to the information already provided by EZH2 and EED.

Since the partial correlation coefficients are related to the precision matrix, a more efficient way to compute them involves the matrix inversion of the covariance matrix

$$p_{x,y|z} = \frac{-\Sigma_{xy}^{-1}}{\sqrt{\Sigma_{xx}^{-1}\Sigma_{yy}^{-1}}}, \tag{4.5}$$

where Σ^{-1} is the inverse of the covariance matrix.

Compared to other applications (e.g., gene regulatory networks), for inferring chromatin signaling networks many more observations than histone

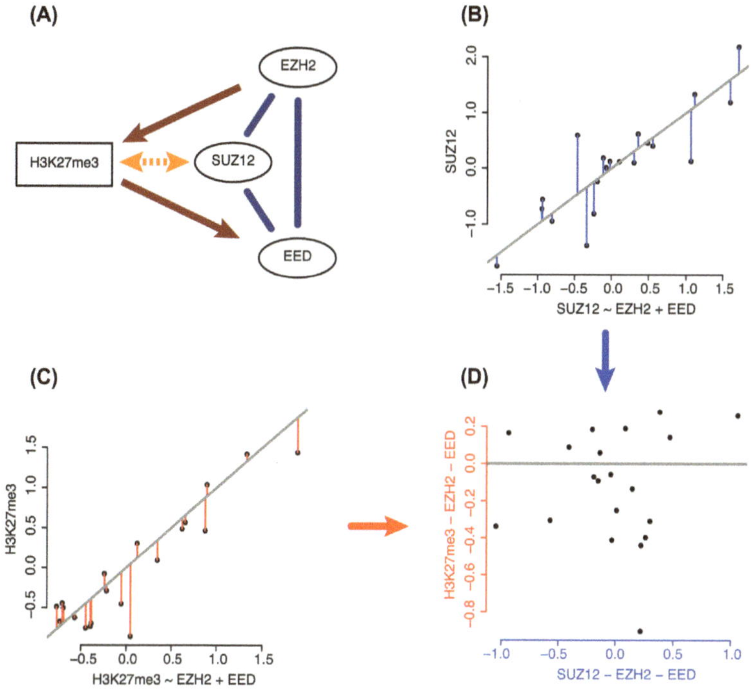

FIGURE 4.3 From correlation to direct interactions.
(A) Model of the interaction of the PRC2 trimeric complex (EZH2, SUZ12, and EED) with H3K27me3. The blue lines indicate protein–protein interactions. The red arrows indicate direct causal interactions, with EZH2 setting H3K27me3 and EED reading H3K27me3. The orange double-headed arrow indicates a correlation between SUZ12 and H3K27me3 induced by either EZH2 and/or EED. (B–D) Toy example of the effect of multivariate regression on the correlation induced by transitivity. Modeling of, e.g., H3K27me3 levels by a linear combination of EZH2, EED, and SUZ12 leads to an estimate of the influence of SUZ12 independent of the influence of EZH2 and EED. This is achieved by modeling SUZ12 (B) and H3K27me3 levels (C) by a linear combination of EZH2 and EED. The predictions of these models are subtracted from the actual SUZ12 and H3K27me3 levels (residuals, depicted by blue (SUZ12) and red (H3K27me3) vertical lines). The residuals of SUZ12 after incorporating the information of EZH2 and EED are used to predict the corresponding residuals of H3K27me3 (D), which in this case fails because there is no information of SUZ12 on H3K27me3 left after considering EZH2 and EED levels. *Reproduced with permission from Perner J, Lasserre J, Kinkley S, Vingron M, Chung H-R. Inference of interactions between chromatin modifiers and histone modifications: from ChIP-Seq data to chromatin-signaling. Nucleic Acids Res. 2014;42(22):13689–13695.*

modifications or chromatin modifiers are available to estimate the partial correlation coefficients. However, many ChIP-seq experiments are highly correlated, which leads to unstable estimates of the coefficients in the regression. As a consequence, small perturbations in the data lead to large changes in the estimated regression coefficients. In addition, often only one highly correlated protein is selected over the other. These two characteristics of ChIP-seq data make the estimation of the partial correlation coefficients rather unstable. The following two-step procedure was proposed to remedy this situation [10].

The first step is based on the observation that partial correlations can be approximated by the coefficients in a linear regression [32]. Using linear regression, each histone modification at a time is predicted by all the chromatin modifiers. Elastic Nets, which is a regularized regression approach, can help in situations where highly correlated features exist [33]. These groups of features are selected or removed all together, i.e., they achieve similar coefficients in the regression instead of selecting one over the other. At the same time, Elastic Nets allow to stress the differences between informative and less informative features.

The coefficients are estimated for each feature in a 10-fold cross-validation where the observations are split into 10 sets and each set is left out iteratively when training the model. The left-out set is then used to test the predictions of the model. Interactions between i and j are called if the feature i is predicted by a feature j with a coefficient that deviates from the average of the coefficients by at least one standard deviation. This approach leads to the discovery of the most informative interactions between chromatin modifiers and histone modifications. However, highly correlated histone modifications tend to interact with similar chromatin modifiers.

In the second step, sparse partial correlation networks [19] are constructed. This procedure is designed to be very conservative in calling an interaction. It is based on a 10-fold cross-validation scheme, where the observations are split into 10 sets, and each set is left out iteratively such that the partial correlations are computed on the basis of the remaining nine sets of the rank-transformed observations. The left-out set is used to mask nonzero partial correlation coefficients that if set to zero would lead to a relatively small loss in accuracy [19]. By design, the sparse partial correlation networks approach is likely to return high-confidence true interactions given all available data but the failure to discover an interaction cannot be interpreted as a truly non-existing interaction. Further, due to statistical associations arising from logical dependencies, which are induced by sharing a common neighbor, some recovered interactions are false positives.

Thus, to retrieve only true interactions, the results of the two steps are combined into a consensus set of interactions. This way the set of interactions is based on only those interactions in which one feature is consistently predictive for another feature. This set is then pruned by the interactions from the sparse correlation network to focus only on those interactions that cannot be explained by any other feature in the dataset.

We have applied this framework to histone modification and chromatin modifier ChIP-seq data for the human immortalized myelogenous cell line K562 [13]. We recovered in total 34 interactions [10] (Fig. 4.4). Among these, 19 (56%) had support from experimental evidence [10]. For example, the interactions between the histone modification H3K27me3 and the chromatin modifiers CBX2, CBX8, SUZ12, and EZH2 are well known. CBX2 and CBX8 are part of the polycomb repressive complex 1 [26,34] (PRC1) and at least, CBX2 has been shown to directly bind to H3K27me3 [35]. EZH2 and SUZ12 are part of PRC2 [26]. EZH2 has been shown to catalyze H3K27me3 [27–30] indicating that the uncovered interaction is direct. SUZ12 is part of the PRC2 core complex, but has not been shown to directly interact with H3K27me3. However, the EED protein has been shown to bind to H3K27me3. The uncovered interaction of SUZ12 with H3K27me3 may be due to an indirect interaction via EED [31], which is not part of the analyzed dataset.

Two of the novel positive interactions involve the monomethylation of lysine 20 of the histone protein H4 (H4K20me1) and CBX2 of PRC1 and EZH2 of PRC2. The enzymatic activities of PRC1 and PRC2 suggest that CBX2 and

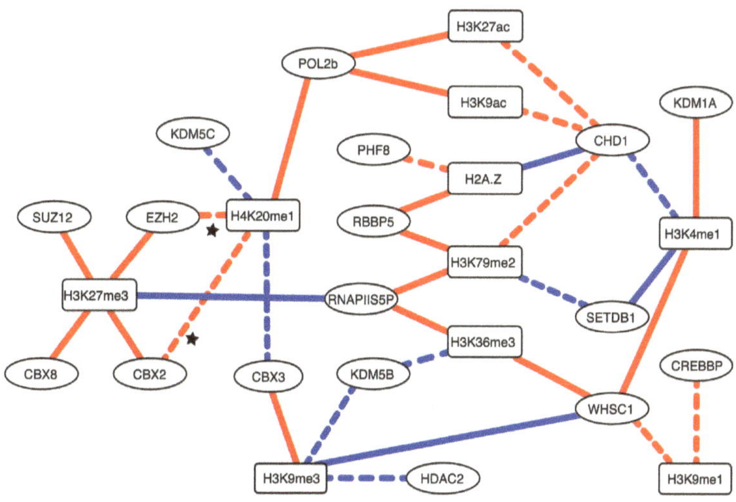

FIGURE 4.4 Chromatin signaling network.
Graphical representation of the interactions between chromatin modifiers (circles) and histone modifications (squares). Shown are the interactions recovered both by the Elastic-Net and the SPCN approach. Red lines indicate colocation and blue lines mutual exclusivity. The continuous lines indicate interactions with supporting evidence in the literature, whereas the dashed lines indicate interactions without supporting evidence. The stars indicate the two interactions confirmed in this study. *Reproduced with permission from Perner J, Lasserre J, Kinkley S, Vingron M, Chung H-R. Inference of interactions between chromatin modifiers and histone modifications: from ChIP-Seq data to chromatin-signaling. Nucleic Acids Res. 2014;42(22):13689–13695.*

EZH2 directly or indirectly bind to H4K20me1 rather than catalyzing or stabilizing H4K20me1. To test this idea, we performed an immunoprecipitation (IP), where a protein/histone modification of interest is precipitated with a specific antibody. Here, we used an antibody against H4K20me1. If, e.g., CBX2 interacted with H4K20me1 then a precipitation of H4K20me1 should lead to the coprecipitation of CBX2 (Fig. 4.5). The presence of CBX2 and EZH2 in the H4K20me1 IP and their absence in the control IgG IP suggests that both proteins interact with H4K20me1. H4K20me1 is a cell-cycle-regulated histone modification. Its levels increase during S-phase and peak during mitosis [36]. Given this pattern, H4K20me1 may be involved in maintaining PRCs at their target sites through the replicative and mitotic challenges [37] by retaining PRC1 and PRC2 complexes in regions with old H3K27me3- and new H4K20me1-bearing nucleosomes.

In summary, the inferred interactions between chromatin modifiers and histone modifications provide testable hypotheses for in vivo and genome-wide

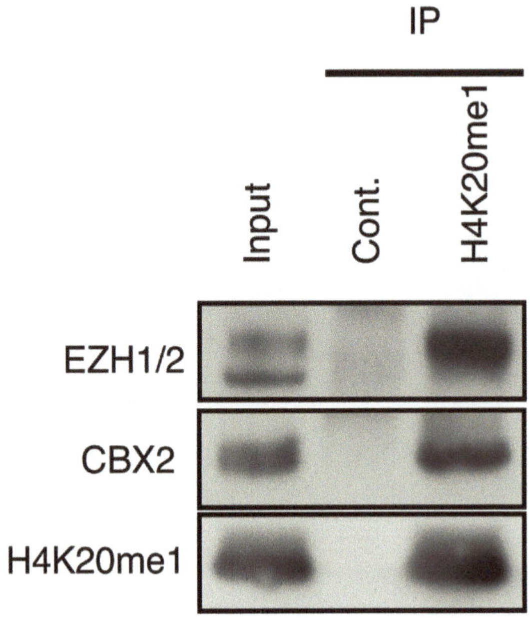

FIGURE 4.5 Verification of two predicted interactions links H4K20me1 to Polycomb-mediated repression.
H4K20me1 coimmunoprecipitation: K562 cells were immunoprecipitated with a control IgG and an H4K20me1 specific antibody and analyzed by immunoblot for the coprecipitation of EZH2, CBX2, and H4K20me1. 10% Input was loaded. *Reproduced with permission from Perner J, Lasserre J, Kinkley S, Vingron M, Chung H-R. Inference of interactions between chromatin modifiers and histone modifications: from ChIP-Seq data to chromatin-signaling. Nucleic Acids Res. 2014;42(22):13689–13695.*

occurring interactions. As such, they complement experimental findings obtained in vitro or by using one or few genes as model system. Thus, the specific predictions by our framework can serve as the starting point to define chromatin signaling and its impact on transcription.

PERTURBING THE SYSTEM

Perturbations of chromatin signaling networks provide a critical test for the existence of an interaction in an in vivo situation. Moreover, perturbations yield information about the directionality of the interactions [38]. If A acts upstream of B, a perturbation of A will lead to an effect on B, while a perturbation on B will lead to no effect on A. However, possible feedback loops and functional redundancies render the interpretation of the results difficult. If A and B are part of a feedback loop, a perturbation of A will lead to an effect on B and a perturbation of B will lead to an effect on A. If redundancies are present, i.e., there is a factor C that can fulfill the function A, a perturbation of A will lead to no effect on B even if A acts upstream of B.

Perturbations, like knockdowns, knockouts, and overexpression, are possible at the level of chromatin modifiers. However, these perturbations may have pleiotropic effects, such as cell death, which again render an interpretation difficult. Another more subtle approach involves the targeted recruitment of chromatin modifiers to model promoter(s) by, e.g., Cas9 [39]. Here, the effect should be localized to a defined set of loci, alleviating the problem of pleiotropic effects. However, the problems associated with feedback loops remain.

Perturbations are much more difficult at the level of histone modifications. One possibility is to deplete or overexpress the enzyme that catalyzes a particular histone modification. However, these enzymes have frequently additional targets such that an interpretation of the results is difficult. Another way is to introduce mutations in the histone genes. This has been demonstrated for yeast [40] and Drosophila [41,42]. Especially in mammals, all four core histones are encoded by multiple genes located at different genomic loci [43], rendering knockouts and rescue strategies difficult. Finally, it remains unclear whether the effect of a substitution of an amino acid residue is due to the loss of a histone modification or the modulation of an interaction surface for another, unrelated protein or complex.

OUTLOOK

Our approach yields a number of high confidence predictions of interactions between histone modifications and chromatin modifiers. Many of them are already known indicating the high sensitivity of the approach. We validated two novel interactions by coimmunoprecipitation: one involving CBX2 and H4K20me1 and another one involving EZH2 and H4K20me1, providing a link between PRC1 as well as PRC2 and H4K20me1. The remaining predicted but yet unknown interactions remain to be tested.

Coimmunoprecipitation is one line of evidence to support a predicted interaction. However, it cannot be viewed as comprehensive proof of a direct physical interaction, as the interaction may still be indirect (e.g., by resulting from both proteins being part of the same complex). Perturbation experiments provide another line of support. However, perturbing the chromatin signaling network is difficult as outlined above. Due to the presence of multiple feedback loops, functional redundancies and pleiotropic effects, the results derived from perturbation experiments are likely to be hard to interpret.

The inference of chromatin signaling networks can be easily applied to other chromatin states, such as enhancers, insulators, etc. In addition more histone modifications, chromatin modifiers, and also other chromatin modifications, such as DNA methylation, can be included. The integration of prior knowledge and the inference of the chromatin signaling networks in distinct chromatin states revealed that in murine embryonic stem cells 5-hydroxycytosine is a central node of this network. 5-hydroxycytosine links DNA demethylation to nucleosome remodeling complexes and to key pluripotency transcription factors. A more detailed analysis of the cooccupancy patterns of 5-hydroxycytosine with its specific interaction partners showed that depending on the colocalization partner different processes, such as epigenetic self-regulation, cell signaling, maintenance of stemness, morphogenesis, and metabolism are affected [21].

Our approach can certainly be improved. For instance, it should be possible to integrate prior knowledge into the estimation of the covariance matrix. Such a possibility is instrumental for the integration of the results of validation experiments into the inference of chromatin signaling. Moreover, additional constraints can be used during the estimation of the covariance matrix, unifying the regularized linear regression by, e.g., Elastic Nets with the partial correlation framework [44]. Thus, further developments are likely to yield a refined view on the topology of chromatin signaling networks.

Finally, chromatin signaling is only a part of the signaling and regulatory events happening in the cell. For example, the activity of chromatin modifier themselves can be modulated by posttranslational modifications and it remains an open question how this affects their interactions with other chromatin modifiers, histones, and the transcriptional machinery. Other unanswered questions include how chromatin signaling is initiated by changes in the regulatory inputs, and how new chromatin states are stabilized. To answer these questions, cell state changes have to be followed by ChIP-seq experiments over time. This dynamic view on cooccupancy patterns of histone modifications and chromatin modifiers provides a source of information for unraveling the intricate (causal) interactions in chromatin signaling.

References

[1] Kornberg RD. Chromatin structure: a repeating unit of histones and DNA. Science 1974;184(4139):868–71.

[2] Allfrey VG, Faulkner R, Mirsky AE. Acetylation and methylation of histones and their possible role in the regulation of RNA synthesis. Proc Natl Acad Sci USA 1964;51:786–94.

[3] Badeaux AI, Shi Y. Emerging roles for chromatin as a signal integration and storage platform. Nat Rev Mol Cell Biol 2013;14(4):211–24.

[4] Ernst J, Kellis M. Discovery and characterization of chromatin states for systematic annotation of the human genome. Nat Biotechnol 2010;28(8):817–25.

[5] Mammana A, Chung H-R. Chromatin segmentation based on a probabilistic model for read counts explains a large portion of the epigenome. Genome Biol 2015;16:151.

[6] Schreiber SL, Bernstein BE. Signaling network model of chromatin. Cell 2002;111(6):771–8.

[7] Perner J, Chung H-R. Chromatin signaling and transcription initiation. Front Life Sci 2013;7(1–2):22–30.

[8] Karlić R, Chung H-R, Lasserre J, Vlahoviček K, Vingron M. Histone modification levels are predictive for gene expression. Proc Natl Acad Sci USA 2010;107(7):2926–31.

[9] Wang C, Tian R, Zhao Q, Xu H, Meyer CA, Li C, et al. Computational inference of mRNA stability from histone modification and transcriptome profiles. Nucleic Acids Res 2012;40(14):6414–23.

[10] Perner J, Lasserre J, Kinkley S, Vingron M, Chung H-R. Inference of interactions between chromatin modifiers and histone modifications: from ChIP-Seq data to chromatin-signaling. Nucleic Acids Res 2014;42(22):13689–95.

[11] Roadmap Epigenomics Consortium Kundaje A, Meuleman W, Ernst J, Bilenky M, Yen A, et al. Integrative analysis of 111 reference human epigenomes. Nature 2015;518(7539):317–30.

[12] ENCODE Project Consortium. An integrated encyclopedia of DNA elements in the human genome. Nature 2012;489(7414):57–74.

[13] Ram O, Goren A, Amit I, Shoresh N, Yosef N, Ernst J, et al. Combinatorial patterning of chromatin regulators uncovered by genome-wide location analysis in human cells. Cell 2011;147(7):1628–39.

[14] Stuart JM, Segal E, Koller D, Kim SK. A gene-coexpression network for global discovery of conserved genetic modules. Science 2003;302(5643):249–55.

[15] Steuer R, Kurths J, Daub CO, Weise J, Selbig J. The mutual information: detecting and evaluating dependencies between variables. Bioinformatics 2002;18(Suppl 2):S231–40.

[16] Margolin AA, Nemenman I, Basso K, Wiggins C, Stolovitzky G, Dalla Favera R, et al. ARACNE: an algorithm for the reconstruction of gene regulatory networks in a mammalian cellular context. BMC Bioinfor 2006;7(Suppl 1):S7.

[17] Zhou J, Troyanskaya OG. Global quantitative modeling of chromatin factor interactions. PLoS Comput Biol 2014;10(3):e1003525.

[18] van Bemmel JG, Filion GJ, Rosado A, Talhout W, de Haas M, van Welsem T, et al. A network model of the molecular organization of chromatin in Drosophila. Mol Cell 2013;49(4):759–71.

[19] Lasserre J, Chung H-R, Vingron M. Finding associations among histone modifications using sparse partial correlation networks. PLoS Comput Biol 2013;9(9):e1003168.

[20] Lundberg SM, Tu WB, Raught B, Penn LZ, Hoffman MM, Lee S-I. ChromNet: learning the human chromatin network from all ENCODE ChIP-seq data. Genome Biol 2016;17(1):82.

[21] Juan D, Perner J, Carrillo de Santa Pau E, Marsili S, Ochoa D, Chung H-R, et al. Epigenomic co-localization and co-evolution reveal a key role for 5hmC as a communication hub in the chromatin network of ESCs. Cell Rep 2016;14(5):1246–57.

[22] Johnson DS, Mortazavi A, Myers RM, Wold B. Genome-wide mapping of in vivo protein–DNA interactions. Science 2007;316(5830):1497–502.

[23] Jain D, Baldi S, Zabel A, Straub T, Becker PB. Active promoters give rise to false positive "Phantom Peaks" in ChIP-seq experiments. Nucleic Acids Res 2015;43(14):6959–68.

[24] Egelhofer TA, Minoda A, Klugman S, Lee K, Kolasinska-Zwierz P, Alekseyenko AA, et al. An assessment of histone-modification antibody quality. Nat Struct Mol Biol 2011;18(1):91–3.

[25] Grzybowski AT, Chen Z, Ruthenburg AJ. Calibrating ChIP-Seq with nucleosomal internal standards to measure histone modification density genome wide. Mol Cell 2015;58(5):886–99.

[26] Margueron R, Reinberg D. The Polycomb complex PRC2 and its mark in life. Nature 2011;469(7330):343–9.

[27] Müller J, Hart CM, Francis NJ, Vargas ML, Sengupta A, Wild B, et al. Histone methyltransferase activity of a Drosophila Polycomb group repressor complex. Cell 2002;111(2):197–208.

[28] Kuzmichev A, Nishioka K, Erdjument-Bromage H, Tempst P, Reinberg D. Histone methyltransferase activity associated with a human multiprotein complex containing the Enhancer of Zeste protein. Genes Dev 2002;16(22):2893–905.

[29] Cao R, Wang L, Wang H, Xia L, Erdjument-Bromage H, Tempst P, et al. Role of histone H3 lysine 27 methylation in Polycomb-group silencing. Science 2002;298(5595):1039–43.

[30] Czermin B, Melfi R, McCabe D, Seitz V, Imhof A, Pirrotta V. Drosophila enhancer of Zeste/ESC complexes have a histone H3 methyltransferase activity that marks chromosomal Polycomb sites. Cell 2002;111(2):185–96.

[31] Hansen KH, Bracken AP, Pasini D, Dietrich N, Gehani SS, Monrad A, et al. A model for transmission of the H3K27me3 epigenetic mark. Nat Cell Biol 2008;10(11):1291–300.

[32] Lauritzen SL. New York: Oxford University Press Graphical models. Oxford: Clarendon Press; 1996.

[33] Zou H, Hastie T. Regularization and variable selection via the elastic net. J Royal Stat Soc B 2005;67:768–768.

[34] Simon JA, Kingston RE. Mechanisms of polycomb gene silencing: knowns and unknowns. Nat Rev Mol Cell Biol 2009;10(10):697–708.

[35] Bernstein E, Duncan EM, Masui O, Gil J, Heard E, Allis CD. Mouse polycomb proteins bind differentially to methylated histone H3 and RNA and are enriched in facultative heterochromatin. Mol Cell Biol 2006;26(7):2560–9.

[36] Pesavento JJ, Yang H, Kelleher NL, Mizzen CA. Certain and progressive methylation of histone H4 at lysine 20 during the cell cycle. Mol Cell Biol 2008;28(1):468–86.

[37] Steffen PA, Ringrose L. What are memories made of? How Polycomb and Trithorax proteins mediate epigenetic memory. Nat Rev Mol Cell Biol 2014;15(5):340–56.

[38] Koller D, Friedman N. Probabilistic graphical models. Cambridge, Massachusetts, MA: MIT Press; 2009. 1 p.

[39] Wang H, La Russa M, Qi LS. CRISPR/Cas9 in genome editing and beyond. Annu Rev Biochem 2016;85:227–64.

[40] Park EC, Szostak JW. Point mutations in the yeast histone H4 gene prevent silencing of the silent mating type locus HML. Mol Cell Biol 1990;10(9):4932–4.

[41] Pengelly AR, Copur Ö, Jäckle H, Herzig A, Müller J. A histone mutant reproduces the phenotype caused by loss of histone-modifying factor Polycomb. Science 2013;339(6120):698–9.

[42] Günesdogan U, Jäckle H, Herzig A. A genetic system to assess in vivo the functions of histones and histone modifications in higher eukaryotes. EMBO Rep 2010;11(10):772–6.

[43] Marzluff WF, Gongidi P, Woods KR, Jin J, Maltais LJ. The human and mouse replication-dependent histone genes. Genomics 2002;80(5):487–98.

[44] Witten DM, Tibshirani R. Covariance-regularized regression and classification for high-dimensional problems. J Royal Stat Soc B 2009;71(3):615–36.

GLOSSARY

Histone modification a chemical group (e.g., methylation or acetylation) added to the tail of a histone protein.

Chromatin modifier a protein that on its own or as part of a complex interacts with histones or DNA and affects the local chromatin environment.

Chromatin signaling the sequence of interactions between histone modifications and chromatin modifiers that leads to the establishment, maintenance or change of chromatin states ultimately affecting the transcriptional machinery.

Transitive/indirect interaction an association between two factors observed in the experimental data that arises from the shared interaction with an underlying unknown factor. In contrast a direct interaction is an association between two factors that cannot be explained by any other factor (e.g., resulting from a protein–protein interaction).

LIST OF ACRONYMS AND ABBREVIATIONS

H3K4me3 trimethylation of lysine 4 of the histone protein H3
H3K4me2 dimethylation of lysine 4 of the histone protein H3
H3K27me3 trimethylation of lysine 27 of the histone protein H3
H4K20me1 monomethylation of lysine 20 of the histone protein H4
ChIP-seq chromatin immunoprecipitation followed by sequencing
DNA deoxyribonucleic acid
EZH2 Enhancer of Zeste Homolog 2
EED Embryonic Ectoderm Development
SUZ12 Suppressor of Zeste 12 Homolog
K562 immortalized human myelogenous leukemia line
CBX2 Chromobox Homolog 2
CBX8 Chromobox Homolog 8
RNF2 Ring Finger Protein 2
PRC1 Polycomb Repressive Complex 1
PRC2 Polycomb Repressive Complex 2

SECTION

Everything's Moving: In Vivo Dynamics of Epigenetic Regulators: Kinetic Models Based on Ordinary Differential Equations

The previous chapters dealt with data that give a "snapshot" of fixed binding events or modifications. However, if we were to be able to take a movie of the individual molecules in the nucleus, we would see that each one of them is actually continuously in motion, occasionally settling on its binding site but also often leaving to sample other places, rather like a flock of birds on a shoreline. Similarly, the modifications on histones that appear as "enrichments" in chromatin profiling data sets are in fact constantly being erased and replaced by new ones. How can a system whose components are in constant flux nevertheless give stable epigenetic regulation? The concept of "stability" is fundamental to epigenetics, but it is more appropriate to think of this as a stable collective group (e.g., a flock of birds of a certain constant size at a certain place) rather than a stable individual interaction (e.g., a flock of birds that never move from their individual spots). Indeed, it is this flux of individual molecules that gives room for change and regulation.

This section focuses on mathematical models that capture this dynamic behavior using ordinary differential equations (ODEs). Such equations describe the changes in the state of a system over time. Kinetic simulations are performed by calculating how many of a given species there are at a given time point (e.g., how many birds are there on the beach?) One then calculates how many there will be at the next time point, depending on how fast they move (e.g., how often do birds leave the flock and how often do new ones join?). A beautifully simple system of ODEs can give profound insights into average population behavior, for example, how will the system be affected if something changes: e.g., a big wave washes over the beach, dispersing the flock of birds. How many of them leave, and how long will it take them to come back?

Chapters 5 and 6 give examples of the use of simple ODE models to investigate the changes in binding of epigenetic regulatory proteins to chromatin (Chapter 5: "In Vivo Biochemistry": Absolute Quantification and Kinetic Modeling Applied to Polycomb and Trithorax Regulation) and the dynamics of histone modifications (Chapter 6: Modeling Distributive Histone Modification by Dot1 Methyltransferases: From Mechanism to Biological Insights) during the cell cycle. These two chapters have in common that they illustrate the need for quantitative in vivo measurements to make the best use of ODE modeling and explain in detail how that was done in *Drosophila* (Chapter 5) and yeast (Chapter 6).

CHAPTER 5

"In Vivo Biochemistry": Absolute Quantification and Kinetic Modeling Applied to Polycomb and Trithorax Regulation

Leonie Ringrose
Humboldt-Universität zu Berlin, Berlin, Germany

CONTENTS

INTRODUCTION

The Journey Begins: DNA Recombinases Yield Their Secrets to Mathematics

In 1996, I was a struggling PhD student. After several years of my PhD, I had generated very little usable data and instead had stumbled upon one curious thing. I had been working on the DNA recombinases, "FLP" (from the yeast *Saccharomyces cerevisiae*) and "Cre" (from bacteriophage P1). These enzymes recognize specific sites on DNA and recombine them by precisely cutting and rejoining the ends. Both enzymes use the same reaction pathway, in which four enzyme molecules recombine two DNA molecules, but I had observed very different in vitro behaviors. Although the FLP enzyme seemed to be able to greedily recombine an ever-increasing amount of DNA, the Cre enzyme very quickly became saturated and actually got more quickly exhausted the more DNA I gave it. After a long detour into my vaguely remembered university textbooks on Michaelis–Menten enzyme kinetics, I became convinced that if I could somehow describe the pathway mathematically, I might be able to understand why their behaviors were so different. It was clear that the Michaelis–Menten equation, which describes one enzyme and one substrate, was not going to do the job, but how should I describe my system of four enzyme molecules and two substrate molecules?

I talked incessantly about this to anyone who would listen, and luckily, Valère Lounnas and Lutz Ehrlich took the bait. They were able to show me how to construct a model of the reaction pathway out of ordinary differential equations, and to simulate the results of the in vitro reactions. We were all set, and I was sure we were near to the answer. Every day I would bring new data to Valère and Lutz, and ask them to make the models fit it. One day, Valère became quite exasperated, and asked me, "but Leonie, what are your units?" I answered smartly: "microliters of DNA, microliters of protein, and percent

Epigenetics and Systems Biology. DOI: http://dx.doi.org/10.1016/B978-0-12-803075-2.00005-2

recombination." He then explained that we needed to know absolute quantities in terms of the actual numbers of molecules per volume, in other words, the micromolar or nanomolar concentrations. And so I spent the next 6 months generating that information, until finally we could describe the system in absolute terms. I could indeed understand why Cre became so quickly exhausted, and the reason was quite counterintuitive. It bound very strongly to DNA, and therefore each molecule stayed faithfully bound for a long time to the first DNA site it encountered. When there were too many DNA molecules, the reaction was essentially blocked, because no two Cre molecules would ever meet on the same piece of DNA (a prerequisite for the next step of the reaction). FLP on the other hand, bound less tightly and so was able to move easily from one DNA molecule to the next [1].

I would say in retrospect that this conversion from relative to absolute quantification was the most important step that I made in my entire PhD and was a turning point in my early career. Ever since then, I have been fascinated by the insights that can be gained from absolute quantification and simple modeling and have been determined to apply these concepts to the more messy world of epigenetic gene regulation in vivo. In contrast to the 6 months I needed in order to quantify two proteins and their DNA substrates in vitro, it has taken me and a large number of my students, 10 years to make a start on this endeavor for my favorite epigenetic regulators: The Polycomb (PC) group (PcG) and Trithorax group (TrxG) proteins. The journey was much longer and tougher than we expected, and is by no means finished, but the views have been well worth the extra effort in climbing this mountain. We have been to places that relative quantification cannot go, and have gained molecular insights that only absolute quantification can give. This is the story of that journey so far.

The Journey Continues: Epigenetics Meets Mathematics

The highly conserved PcG and TrxG of proteins are epigenetic gene regulatory proteins. These proteins can provide a "memory" system that is essential for maintaining the correct identity of both stem cells and differentiated cells, and for orchestrating transitions between them. Aberrant expression of these proteins can lead to developmental defects and cancer [2]. The PcG and TrxG proteins work antagonistically on several hundred developmentally important target genes, to maintain repressed (PcG) or active (TrxG) transcription states [3].

The biochemistry of the PcG and TrxG proteins and the DNA-binding proteins that recruit them has been intensely studied over the last decades, showing that these proteins bind to their target genes in many different ways, including direct DNA binding, and binding of modified histones and noncoding RNAs. In addition, many of these proteins have activities that can modify the chromatin template itself, including enzymatic activities that modify histone tails and the transcriptional apparatus, and those that promote nucleosome remodeling or compaction of chromatin (for reviews, see Refs. [3,4]). The emerging picture is of a "bi-stable" system, in which one or other group of proteins (the activators or the silencers) takes possession of a given target gene locus, and modifies its local chromatin environment in a manner that favors activation or silencing and is stable and heritable across cell generations (see Chapter 8: Quantitative Environmentally Triggered Switching Between Stable Epigenetic States; Chapter 7: Modeling Bistable Chromatin States for discussions of chromatin mediated heritable bistability).

Because the PcG/TrxG proteins can maintain stable states of gene expression over many rounds of cell division, it was long believed that they must form stable static structures on their chromatin target sites. However in recent years, we and others have shown that this system is extraordinarily dynamic: All of the PcG and TrxG proteins so far tested exchange rapidly on chromatin within seconds [5–8] (Fig. 5.1). Furthermore, it has become clear that the histone modifications that they catalyze, long thought to be stable and irreversible, are in fact reversible and therefore also highly dynamic [9]. Together these dynamic properties offer opportunities for regulation, allowing switching or modulation of output in response to cell cycle, developmental, environmental, or metabolic signaling.

The PcG/TrxG system is both robust and dynamic. The dynamic nature of the system raises some fascinating questions about how it works, for example, how can a system whose components are in constant flux nevertheless maintain stable states of gene expression? How does this flux allow change and what causes that change to occur? Addressing these questions requires quantitative experimental analysis in combination with a coherent theoretical framework that recapitulates the dynamic nature of the system. In recent years, we have focused on using live imaging to quantify the molecules involved in absolute terms and to measure their kinetic behavior at different stages of the cell cycle. Simple mathematical models describe this dynamic behavior in formal terms and provide predictions for further experimental testing. This has led to insights into molecular mechanisms of chromatin binding that are key points for regulating the stability and flexibility of gene expression states. In this chapter, I describe our efforts to bring the study of epigenetic gene regulation into the era of molecular systems biology.

ABSOLUTE QUANTIFICATION IN VIVO: A TECHNICAL CHALLENGE

In 1996, it took about 6 months to quantify all the components of my in vitro system. Measuring the absolute concentrations of purified DNA and protein

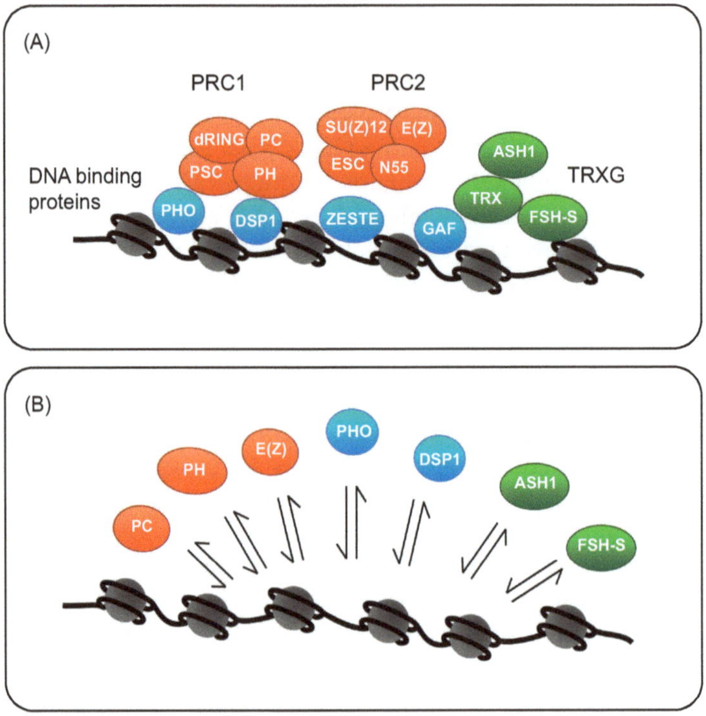

FIGURE 5.1 Polycomb and Trithorax group proteins bind dynamically to chromatin.
(A) Selected fly complexes and proteins are shown (for comprehensive reviews, see Refs. [3,4]). Red: Polycomb group silencing complexes PRC1 and PRC2. Green: Trithorax group activating proteins. Blue: DNA binding proteins that can recruit Polycomb and Trithorax group proteins. (B) Chromatin binding is dynamic, not static. Fly proteins whose mobility has been demonstrated experimentally are shown. PC and PH: [5,6,8]. PHO, E(Z), and ASH1: [8]. FSH-S and DSP1 (Philipp Steffen, unpublished).

is relatively simple. To determine binding rates, I stood in the cold room, loading samples of an ongoing reaction onto a running gel at 15-second intervals. I felt as if I was a classical musician giving a live performance: woebetide anybody who dared talk to me during those critical 2 minutes. I took to working at night to avoid any human intervention. The result, apart from a rather strange state of mind induced by lack of daylight, was a very useful set of numbers [1].

How can we do this type of analysis in a living organism? For DNA and chromatin, we have a good starting point: Given a known genome size and nuclear volume it is possible to estimate the expected concentrations of DNA and histones (e.g., the concentration of core histones assembled in DNA is estimated between 10 and 150 μM depending on cell type and organism [7]).

Proteins or histone modifications are more tricky. One could extract and detect proteins from a known number of cells, with known volumes, to get an average concentration. This is useful for ubiquitously expressed proteins in tissues that are relatively homogeneous and can be isolated in sufficient quantities for analysis, such as early embryos or cells in culture [10,11], for more examples, see Ref. [7]. Alternatively, one can fix the tissues and examine individual cells by immunofluorescence using an antibody to the protein of interest, calibrated via a standard [12]. However, both of the above approaches have the disadvantage that they rely on extracting or fixing the molecules we want to quantify. Thus, these techniques cannot address the dynamic aspect of the processes we are interested in, and they lack the developmental dimension. Live imaging offers a solution.

Fluorescent Tagging: A Powerful Tool for Live Imaging

In 2008, the Nobel Prize in Chemistry was awarded for the discovery and development of the Green fluorescent protein (GFP) (http://www.nobelprize.org/nobel_prizes/chemistry/laureates/2008/press.html). This small protein (238 amino acids, 26.9 kDa) has revolutionized modern biology. GFP glows bright green when exposed to light in the blue to ultraviolet range, and can be fused by genetic engineering to any protein of interest, making that protein visible in living cells under the fluorescence microscope. Since the discovery of GFP, other variants with other colors have been developed, enabling several different proteins to be tagged and imaged simultaneously [13].

Although GFP is a small protein, it may still interfere with the function of the protein it is fused to. Genetic rescue experiments can help determine whether your GFP fusion protein is still functional (Fig. 5.2A). The gene encoding the protein of interest (protein X) is first fused in frame to a gene encoding the fluorescent tag and introduced into the genome (e.g., of *Drosophila*). Transgenic animals are tested for the ability of the tagged protein to substitute for the endogenous protein, by checking whether they give "genetic rescue" of lethal mutations in the endogenous allele. In the example shown (Fig. 5.2A, left), an N-terminal GFP tag interferes with the function of protein X, and it cannot assemble into the right complex. Animals carrying only this tagged version of protein X do not survive, so there is no genetic rescue. On the right (Fig. 5.2A), the C-terminal tag does not interfere with function. Both tagging positions should be tested. Other deleterious effects of GFP could impinge on the correct folding of the protein or its stability.

The GFP tag can be introduced either into the endogenous gene encoding the protein of interest, or as an additional "transgenic" copy of the gene, placed elsewhere in the genome. These two strategies have different consequences for live imaging (Fig. 5.2B). If the tagged protein is provided as a transgenic

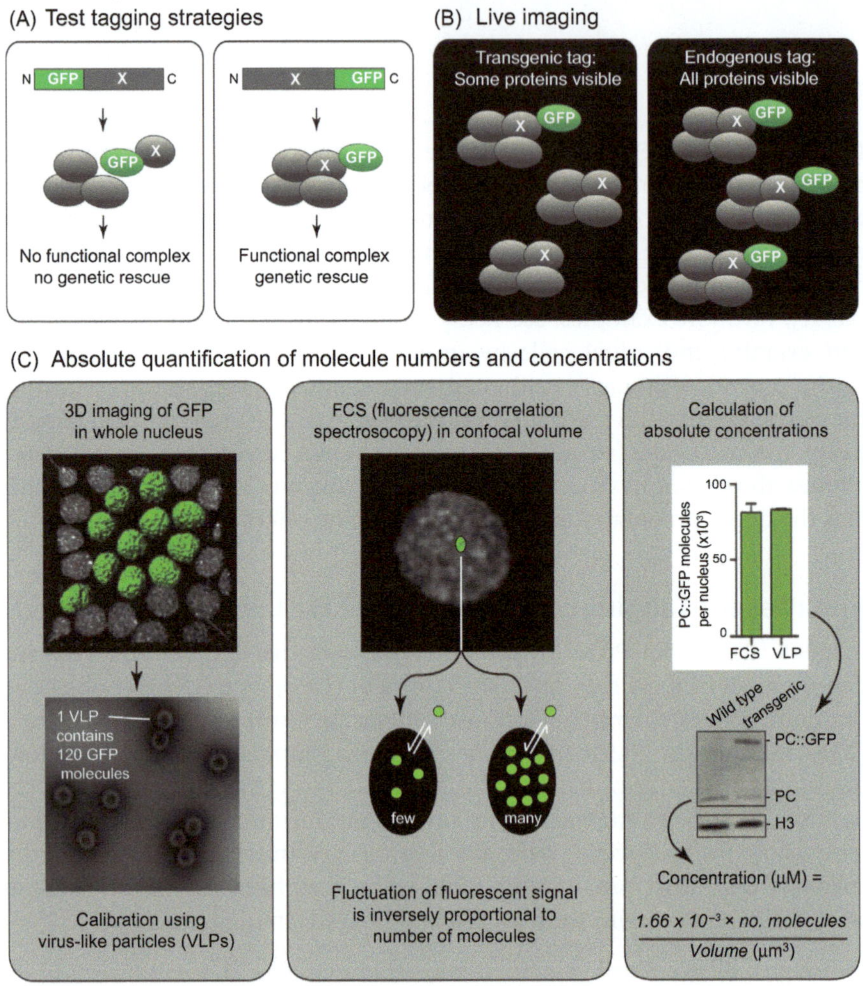

FIGURE 5.2 Imaging-based methods for quantifying molecule numbers in vivo.
(A) Testing fusion tags by genetic rescue. Top: transgenic construct expressing a fusion of protein X (gray) with GFP (green). Middle: complexes (gray) that would normally contain protein X are shown. See main text for details. (B) Different strategies for tagging give different live imaging situations. See main text for details. (C) Different methods allow quantification of molecule numbers in living animals or cells. Left panel: 3D imaging and calibration using virus-like particles (VLPs). Top: nuclei of a *Drosophila* embryo at cycle 12 expressing PC GFP. Bottom: VLPs are used to calibrate the GFP signal. See main text for details. Middle panel: Fluorescence correlation spectroscopy (FCS). Top: a confocal volume is defined. Bottom: fluctuations of fluorescence intensity in the confocal volume are caused by single molecules moving in and out of that volume. If there are few molecules (left), then the exit or entry of a single molecule will cause a large change in intensity. If there are many molecules (right) then the gain or loss of a single molecule will not make a big difference to the total intensity. In this way the fluctuation compared to the mean total intensity is inversely proportional to the total average number of molecules in the confocal volume [19]. For FCS, it is essential to have a low average number of molecules in the confocal volume, so that fluctuations of a few molecules cause a measurable change in the intensity of fluorescence. In our lab we typically measure between 5 and 100 molecules in the confocal volume, depending on the expression level of the tagged protein [8]. Right panel: calculation of total GFP molecules per nucleus and conversion to absolute concentrations of endogenous protein. Top: total GFP molecules measured by VLP and FCS. Middle: Western blot of embryo nuclear extract probed with α-Polycomb antibody (top panel) or α-H3 (bottom panel) [6]. See main text for details.

copy then the endogenous protein is still present and only the tagged copies are visible in live imaging (Fig. 5.2B, left). This strategy allows expression of the tagged protein at levels that are independent of those of the endogenous protein, which may be essential for some applications (see Fig. 5.3 and Box 5.1). If, on the other hand, the endogenous locus is tagged, then all copies of the protein will be visible in live imaging (Fig. 5.2B, right). This strategy has the advantage that the gene encoding the tagged protein is under the correct endogenous regulation but does not allow any fine-tuning of expression levels of the tagged version. Depending on the model organism, transgenic strategies may be faster and more tractable than those that require engineering of the endogenous locus. For this reason, we used a transgenic strategy for our work on PcG and TrxG proteins in *Drosophila* [6,8], However, the advent and rapid development of genome editing technologies based on the Crispr–Cas9 system may enable engineering at a locus of choice in a wide range of organisms in the near future [14].

Absolute Quantification of Molecule Numbers: Double Check!

With a functional GFP fusion protein in hand, we can see where it is in the cell, but we can also do much more. For example, we can also count how many GFP molecules there are in a defined volume, an essential step toward absolute quantification of endogenous molecule numbers and concentrations. In my experience, the idea of absolute rather than relative quantification has raised quite some opposition from reviewers and critical colleagues. How dare I claim to know an absolute number? The critics are right—if I am going to stick my neck out and say, "this is how many there are" then it is important to have at least a double check on that number. In Fig. 5.2C, I show two independent ways of estimating the number of GFP molecules in a given volume and show that the two methods agree very well with each other. For alternative methods, see for example, Refs. [5,15,16].

The first method is based simply on imaging of the entire nucleus (Fig. 5.2C, left panel). GFP intensities are measured by 3D imaging of the entire nuclear volume (in this example of *Drosophila* embryos at cycle 12, the nuclei have an average volume of $222.6 \pm 7.5\,\mu m^3$) [8]. 3D imaging gives us a measure of total intensity but does not tell us what that represents in real molecule numbers. To address this, the GFP intensity is calibrated by comparing it to "virus-like particles" (VLPs) imaged under the same conditions. Each of these VLPs comprises 120 GFP molecules fused to the viral coat protein, and thus they serve as a "molecular ruler" to allow estimation of absolute GFP numbers [6,17,18].

A second method, called fluorescence correlation spectroscopy (FCS) [19], gives a more direct measure of molecule numbers (Fig. 5.2C, middle panel;

see legend for more detail). FCS measures the fluctuations in fluorescence intensity within a very small volume, defined by the optical properties of the confocal microscope. The amount of fluctuation, caused by single molecules moving in and out of the volume, is inversely proportional to the average number of molecules in the volume. This volume is typically less than 1 μm^3 or 1 femtoliter (1 fl). In our lab, the volume is typically 0.1 fl, and thus it is over 2000 times smaller than that of the entire nucleus mentioned above [8]. Calibration of the confocal volume is absolutely essential if we want to convert molecule numbers measured by FCS to molar concentrations and can be done by performing FCS on standard solutions of fluorophores (small synthetic fluorescent molecules) with known concentrations and diffusion coefficients [20].

Once we have our estimate of the number of molecules in the tiny volume measured in FCS, we then convert this into an estimated number for the whole nucleus, using the nuclear volume. The huge difference in volume between the confocal volume and the nucleus itself, illustrates why it is so important to accurately calibrate the confocal volume used in FCS, as any error in this measurement will be amplified by the conversion to total nuclear volume. The estimate of total GFP molecules per nucleus then enables the first (VLP calibration) and second (FCS) techniques to be compared to each other (Fig. 5.2C, right panel). The calculation of absolute molecule numbers of PC::GFP using VLPs and FCS is shown for *Drosophila* embryos, showing good agreement between the two methods, and estimating approximately 80,000 GFP molecules per nucleus in both cases [6,8].

If, as in this example, a transgenic copy of the GFP fusion protein is used, then the endogenous molecule numbers are calculated by comparing the transgenic animals and wild-type animals with a standard (here, Histone H3) in a quantitative Western blot (Fig. 5.2C, right panel, top). The use of the standard allows correction for any effect of the transgene on the levels of endogenous protein when the two are coexpressed. In this example, endogenous PC is downregulated in the presence of the transgene. This effect has been observed by several authors for different proteins and should be accounted for [5,8]. Once the total number of endogenous molecules is known, then the micromolar concentration can be calculated via the Avogadro constant ($=6.022 \times 10^{23}$ which gives the number of molecules in one mole), and the known total nuclear volume, as shown (Fig. 5.2C, right panel, bottom). Concentrations of PcG and TrxG proteins measured by this method are typically between 50 and 400 nM, depending on cell type and developmental stage [6,8].

Using this approach, we have generated a catalog of concentrations of PcG and TrxG proteins in different cell types of living *Drosophila* [6,8]. By using absolute quantification, we were able to compare multiple proteins to one another without needing to tag them in the same animal. So we have our units, which is the first step on the journey toward a quantitative model of the system. But what else do we need? The next step is to measure the kinetic properties of our system.

EVERYTHING'S MOVING: METHODS FOR MEASURING KINETIC PARAMETERS IN VIVO

Once we know how many molecules we have, we next want to know how fast they move, and how long they stay in one place. It came as a big surprise to the field of epigenetics to learn that even molecules that mediate very stable long-term gene silencing are constantly in motion, hopping on and off their chromatin binding sites in a matter of seconds [5,21,22]. This demanded that we rethink our ideas of how silencing works. We may now envisage silent chromatin, not as a dense forest of interlocked branches, but rather as a noisy flock of birds, constantly visiting and leaving, but always filling the trees. There are several ways to measure and quantify these movements, based on fluorescent live imaging. Three of these, namely FRAP (fluorescence recovery after photobleaching), FCS, and SMT (single molecule tracking) are briefly described in Fig. 5.3 and box 5.1. For more information on these techniques see Refs. [19,23–25].

IN VIVO BIOCHEMISTRY OF AN EPIGENETIC SYSTEM: WHAT DID WE LEARN?

We have used both FRAP and FCS to extract information about the kinetic behaviors of the PcG and TrxG proteins in different cell types of living *Drosophila* [6,8]. These experiments, combined with absolute quantification for several of our favorite proteins, gave insights into the "in vivo biochemistry" of our system. What did we learn from this painstaking analysis? One question that we were particularly interested in was what happens to these properties when the cell enters mitosis? The molecular disruptions of DNA replication and mitosis are important challenges to the stability of memory, because the information that was present in the previous cell generation must be copied at replication and faithfully transmitted to daughter cells after mitosis [3]. Consequently, these phases of the cell cycle also offer the most important windows of opportunity for change. Because we were working in living, developing *Drosophila*, we were able to examine mitosis in several different cell types, at different stages of the differentiation process. For example, in early embryos, and in stem cells of the larval brain, the final identity of the cell has

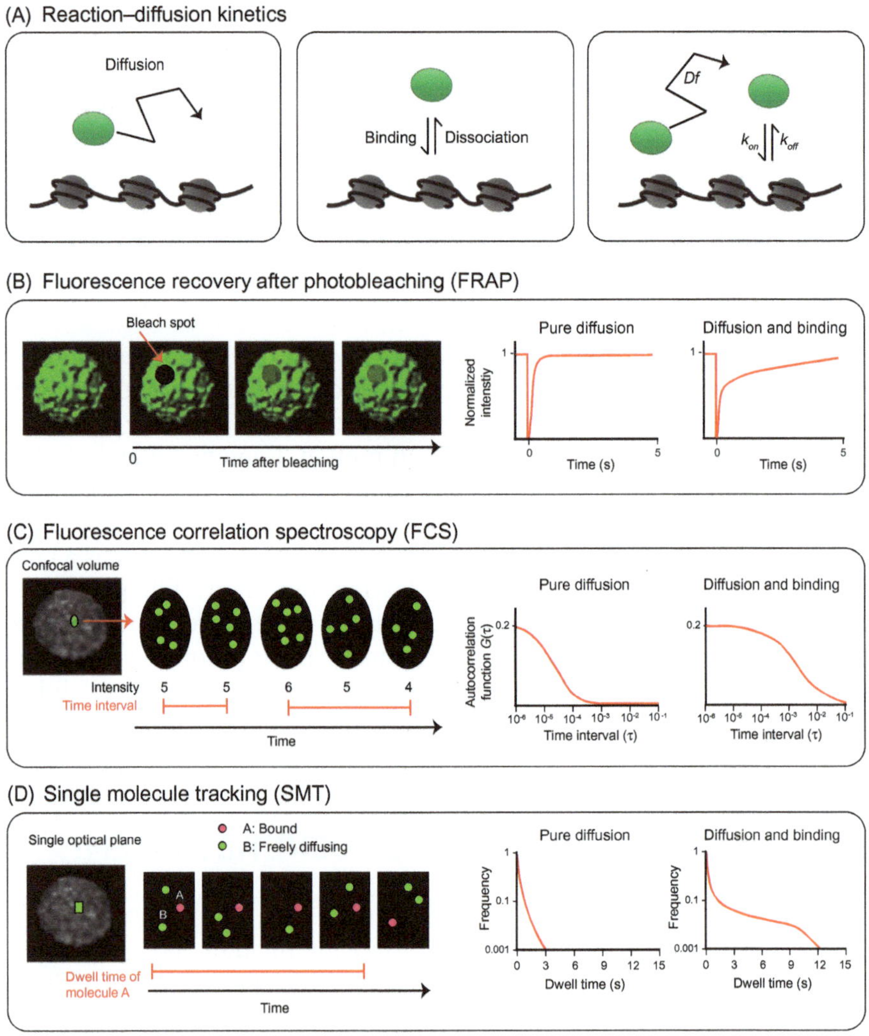

FIGURE 5.3 "In vivo biochemistry": Imaging-based methods for measuring kinetic behavior in living cells. See box 5.1 for detailed explanation.

BOX 5.1 FIGURE 5.3 DETAILED EXPLANATION

(A) Reaction and diffusion kinetics are different. Fig. 5.3, Left panel: Molecules that bind chromatin do not always do so. For much of the time, the molecules are merely randomly diffusing through nuclear space. Middle panel: The same molecules sometimes bind to chromatin or DNA, they stay for a certain time ("dwell time" or "residence time") and then dissociate. Right panel: In live imaging experiments, we aim to distinguish between diffusion [defined by the diffusion coefficient *Df* (units: $\mu m^2 s^{-1}$)], binding [defined by the association rate, k_{on} (units: $M^{-1} s^{-1}$)] and dissociation [defined by the dissociation rate k_{off} (units: s^{-1})]

[26]. If there is a sufficient difference in the speed of these events (i.e., diffusion is fast and binding is slow), then they can be distinguished from each other by fitting models to data generated by the experiments shown in (B) to (D). Recent models allow extraction of both diffusion and binding parameters, and have been developed for both FRAP [27] and FCS [28].

(B) Fluorescence recovery after photobleaching (FRAP). FRAP measures mobilities of fluorescent molecules by first creating a dark space, and then observing how fast that space gets filled with fluorescence. Left: To create the dark space, a spot within the nucleus of a cell expressing a fluorescently tagged protein is bleached with a laser. All the molecules in this volume become irreversibly invisible. The recovery of fluorescence in the same spot, caused by other unbleached molecules moving in as the bleached ones move out, is measured over time. This gives information about the mobility of the molecules in the bleached region. The plots show (left) a typical curve that might be observed if there is only diffusion: A very fast and total recovery of fluorescence (note that an identical curve would be obtained if there is diffusion and equally fast binding). The right plot shows a typical curve that might be observed if there is both fast diffusion and slow binding. If these occur on sufficiently different time scales, then the fast and slow phases of the recovery are clearly distinguishable. The diffusion coefficient and the binding constants can be extracted from such curves by fitting suitable models to the data [27]. Not shown in the figure: If a fraction of molecules does not move at all, this would result in an incomplete recovery. Other photobleaching based methods have been developed for different purposes. In fluorescence loss in photobleaching (FLIP), a region is repeatedly bleached and the rate of fluorescence loss at a region of interest outside the bleached region is measured [29]. In inverse FRAP (iFRAP), the whole cell is bleached with the exception of the region of interest, and the rate of fluorescence loss in that region is measured [5]. One disadvantage of FRAP and related techniques is that they require photobleaching of a relatively large area of the cell or nucleus, which can cause unseen damage. An alternative method that does not require intentional bleaching is FCS.

(C) Fluorescence correlation spectroscopy (FCS). In addition to giving information about molecule numbers (Fig. 5.2), FCS can also give information about mobilities [19,25]. Fluctuations in the total fluorescence intensity in the confocal measurement volume give information about how fast molecules are moving. Fast moving molecules (e.g., those that are diffusing) will move out of the volume and will immediately be replaced by those coming in from outside. Thus, over very short time scales, consecutive images will have very similar intensities despite movement of molecules in and out of and through the volume. Compare the first two panels of the time course, each containing five molecules but differently positioned within the space. Over longer time scales (compare the third and fifth images) this correlation breaks down, as random movements generate a range of images with different intensities. For slower moving molecules, the time interval (τ) over which any two images are still correlated becomes longer. The plots on the right of Fig. 5.3C show the autocorrelation function $G(\tau)$ plotted against τ for our two different scenarios. On the left, a rapidly diffusing molecule shows correlation between pairs of images only at very short time scales. On the right, a molecule that engages in slow binding as well as diffusion shows a typical increase in the time scales over which the intensities of pairs of images are correlated. If the rates of diffusion and binding are sufficiently different, then the diffusion coefficient and the binding constants can be extracted from such curves by fitting suitable models to the data [28]. An extension of FCS is FCCS (fluorescence cross-correlation spectroscopy), in which two differently labeled molecules are examined to address whether they move together or separately [30].

When combined with suitable models, both FRAP and FCS can give information about the average behavior of molecules in the volume of interest. For a comparative discussion see Refs. [7,25]. The key parameters that relate to biological properties are the diffusion coefficient and the on and off rates [31]. The diffusion coefficient can give an indication of whether the protein is diffusing as a monomer (if the diffusion coefficient is as expected for a protein of that size) or in a complex (if the diffusion coefficient is slower than expected). The inverse of the off rate ($1/k_{off}$) gives the average residence time of the protein of interest. The on rate is a little more cryptic: the parameter k^*_{on} extracted from FRAP or FCS data is defined as $k_{on} \times [B]$, where [B] is the concentration of free binding sites at equilibrium (see Fig. 5.5A). As [B] is usually unknown, it is extremely difficult to determine true binding rates from this kind of data. However k^*_{on} does have its uses: The inverse ($1/k^*_{on}$) gives the average time a molecule needs to find its binding site. The simple relationship $k^*_{on}/(k^*_{on}+k_{off})$ gives the average bound fraction of total molecules. These operations are based on parameters extracted from fitting models to data that indirectly measures movement. A further recent technique, shown in (D), allows tracking of single molecules, and therefore, enables direct observation of diffusion and binding events.

(D) Single molecule tracking (SMT). Individual molecules are visualized as they move (or do not move) in a single optical plane [23,25]. In contrast to FCS and FRAP, which measure total intensity and not individual molecules, SMT tracks the movement of each molecule. Prerequisites are a very low number of fluorescent molecules in the optical space, a very bright and robust fluorescence that is not bleached during the time of imaging, and an imaging setup that enables a single sheet and not a volume to be illuminated (in order to exclude signal from above and below the region of interest in the Z direction). Different setups are possible [23] but the principle is similar: The speed of movement of diffusing molecules (green on the figure), the frequency of events in which a molecule binds for longer (pink) and the dwell time of those molecules, can all be measured directly. The plots (right) show typical data for our two scenarios. On the left, if all molecules are diffusing, then the vast majority will have a short dwell time. On the right, if both diffusion and binding are occurring then there will be both short and longer dwell times. SMT has the advantage that it measures single events, rather than an average of events. It has been used recently to measure behavior of transcription factors as they visit specific places in the cell, by colocalizing with a second tag [32]. SMT has also been used to image calcium channels in whole worms, thus opening the way for its application to developmental questions [33].

not yet been determined, and we may expect chromatin to be in a rather flexible state, allowing for different gene expression patterns to be established later. This, in turn, might be reflected in the way that the PcG and TrxG proteins interact with chromatin. In contrast, in a more differentiated cell lineage, in which many irreversible decisions have already been taken, we might expect the PcG and TrxG proteins to be more tightly bound to chromatin.

Indeed, these expectations were partly borne out by the experiments, but there were also some surprises (summarized in Fig. 5.4) [6,8]. For simplicity, I show data for only the PcG protein, PC, and the TrxG protein, Absent, small, or homeotic discs 1 (ASH1), as representative proteins that are involved in silencing (PC) and activation (ASH1). In stem cells and embryos, the PC

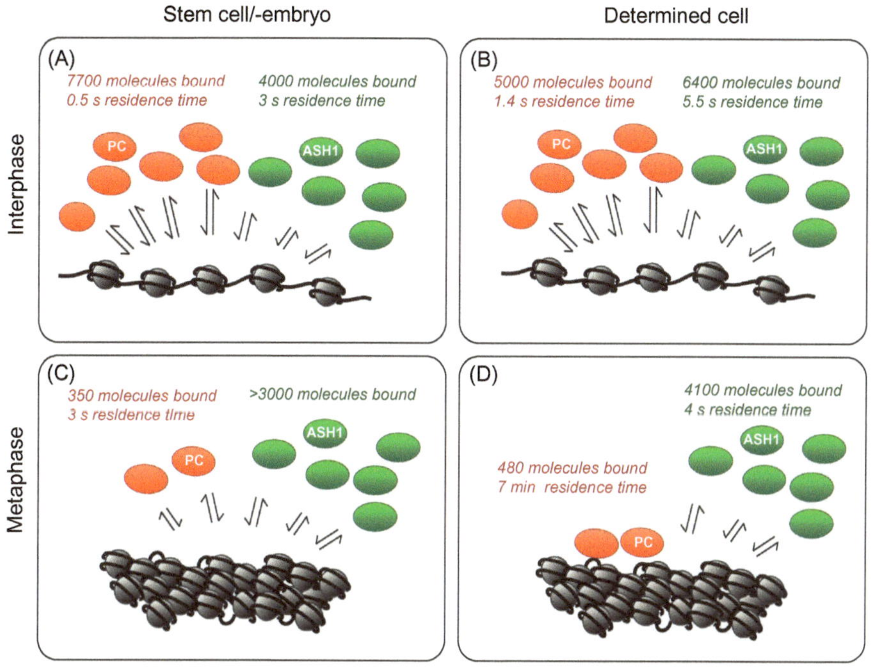

FIGURE 5.4 Insights gained from in vivo biochemistry on the Polycomb and Trithorax group proteins.

A combination of absolute quantification (see Fig. 5.2) and kinetic analysis (Fig. 5.3) gives information about the average interaction of Polycomb (red) and Trithorax group (green) proteins with chromatin in different cell types and at different stages during the cell cycle. (A) Interphase binding kinetics in stem cells/embryos; (B) Interphase in determined cells; (C) Metaphase binding in stem cells/embryos; (D) Metaphase in determined cells. The values shown were measured using FRAP in preblastoderm embryos, in stem cells (neuroblasts of third instar larvae) and determined cells (sensory organ precursor cells of pupae) [6,8]. See main text for details.

and ASH1 proteins bind chromatin with short residence times of a few seconds, and in similar abundance to each other (Fig. 5.2A). In the more determined sensory organ precursor cell lineage, found in pupae, both proteins bind for a little longer, suggesting a reduction in plasticity of the system as differentiation proceeds (Fig. 5.2B). When all three cell types enter mitosis, the PC protein largely dissociates from mitotic chromatin (Fig. 5.2C and D). Surprisingly, a large proportion of the ASH1 protein remains bound. This may reflect a requirement for ASH1 in "bookmarking" genes for reactivation after mitosis [34]. This idea is currently under investigation. What happens to the mobility of the proteins as cells enter mitosis? In stem cells, we measured a three fold increase in PC residence time at mitosis compared to interphase (Fig. 5.2C), whereas in determined cells (Fig. 5.2D), the residence time was dramatically increased in mitosis, to 280-fold longer than in interphase. This suggests that the mitotic binding properties of PC are highly regulated and may indicate a more important role for PC binding in propagating epigenetic memory through mitosis in more determined cells.

In summary, these experiments show that by combining absolute quantification and kinetic analysis for different proteins in living animals, very different characteristics emerge, both in terms of the "personality" of each protein, and in terms of their behavior at different stages of development and during the cell cycle. It was possible to detect these differences only because we used quantitative analysis, and because we studied living and developing animals. The next challenge is to understand what causes these differences in molecular terms, and whether they are important for the biology of our system. Although we have pursued several different angles [6,8], I will focus here on the question: Why does ASH1 bind so much more abundantly to mitotic chromatin than PC, and how is mitotic binding regulated for both proteins? To address this we turned to modeling and perturbation.

MATHEMATICAL MODELING OF THE SYSTEM: DEFINING WHAT WE DON'T KNOW

We want to understand why ASH1 stays on chromatin and why PC comes off during the transition between interphase and mitosis. However, in the above experiments, we were not able to measure this transition itself, because it is too dynamic for different stages to be reliably identifiable. Instead, we took a "snapshot" of our proteins binding to chromatin at only two stages of the cell cycle (interphase and mitosis). I will show how modeling helps to understand what could be happening in those missing moments and makes concrete testable predictions about mechanisms.

We used kinetic modeling based on ordinary differential equations (ODEs, described in more detail below and in Fig. 5.5). ODEs express changes in the quantities of components of a system over time, in terms of the concentrations of those components themselves, and their rates of gain and loss from the system. We used ODEs to simulate the whole time course of the cell cycle, taking into account the measured changes in volume, and in the concentrations, bound fractions and kinetic properties of each of our two proteins. ODE modeling is appropriate for modeling changes over time, and in our case is very useful because we know the exact duration of each phase of the cell cycle, which places stringent constraints on fitting the model to the data, as we shall see below.

Basic Principles of Kinetic Modeling

The basic principles of kinetic modeling are explained in Fig. 5.5A–C and their legends. We model proteins binding to chromatin as the interaction between two components, "A" and "B," which associate to give the complex "C" (Fig. 5.5A). Although we know that in "real life" there are many more components involved, this model illustrates an important principle in mechanistic modeling: The model should not be much more complex than the things you can actually measure. As we can measure only one protein at a time and its mobility, and from that we can infer the number of molecules bound to chromatin, we limit our model to only those components. At equilibrium, the concentration of the complex "C" is given by the equation shown in panel A of Fig. 5.5 (bottom). In the cell, the most important event for our modeling of cell cycle transitions is nuclear envelope breakdown (NEBD), which occurs as the cell enters mitosis (Fig. 5.5B). At this moment, the proteins that were previously confined to the nucleus are released into the entire cytoplasmic volume, so their concentration is rapidly reduced. The example shows a *Drosophila* sensory organ precursor cell, in which the volume after NEBD is 3.5-fold increased compared to that of the intact nucleus [6]. The graph shows the effect on the number of complexes "C" of this 3.5-fold dilution, assuming the system reaches equilibrium and assuming there is no change in the concentration of binding sites. Because the number of complexes at equilibrium depends on the concentration of free components (see the equation in panel A), the total amount of complexes is reduced 3.5-fold by this dilution. How do we use this information to simulate a time course? Fig. 5.5C shows how the single kinetic equation on the left (describing the reaction) is converted into the three ODEs on the right (describing the rate of change of each of the components A, B, and C over time). The ODE model was implemented in Matlab. See the legend to Fig. 5.5 for alternative implementations.

Kinetic Modeling of Polycomb and ASH1 Makes Mechanistic Predictions

The ODE model described above was used to examine the effects of protein dilution at NEBD, and the kinetic changes we have measured, to ask what mechanisms might account for the observed differences between PC and ASH1 (Fig. 5.5D–G). We tested three different models.

Model 1: Wrong But Useful

The first model examines the contribution of dilution alone to mitotic dissociation (Fig. 5.5D). The on and off rates measured in interphase for each protein were used throughout the simulation, and the only change was the 3.5-fold dilution of the system at NEBD. In the case of PC, dilution can account for some but not all of the observed mitotic dissociation. In contrast, in the case of ASH1, dilution is predicted to remove more protein than is actually observed to dissociate. This was already a very interesting result, and

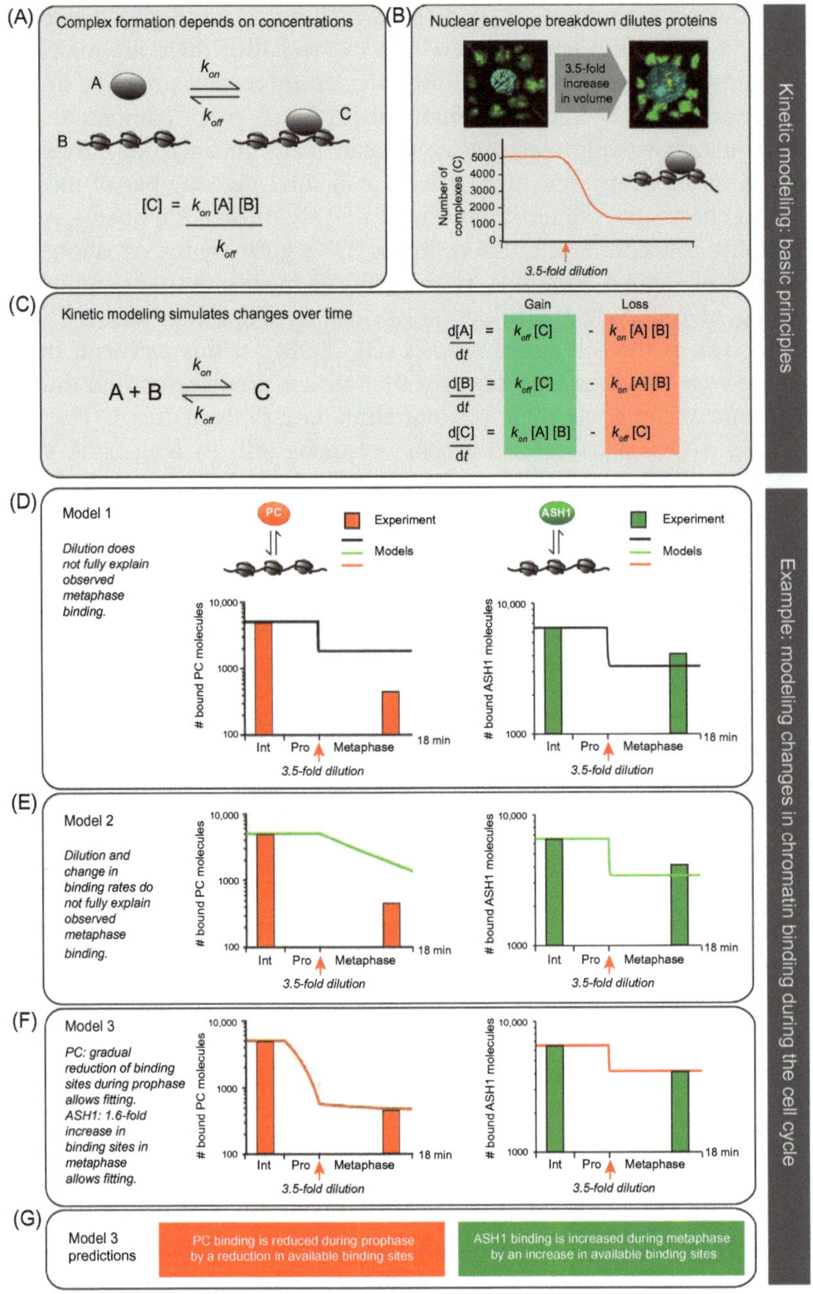

FIGURE 5.5 Kinetic modeling of cell cycle transitions using ordinary differential equations (ODEs).

(A–C) Basic principles of kinetic modeling. (A) Top: We model proteins binding to chromatin as the interaction between two components, "A" and "B," which associate to give the complex "C." Bottom: at equilibrium, the concentration of the complex "C" is given by the equation shown [26]. (B) At nuclear illustrates very nicely another fundamental principle of modeling, namely that "essentially, all models are wrong, but some are useful" [36]. The model we used here was intentionally wrong, because it ignored the real kinetic changes that we had measured for both proteins in metaphase. However, that oversimplification turned out to be very useful, because we now have to explain, what helps to remove PC from mitotic chromatin, and what helps to keep ASH1 there?

Model 2: Still Wrong But More Useful

The next most obvious thing to do was to include the real measured kinetic parameters for metaphase in the model, and ask whether this would be sufficient to explain the observed binding. This is shown in model 2 (Fig. 5.5E). The k^*_{on} and k_{off} values measured at metaphase were introduced into both models in addition to the dilution, with all changes occurring simultaneously at NEBD. The number of available binding sites (which is the only parameter not measured directly in the experiments) was assumed to remain unchanged throughout the cell cycle. In the case of PC, the on and off rates measured in metaphase are much slower than in interphase, with the result that the simulated system does not reach equilibrium within the time constraint imposed

➤ envelope breakdown, the proteins previously confined to the nucleus are released into the whole cell volume, causing a dilution of the system, which should decrease the number of complexes (see main text for detail) (C) Kinetic modeling using ordinary differential equations (ODEs). The kinetic equation (left) is described by a system of three ODEs (right). Changes in concentration of A, B, and C over time (t) depend on the concentrations of each of the three components, and on the rates of dissociation (k_{off}) and association (k_{on}). For example, A is gained by dissociation of C, and it is lost from the system when A and B combine to make C. Similar rules apply to B, whereas C is gained at the expense of A and B and is lost when it itself dissociates. ODE models can be implemented using standard mathematical tools such as Mathematica, Matlab, or R. There also exist open-source modeling packages that allow the user to bypass the mathematics, and simply plug in species and rate constants (such as Celldesigner www.celldesigner.org or Copasi, www.copasi.org). On the other hand, for the simple example shown, it is entirely feasible if you have no background in mathematics, to simulate this system manually using a pencil and paper and a calculator, or an Excel sheet. This exercise can be very fruitful for the nonmodeler and gives a "feel" for how such a system evolves over time. (D–G) The model described in (A–C) was used to examine the effects of protein dilution at nuclear envelope breakdown. The absolute numbers of observed bound molecules of endogenous PC (red) and ASH1 (green) are shown for *Drosophila* sensory organ precursor cells at interphase and metaphase. The *x*-axis shows the known duration of each of the cell cycle phases [35]. Different models were used to simulate the interphase–metaphase transition. As inputs for the models, the measured total numbers of bound and free proteins and the measured k^*_{on} and k_{off} rates for the two different proteins were used. The parameter measured in FRAP and FCS is k^*_{on} (units: s^{-1}), which is defined as $k_{on} \times$ [B]. The total number of binding sites ([B]+[C]) was assumed to be identical in both cases [6–8]. See main text for details.

by the cell cycle (given a longer simulation time, the system would eventually reach the observed level of binding). For ASH1, the kinetics during metaphase are little different to those of interphase, and thus there is little effect of introducing them into the model. So, model 2 is still wrong, but again it is useful in telling us what we might change next.

Model 3: This One Might Be Right!

The only remaining parameter that is unknown and may be changed to fit the model to the data is the number of available binding sites. It is reasonable to imagine that the effective concentration, availability and nature of binding sites may change during prophase (when chromatin condenses) and metaphase (when it is fully compacted), and that these structural changes in chromatin may be perceived differently by different proteins, because they bind through different mechanisms. For PC, we found that a gradual reduction of the number of available binding sites during prophase was required to give the system enough time to reach the observed level of dissociation [6]. For ASH1 on the other hand, it was sufficient to increase the number of available sites by 1.6-fold at the moment of NEBD, to fit the model to the data. In "real life" it is of course more realistic to imagine that the chromatin changes during prophase should affect the available binding sites for ASH1 in a gradual way, as we have used for the PC model, and not in the sudden manner that we have used for the ASH1 model. However, the fact that we did not need to include this feature in the model for ASH1 to get a good fit illustrates a third important principle of modeling: Only include the minimum number of things in your model that are needed to fit it to the data.

Model 3 might be right, as it is consistent with the data, but it might still be wrong, as there may be other explanations for the data that we have not thought about. There is only one way to find out, and this illustrates a fourth principle of modeling: Once we have a model that fits, it is time to go back to the bench and ask whether the predictions of the model are borne out by experiments. Importantly, for both PC and ASH1 the models make clear predictions (Fig. 5.5G), which can be tested experimentally, by specifically perturbing the system. In the case of PC, the model predicts that something accumulates during prophase and causes a reduction in the number of available binding sites for PC. For ASH1, the model predicts the opposite: That there is some kind of mechanism that could increase the number of available binding sites during mitosis. The beauty of this interplay between modeling and experimentation is that the model can guide our experiments so we know exactly where to look. In the next section, I look in detail at what these mechanisms could be and explain how we tested some of them.

BACK TO THE BENCH: TESTING THE MODEL BY PERTURBING THE SYSTEM

Different molecular mechanisms might cause the changes in binding properties we have observed during cell cycle transitions (Fig. 5.6A). For example, the properties of the protein itself might be changed by posttranslational or conformational changes, or other molecules may interfere with the process of binding, or with the chromatin template itself. These hypotheses can be tested experimentally in living cells, for example, by mutating the GFP fusion protein to discover which domains, binding surfaces, or amino acid residues contribute the observed binding behavior (Fig. 5.6B, top). Alternatively, other candidate molecules that help or prevent binding, or that modify the chromatin template, can be perturbed to ask whether this has any effect on the observed behavior (Fig. 5.6B, bottom).

We used different strategies to perturb the system for PC and ASH1 (Fig. 5.7). For PC, to identify molecules that could reduce the available number of binding sites during prophase, we used tissue-specific RNAi knockdown of a panel of candidate molecules (Fig. 5.7A). This enables an induced knockdown in specific cells, of transcripts of genes for which mutations would otherwise be lethal. We measured PC-GFP kinetics upon knockdown of other molecules and found that knockdown of the Chromosomal serine/threonine-protein kinase JIL-1 [6] or ASH1 itself [8] caused an increase in the residence time of PC, implying that they or their action would both normally reduce or prevent PC binding. Both ASH1 and the chromatin modification that is catalyzed by JIL-1 accumulate on chromosomes during prophase, making them good candidates for regulators of PC dissociation [6,8].

JIL-1 is a kinase that phosphorylates the tail of histone H3 on serines 10 and 28 (H3S28) [37–39]. Importantly, the residue adjacent to serine 28, which is lysine 27 (H3K27), can carry a trimethylation, which forms a binding site for the PC protein. The H3 tail modified at both S28 and K27 is a poor fit for the PC binding pocket, so the data are consistent with the idea that this additional phosphorylation would act to remove PC from these sites. Such an effect has been observed in interphase by other studies [38,39], and our results imply a major role for this "phospho-methyl switch" in regulating mitotic displacement of PC. In the case of ASH1, genetic data show that ASH1 and PC compete for the same target genes (reviewed in Ref. [3]). Our kinetic data show for the first time that ASH1 also affects the binding kinetics of PC. This may be directly, via physical competition, or indirectly via modification of PC or the chromatin template (ASH1 has histone methyltransferase

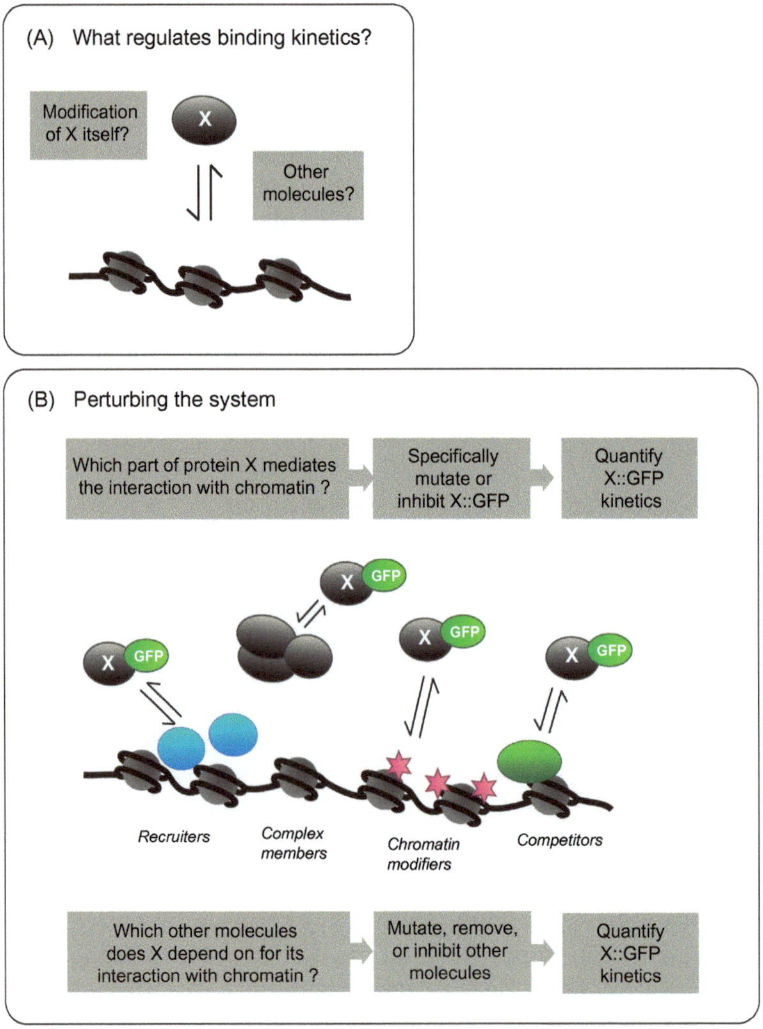

FIGURE 5.6 Perturbing the system experimentally.
(A) Different molecular mechanisms might cause the changes in kinetic properties we have observed during cell cycle transitions. For example, the properties of protein X itself might be changed by posttranslational or conformational changes, or other molecules may interfere with the process of binding or with the chromatin template. (B) These hypotheses can be tested experimentally in living cells, for example (top) by mutating the GFP fusion protein to discover which domains, binding surfaces or amino acid residues contribute the observed kinetics. Alternatively (bottom), other candidate molecules that help or prevent binding, or that modify the chromatin template, can be perturbed to ask whether this has any effect on the observed kinetics.

activity toward H3K36, which may also interfere directly or indirectly with PC binding [40–42]).

To perturb the binding of the ASH1 protein itself, we took a different approach. Because the mitotic binding of ASH1 is so robust, we were able to use direct imaging to screen for defects in binding for different mutant versions of the protein. Thus, we focused on finding out whether ASH1 uses different domains to bind to mitotic and interphase chromatin (Fig. 5.7B). To this end, we made a series of mutations and deletions in the ASH1::GFP fusion protein and measured their kinetic properties. We found that ASH1 binds to chromatin via different domains in interphase and mitosis. We propose that this change in binding mode may allow ASH1 to access a different and increased set of binding sites in mitotic chromatin. Most importantly, we found that impairment of the domains required for mitotic binding led to developmental defects and in specific cases, to lethality. This shows that the mitotic binding of ASH1 is most likely very important to the well-being of the fly.

In summary, this comparative analysis of two proteins using a combination of simple modeling, quantitative experiments and perturbation was very fruitful for unraveling mechanisms that may be key points for regulation. The models told us what to look for and the experiments told us what is and isn't important for binding, and whether this matters in a living animal. The power of *Drosophila* genetics was essential for thorough testing of the models, but essentially this approach can be applied to any system in which quantitative analysis is possible and controlled perturbation of some kind can be applied. In conclusion, I would like to point out that this is a strictly reductionist approach, and that there may be many more mechanisms of regulation in our system than those we have identified here. However, those that we have identified were only found because we combined rigorous quantitative analysis with predictive, reductionist modeling. This led to the identification of specific molecular mechanisms that may be key points for regulation and for misregulation in disease. The "wet-dry cycle" will continue with the future refinement of our models to account for these mechanisms.

OUTLOOK

We have come to the end of our journey for the purposes of this chapter, but for understanding dynamic epigenetic regulation, I suspect it is just the beginning. I hope I have given you some insights into how simple models and absolute quantification can be combined to answer some fundamental biological questions. The analysis we have done so far gives information about average behavior of all the molecules in the cell or nucleus we are measuring.

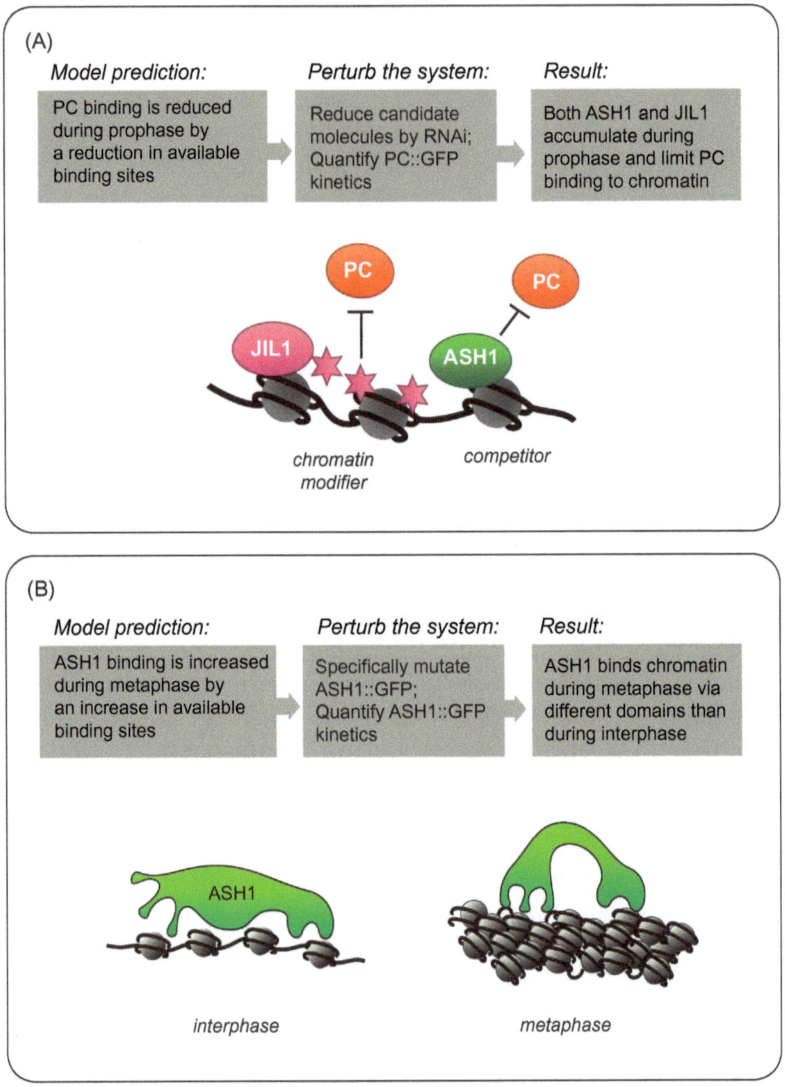

FIGURE 5.7 Perturbing the system: What did we learn?

(A) PC. To identify molecules that could reduce the available number of binding sites during prophase, we used tissue-specific RNAi knockdown. We found that knockdown of the chromatin kinase JIL-1 or the TrxG protein ASH1 caused an increase in the residence time of PC [6,8]. (B) ASH1. To identify domains of ASH1 that are required for interphase and mitotic binding, we made a series of mutations and deletions in the ASH1::GFP fusion protein and measured their kinetic properties using quantitative live imaging and FCS. We found that ASH1 binds to chromatin via different domains in interphase and mitosis. See main text for more details (Steffen et al., unpublished results).

This has been extremely useful for identifying "personality differences" between proteins, for understanding the effect of mutations and for identifying big global developmental changes. In future, it would be very interesting to extend these studies and to find a way to measure dynamics at single loci, with defined transcription status, in living animals. The combination of SMT with in vivo visualization of DNA and RNA [43] and genome editing [14] offers great promise. To model single binding events at single genes, we will need to move from the type of modeling I have described here, which represents an average system behavior, toward stochastic modeling (described in chapters 7 and 8) which describes single events for individual molecules. This brings me to a fifth principle of modeling: The model you choose must be appropriate for the question you want to answer.

Modeling and experimental manipulation of single defined loci may open the door to whole new world of insights into what the PcG and TrxG proteins are doing on their targets genome-wide. Indeed, it is becoming clear that stable memory of "on" or "off" transcriptional states is only one of a rich repertoire of potential outputs of PcG/TrxG regulation (reviewed in Ref. [3]). It has been proposed that this regulatory system may also stabilize gene expression states, amplify transcriptional oscillations, reduce noise by thresholding, and provide robustness to existing transcriptional networks [3,44]. Which of these outputs the system produces will depend on parameters such as transcriptional activity of the target gene, DNA sequence of the target site, genomic context, and fine-tuning, both by developmental signals and by noncoding RNAs [45]. I believe that in future, it will be essential to combine quantitative experimentation with a coherent theoretical framework to fully understand this fascinating dynamic regulatory system.

Acknowledgments

The author would like to thank Valère Lounnas and Lutz Ehrlich for those early insights into ODE modeling, and former members of the Ringrose lab at the IMBA, Vienna, for their contributions to quantification and modeling, in particular Philipp Steffen, Joao Fonseca, Cornelia Gänger, and Eva Dworschak.

References

[1] Ringrose L, Lounnas V, Ehrlich L, Buchholz F, Wade R, Stewart AF. Comparative kinetic analysis of FLP and Cre recombinases: mathematical models for DNA binding and recombination. J Mol Biol 1998;284(2):363–84.

[2] Mills AA. Throwing the cancer switch: reciprocal roles of polycomb and trithorax proteins. Nat Rev Cancer 2010;10(10):669–82.

[3] Steffen PA, Ringrose L. What are memories made of? How Polycomb and Trithorax proteins mediate epigenetic memory. Nat Rev Mol Cell Biol 2014;15(5):340–56.

[4] Bauer M, Trupke J, Ringrose L. The quest for mammalian Polycomb response elements: are we there yet? Chromosoma 2016;125(3):471–96.

[5] Ficz G, Heintzmann R, Arndt-Jovin DJ. Polycomb group protein complexes exchange rapidly in living *Drosophila*. Development 2005;132(17):3963–76.

[6] Fonseca JP, Steffen PA, Müller S, Lu J, Sawicka A, Seiser C, et al. In vivo Polycomb kinetics and mitotic chromatin binding distinguish stem cells from differentiated cells. Genes Dev 2012;26(8):857–71.

[7] Steffen PA, Fonseca JP, Ringrose L. Epigenetics meets mathematics: towards a quantitative understanding of chromatin biology. Bioessays 2012;34(10):901–13.

[8] Steffen PA, Fonseca JP, Gänger C, Dworschak E, Kockmann T, Beisel C, et al. Quantitative in vivo analysis of chromatin binding of Polycomb and Trithorax group proteins reveals retention of ASH1 on mitotic chromatin. Nucleic Acids Res 2013;41(10):5235–50.

[9] Nowak RP, Tumber A, Johansson C, Che KH, Brennan P, Owen D, et al. Advances and challenges in understanding histone demethylase biology. Curr Opin Chem Biol 2016;33:151–9.

[10] Wilkins RC, Lis JT. GAGA factor binding to DNA via a single trinucleotide sequence element. Nucleic Acids Res 1998;26(11):2672–8.

[11] Bagshaw JC, Bernstein RS, Bond BH. DNA-dependent RNA polymerases from *Artemia salina*: decreasing polymerase activities and number of polymerase II molecules in developing larvae. Differentiation 1978;10(1):13–21.

[12] Dubuis JO, Samanta R, Gregor T. Accurate measurements of dynamics and reproducibility in small genetic networks. Mol Syst Biol 2013;9:639.

[13] Wiedenmann J, Oswald F, Nienhaus GU. Fluorescent proteins for live cell imaging: opportunities, limitations, and challenges. IUBMB Life 2009;61(11):1029–42.

[14] Thakore PI, Black JB, Hilton IB, Gersbach CA. Editing the epigenome: technologies for programmable transcription and epigenetic modulation. Nat Methods 2016;13(2):127–37.

[15] Morrison AH, Scheeler M, Dubuis J, Gregor T. Quantifying the Bicoid morphogen gradient in living fly embryos. Cold Spring Harb Protoc 2012;2012(4):398–406.

[16] David DJ, McGill MA, McKinley RF, Harris TJ. Live imaging of *Drosophila* embryos: quantifying protein numbers and dynamics at subcellular locations. Methods Mol Biol 2012;839:1–17.

[17] Dundr M, McNally JG, Cohen J, Misteli T. Quantitation of GFP-fusion proteins in single living cells. J Struct Biol 2002;140(1–3):92–9.

[18] Rabut G, Doye V, Ellenberg J. Mapping the dynamic organization of the nuclear pore complex inside single living cells. Nat Cell Biol 2004;6(11):1114–21.

[19] Medina MA, Schwille P. Fluorescence correlation spectroscopy for the detection and study of single molecules in biology. Bioessays 2002;24(8):758–64.

[20] Culbertson CT, Jacobson SC, Michael Ramsey J. Diffusion coefficient measurements in microfluidic devices. Talanta 2002;56(2):365–73.

[21] Cheutin T, McNairn AJ, Jenuwein T, Gilbert DM, Singh PB, Misteli T. Maintenance of stable heterochromatin domains by dynamic HP1 binding. Science 2003;299(5607):721–5.

[22] Festenstein R, Pagakis SN, Hiragami K, Lyon D, Verreault A, Sekkali B, et al. Modulation of heterochromatin protein 1 dynamics in primary Mammalian cells. Science 2003;299(5607):719–21.

[23] Levi V, Gratton E. Exploring dynamics in living cells by tracking single particles. Cell Biochem Biophys 2007;48(1):1–15.

[24] Mueller F, Mazza D, Stasevich TJ, McNally JG. FRAP and kinetic modeling in the analysis of nuclear protein dynamics: what do we really know? Curr Opin Cell Biol 2010;22(3):403–11.

[25] Mueller F, Stasevich TJ, Mazza D, McNally JG. Quantifying transcription factor kinetics: at work or at play? Crit Rev Biochem Mol Biol 2013;48(5):492–514.

[26] Phillips R, Kondev J, Theriot J, Orme N. Physical Biology of the Cell. New York: Garland Science; 2009.

[27] Mueller F, Wach P, McNally JG. Evidence for a common mode of transcription factor interaction with chromatin as revealed by improved quantitative fluorescence recovery after photobleaching. Biophys J 2008;94(8):3323–39.

[28] Michelman-Ribeiro A, Mazza D, Rosales T, Stasevich TJ, Boukari H, Rishi V, et al. Direct measurement of association and dissociation rates of DNA binding in live cells by fluorescence correlation spectroscopy. Biophys J 2009;97(1):337–46.

[29] Lippincott-Schwartz J, Altan-Bonnet N, Patterson GH. Photobleaching and photoactivation: following protein dynamics in living cells. Nat Cell Biol 2003(Suppl):S7–S14.

[30] Bacia K, Kim SA, Schwille P. Fluorescence cross-correlation spectroscopy in living cells. Nat Methods 2006;3(2):83–9.

[31] McNally JG. Quantitative FRAP in analysis of molecular binding dynamics in vivo. Methods Cell Biol 2008;85:329–51.

[32] Morisaki T, Muller WG, Golob N, Mazza D, McNally JG. Single-molecule analysis of transcription factor binding at transcription sites in live cells. Nat Commun 2014;5:4456.

[33] Zhan H, Stanciauskas R, Stigloher C, Dizon KK, Jospin M, Bessereau JL, et al. In vivo single-molecule imaging identifies altered dynamics of calcium channels in dystrophin-mutant *C. elegans*. Nat Commun 2014;5:4974.

[34] Kadauke S, Blobel GA. Mitotic bookmarking by transcription factors. Epigenetics Chromatin 2013;6(1):6.

[35] Mayer B, Emery G, Berdnik D, Wirtz-Peitz F, Knoblich JA. Quantitative analysis of protein dynamics during asymmetric cell division. Curr Biol 2005;15(20):1847–54.

[36] Box GEP, Draper NR. Empirical model-building and response surfaces. New York: John Wiley and Sons; 1987.

[37] Jin Y, Wang Y, Walker DL, Dong H, Conley C, Johansen J, et al. JIL-1: a novel chromosomal tandem kinase implicated in transcriptional regulation in *Drosophila*. Mol Cell 1999;4(1):129–35.

[38] Gehani SS, Agrawal-Singh S, Dietrich N, Christophersen NS, Helin K, Hansen K. Polycomb group protein displacement and gene activation through MSK-dependent H3K27me3S28 phosphorylation. Mol Cell 2010;39(6):886–900.

[39] Lau PN, Cheung P. Histone code pathway involving H3 S28 phosphorylation and K27 acetylation activates transcription and antagonizes polycomb silencing. Proc Natl Acad Sci USA 2011;108(7):2801–6.

[40] Yuan W, Xu M, Huang C, Liu N, Chen S, Zhu B. H3K36 methylation antagonizes PRC2-mediated H3K27 methylation. J Biol Chem 2011;286(10):7983–9.

[41] Tanaka Y, Katagiri Z, Kawahashi K, Kioussis D, Kitajima S. Trithorax-group protein ASH1 methylates histone H3 lysine 36. Gene 2007;397(1–2):161–8.

[42] An S, Yeo KJ, Jeon YH, Song JJ. Crystal structure of the human histone methyltransferase ASH1L catalytic domain and its implications for the regulatory mechanism. J Biol Chem 2011;286(10):8369–74.

[43] Brody Y, Shav-Tal Y. Measuring the kinetics of mRNA transcription in single living cells. J Vis Exp 2011;54:e2898.

[44] Ringrose L, Paro R. Polycomb/Trithorax response elements and epigenetic memory of cell identity. Development 2007;134(2):223–32.

[45] Herzog VA, Lempradl A, Trupke J, Okulski H, Altmutter C, Ruge F, et al. A strand-specific switch in noncoding transcription switches the function of a Polycomb/Trithorax response element. Nat Genet 2014;46(9):973–81.

GLOSSARY

Polycomb group proteins a group of proteins, originally defined genetically, but now well characterized biochemically. Several PcG proteins are in large complexes that act on chromatin to silence their target genes.

Trithorax group proteins a group of proteins, originally defined genetically, but now well characterized biochemically. The TrxG proteins act antagonistically on chromatin to counteract PcG-mediated silencing and keep their target genes active.

Epigenome chemical modifications to DNA and histone proteins, and other molecules bound to chromatin (such as PcG and TrxG proteins).

Epigenetic memory the transmission of a gene expression state through cell generations in the absence of the initiating signal that switched the gene on or off.

Transgene a copy of a gene (usually tagged or modified) that is added to the host genome. The transgene may originate from the same species as the host or it may be from a different species.

Genetic rescue a transgenic copy of a gene is tested for its ability to substitute for a mutated copy of the endogenous gene, that would normally be lethal. If transgenic animals survive they are said to be "rescued" by the transgene.

Dwell time (or residence time) the amount of time a protein spends bound to its target site. Given by $1/k_{off}$.

Avogadro constant the number of molecules in one mole of any substance, $= 6.022 \times 10^{23}$.

Replication the process of copying each chromosome which occurs once each cell cycle.

Mitosis the process of partitioning the two copies of the chromosomes correctly to two daughter cells.

Histone methyltransferase an enzyme that adds one, two or three methyl groups to an amino acid side chain (usually lysine) in a histone.

Histone kinase an enzyme that phosphorylates a histone.

LIST OF ACRONYMS AND ABBREVIATIONS

PcG Polycomb group

TrxG Trithorax group

GFP Green fluorescent protein

VLP virus-like particle

FCS fluorescence correlation spectroscopy

FRAP fluorescence recovery after photobleaching

SMT single molecule tracking

Df **(units: $\mu m^2/s^{-1}$)** diffusion coefficient

k_{on} **(units: $/(M^{-1}\ s^{-1})$)** association rate constant

k_{off} **(units: s^{-1})** dissociation rate constant

k^*_{on} (units: s^{-1}) pseudo-first-order association rate constant, defined as $k_{on} \times [B]$ at equilibrium
[A] (units: M) concentration of free protein
[B] (units: M) concentration of free binding sites
[C] (units: M) concentration of complex of A bound to B
NEBD nuclear envelope breakdown
ASH1 Absent, small, or homeotic discs 1.
PC Polycomb
ODE ordinary differential equation
H3K27 lysine 27 of histone H3
H3K36 lysine 36 of histone H3
H3S28 serine 28 of histone H3

Modeling Distributive Histone Modification by Dot1 Methyltransferases: From Mechanism to Biological Insights

Dirk De Vos[1], Hanneke Vlaming[2], Barbara M. Bakker[3] and Fred van Leeuwen[2]

[1]University of Antwerp, Antwerp, Belgium [2]Netherlands Cancer Institute, Amsterdam, The Netherlands [3]University of Groningen, University Medical Center Groningen, Groningen, The Netherlands

CONTENTS

INTRODUCTION

Histone protein methylation is intimately involved in many chromatin-mediated regulatory functions of the genome [1]. Methylation of histones predominantly occurs on lysines and arginines, and the exact site of modification determines the downstream consequences. Methylation of histone H3K79 by Dot1 (in mammals called DOT1-Like or DOT1L) is an evolutionary conserved modification system with important functions [2–6]. Dot1 was originally discovered in budding yeast where it affects gene silencing, DNA damage response and repair, and meiotic checkpoint signaling. In flies, worms and mammals DOT1L is known to be involved in gene transcription. In mammals DOT1L and H3K79 methylation have been shown to act as a barrier against reprogramming [7] and to mediate tumorigenesis by certain oncogenic fusions of the mixed lineage leukemia (MLL) 1 gene (*MLL1*) [8]. Importantly, DOT1L inhibitors are currently in clinical trials for treatment of human MLL-rearranged leukemia, in which DOT1L misregulation is responsible for oncogenic transformation [8]. Finally, in African trypanosomes two Dot1 proteins together set up a dynamic methylation pattern that ensures proper cell cycle progression and antigenic variation [9–13]. Although a few proteins have been identified that can mediate the effect of methylation of H3K79 or "read" its methylation state, how Dot1 and H3K79 methylation control chromatin structure and function is still poorly understood. More

Epigenetics and Systems Biology. DOI: http://dx.doi.org/10.1016/B978-0-12-803075-2.00006-4

information on the function and regulation of Dot1/DOT1L can be found in several review papers [2–6].

H3K79 methylation by Dot1/DOT1L has also received special attention because of its uncommon properties [2,4]. One important feature is the location of H3K79 on the globular surface of the nucleosome core (Fig. 6.1), which contrasts with most other methylation sites, which are situated on the unstructured histone tails [16–18]. Indeed, Dot1 methylates H3K79 only in the context of a nucleosome, which precludes the use of simple peptide substrates in methyltransferase assays [16,19,20]. Therefore, analyzing Dot1's binding and activity toward synthetic peptides with defined methylation states is not possible. Another interesting feature that distinguishes Dot1 from other histone lysine methyltransferases is that it does not have a SET (Su(var)3-9, Enhancer-of-zeste, Trithorax) domain. This is unusual because the SET-domain constitutes the catalytic part of the vast majority of histone lysine methyltransferases [21–23]. Instead Dot1 contains a type-I methyltransferase domain [19,24,25]. This results in distinct mechanisms of methylation and regulation between methyltransferases [21,26,27]. Finally, a single Dot1 protein is responsible for mono-, di-, and trimethylation of H3K79, although in some species this task is split between multiple Dot1 proteins, as will be discussed below. Generally, the control of the different H3K79 methylation states and their specific functions are still poorly understood. A few mechanisms have been identified that promote H3K79 methylation, including ubiquitination of the C-terminal tail of histone H2B (H2Bub) along transcribed regions and recruitment of DOT1L to genes via its interaction with AF9, AF10 and other protein partners [2]. However, how H2Bub influences the activity of Dot1 at a molecular level is not well understood [2,28–31]. An important first step in explaining the changes in methylation seen when regulatory pathways are compromised is to understand how

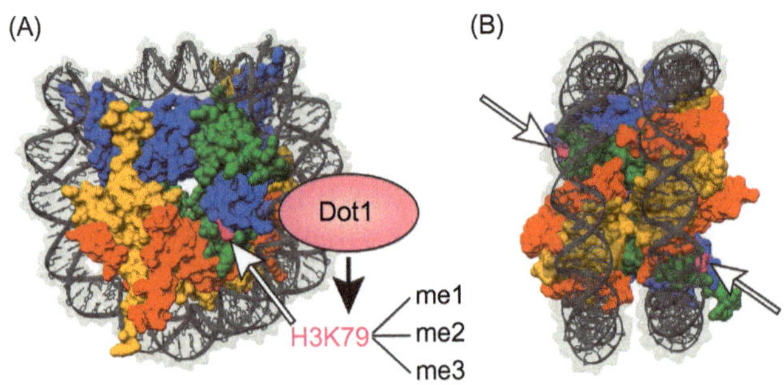

FIGURE 6.1 Dot1 methylates histone H3K79 on the surface of the nucleosome core.
Structure of the yeast nucleosome (PDB 1ID3 [14], visualized using Chimera [15]) in front view (A) and side view (B). Histone tails are absent from the crystal structure.

the (multiple) methylation states are established, i.e., knowing the mode of methylation and the kinetic parameters.

Another outstanding issue is how dynamic the cellular H3K79 methylation landscape is. Histone modifications together with other mechanisms define the epigenetic identity of a cell. This does not mean, however, that histone modifications are static. Some modifications are subject to dynamic addition and removal by modifying and demodifying enzymes, providing opportunities for adjustments by signaling to these opposing activities [1]. To date, demethylases for H3K79me have not been reported [32], suggesting that other mechanisms may be at play to counteract the Dot1 enzymes. DNA replication and the concomitant duplication of chromatin is by definition a process that disrupts the methylation pattern by the dilution of modified chromatin-bound histones with newly synthesized histones that are unmodified, or modified by so-called ground-state modifications [33–35]. Although some histone modifications such as histone acetylation and ubiquitination show high turnover and can be rapidly reestablished following S phase, histone methylation is generally more stable and only slowly reestablished [36–41]. Whether and how cells reestablish the modification pattern that existed prior to S phase is a fundamental question that lies at the heart of epigenetics as determinant of cell identity. Understanding the emergence of histone methylation patterns throughout the cell cycle requires a detailed understanding of the intrinsic biochemical properties of methyltransferases like Dot1.

Given the limitations of and challenges with in vitro methylation assays we have undertaken a combination of genetic engineering in yeast, quantitative measurements, and mathematical modeling to extract information on the basic properties of Dot1 in the living cell (Fig. 6.2). Here we describe our efforts to determine the mechanism of methylation by Dot1 and how this feeds into the regulation of the H3K79me landscape in the cell. This question is highly relevant because of the role of Dot1 in very fundamental cellular processes [2–5,42].

MODELING HISTONE MODIFICATION BY DOT1

Dot1 Action: Distributive Versus Processive

The first requirement for understanding the mechanisms of regulation of H3K79 methylation is knowing the intrinsic mechanism of methylation. Dot1 mono-, di-, and trimethylates H3K79. In general, such a multiple methylation reaction can occur processively or distributively (Fig. 6.3). In processive methylation, the methyltransferase remains bound to the substrate histone and performs its subsequent methylation reactions before it releases the final methylation product. A new round of methylation requires that the

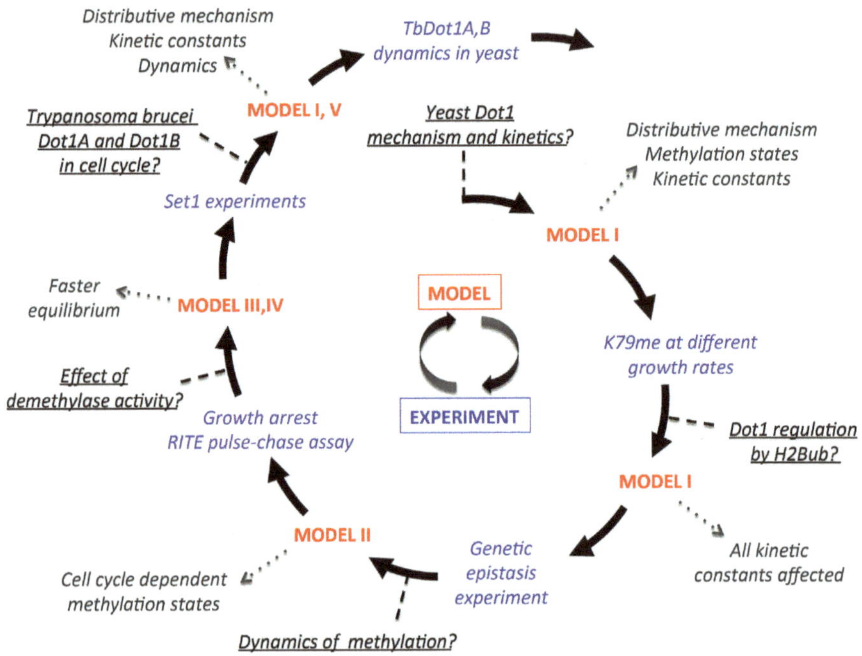

FIGURE 6.2 Schematic representation of the alternating model-experiment interactions throughout the described studies.
Models I–V (red color) correspond to distinct systems of differential equations (not just parameter differences). Model I represents the population (steady state) models with distributive kinetics. Model II represents the so-called single cell (cycle) models. Models III and IV represent the steady state and cell cycle models, respectively, with reversible reactions. Finally, Model V is the single cell model adapted for Trypanosoma (i.e., with three distinct cell cycle phases). Along the systems biology cycle each model(ing) step is preceded by a (symbolic) research question (underlined). Outgoing arrows point to model output. In blue color fonts experimental validation of the preceding model is indicated. Detailed explanations can be found in the main text.

reaction by-product *S*-adenosyl-L-homocysteine (SAH) is exchanged for the methyl donor *S*-adenosyl-L-methionine (SAM), without releasing the substrate histone. In this situation, the concentration of the intermediate states never exceeds that of the enzyme. In distributive methylation, the intermediate methylation states and SAH are released after each addition of a methyl group. Binding of a SAM-bound enzyme to intermediate states is needed for subsequent methylation events. The methylation mechanism is important for understanding the regulation and function of histone methylation states because the two mechanisms result in a different relation between the amount of enzyme activity and the absolute and relative abundance of the methylation states. Although in processive methylation all methylation states are expected to increase proportionally with increasing amounts of enzyme,

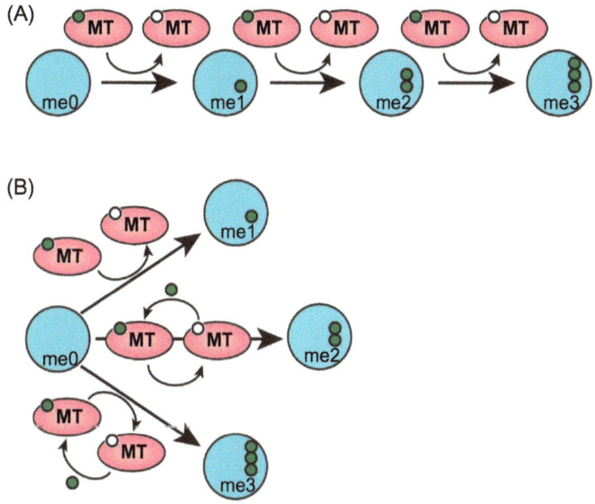

FIGURE 6.3 Schematic representation of distributive and processive methylation.
(A) In distributive methylation the methyltransferase (MT) releases from the substrate and a new binding of MT is required for a next methylation reaction. (B) In processive methylation, the MT remains bound to its substrate and SAH (white circle) is exchanged for SAM (green circle).

in distributive methylation the relative amounts of the methylation states show characteristic changes. For example, when starting from an unmethylated substrate, monomethylation has to accumulate to sufficient levels before a distributive enzyme is likely to encounter it among the many unmethylated substrates and convert it into a dimethylated state and so on. Eventually, the buildup of di- and trimethylated lysines will be at the cost of un-, and monomethylated states.

To determine the mechanism of methylation by Dot1, we measured the changes in H3K79 methylation as a function of changes in Dot1 protein levels in yeast. To that end, Dot1 was expressed from different promoters and the four methylation states of H3K79 (no, mono-, di-, and trimethylation, or me0, me1, me2, and me3, respectively) were quantified by immunoblot or mass spectrometry. As described in Frederiks et al. [43], measurements of methylation fractions as a function of Dot1 levels showed peaks of the fractions of methylation states, as outlined in Fig. 6.4. For a distributive mechanism with equally probable binding of different methylation products, one would expect such a succession of characteristic peaks as the Dot1 amount increases. For a processive model, the unmethylated form alone would enter the processing or production line and exit with a specific probability for each methylation state (see Fig. 6.4). Consequently, increasing the probability of enzyme (Dot1) encounters would only increase the number of events and not the exit probabilities. Briefly, a simple processive mechanism should lead

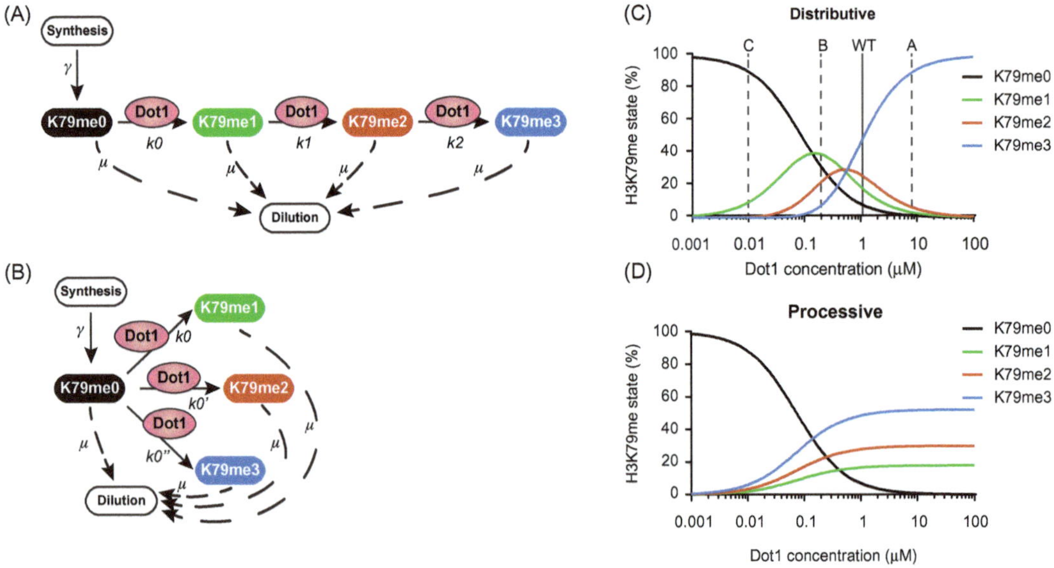

FIGURE 6.4 Models for in vivo H3K79 methylation.

(A) Distributive model. Dot1 places the first, second, or third methyl group on H3K79 with rate constants k_0, k_1, and k_2. Unmodified histones (K79me0) are synthesized at a rate γ and all forms of H3K79 are diluted with a first-order rate constant μ because of an increase in cell volume caused by cell growth. (B) Processive model. Dot1 mono-, di-, and trimethylates unmodified H3K79 with rate constants k_0, $k_{0'}$, and $k_{0''}$. Synthesis and dilution are as in part (A). (C/D) Simulation of distributive and processive H3K79 methylation over a range of Dot1 concentrations. The approximate WT methylation level are indicated with a vertical line. Parts A–C represent high, low, and very low levels of Dot1 activity. *Panels A–C were modified from De Vos D, Frederiks F, Terweij M, van Welsem T, Verzijlbergen KF, Iachina E, et al. Progressive methylation of aging histones by Dot1 functions as a timer. EMBO Rep. 2011;12(9):956–962.*

to characteristic methylation fraction ratios that should be identical for each different amount of Dot1. The distributive mechanism was supported with a basic statistical model in Frederiks et al. [43]. In this model, H3K79 methylation is represented as ($N = 3$) discrete reaction trials (so-called Bernouilli trials) with a certain methylation probability P that is equal for all methylation states in accordance with such a distributive model. The number of successful methylation events K after N trials then represents the actual methylation state (0, 1, 2, or 3). The probability of a methylation state for a certain probability of methylation P (equivalent to the total fraction of methylation) is then given by the binomial probability distribution:

$$f(K;N,P) = \binom{N}{K} P^K (1-P)^{N-K} \tag{6.1}$$

Plotting that function illustrated the expected succession of peaks of methylation fractions. The lack of a quantitative fit to the data was attributed to a potential difference in binding probabilities of the different methylation states.

Subsequently, with in vitro experimental data under controlled conditions, available kinetic models were set up to describe how a fixed amount of histone H3 would be gradually methylated in a processive or distributive way (Appendix). However, contrary to model simulations, a large range (1000-fold) of added Dot1 only produced relatively small differences in overall methylation. Available data suggested that Dot1 activity was higher in vivo and could result in pronounced differences in methylation.

Because a more classical and controlled in vitro assay did not deliver data of sufficient quality for fitting to a kinetic model, we applied a modeling approach using quantitative information on in vivo parameters [44]. Mathematical models for processive and distributive methyltransferases were developed that simulate histone H3K79 methylation in cells to determine the kinetic mechanism and obtain estimates of the rate constants of the different methylation steps (Fig. 6.4). Data on different steady-state expression levels of Dot1 in yeast were obtained using an inducible and tunable promoter. The different Dot1 expression levels were estimated in a semiquantitative manner from immunoblots by comparing them to the endogenous expression level, which had previously been estimated to be 2000 copies per cell [45]. The methylation states were measured by sensitive and accurate quantitative mass spectrometry methods. The substrate histone concentration in yeast was set to 200,000 copies per cell based on the size of the yeast genome and semiquantitative protein abundance measurements [44–46] and for each strain and experimental condition the growth rate was determined. This set of data was used as input for modeling the in vivo methylation of H3K79.

Living cells represent a much greater complexity than in vitro conditions, with homeostasis and balanced processes playing an important role. The regulation of Dot1 activity in vivo is likely to be complex and nonuniform across the genome. However, building a mathematical model requires simplification and abstraction. Several assumptions were made to build the initial model. First, we assumed that the in vivo H3K79 methylation state was entirely controlled by Dot1 activity, histone synthesis rate, and cell growth, without coregulation by other factors. Second, the influence of the substrate/cofactor SAM was assumed to be constant. Third, we considered all histones in the cell as equal substrates, i.e., we assumed the methylation to occur in one compartment without specific chromatin domains with altered properties such as Dot1 accessibility or Dot1 targeting. Finally, a growing population of yeast cells was represented by a single model compartment. At steady state, the assumption of a constant concentration of the histone H3 pool then holds as a result of the production of new histones balanced by continuous growth. The cells were supposed to be asynchronous, and the methylation states were assumed to be in a (population-scale) steady state, as was confirmed repeatedly by temporal sampling.

Ordinary differential equations (ODEs; Appendix) were used to describe dynamic changes in the four methylation states (represented by the variables *me*0, *me*1, *me*2, *me*3) based on the two main mechanisms (Fig. 6.3). In all cases, the histone production rate γ is a constant term that, together with the growth dilution process (represented by the dilution rate constant μ) leads to a constant, steady-state total histone concentration. The dilution process acts on all states proportionally to their abundance. For a processive mechanism (Appendix: Eq. (6.5)), the unmethylated state can undergo three different fates corresponding to mono-, di-, and trimethylation (followed by product release). In a distributive mechanism each methylation state has the potential to interact with the methylase and a methylation event is initiated from any free substrate (Eq. (6.2): v_i terms represents the methylation rates from the respective states i; details can be found in the Appendix). Upon that basic reaction stoichiometry, different types of kinetic expressions can be superimposed. Importantly, the level of kinetic detail (and number of associated parameters) needs to be tailored to the available data. A general assumption was that reaction rates were proportional to the total Dot1 concentration (*dot* in Eq. (6.3)), thus assuming that it reflects free Dot1 concentration. Comparing the most simple (1-parameter) processive and distributive models (Appendix: Eqs. (6.6) and (6.8)) confirmed that the distributive model more successfully predicted the in vivo data from Frederiks et al. [43]. In De Vos et al. [44], this model selection approach of comparing the quality of fit (visually and quantitatively using the sum of the squared errors as a measure) was extended to more complicated kinetic expressions (such as Appendix: Eqs. (6.9)–(6.11)). A distributive model based on simple mass-action (proportional) kinetics was found to give the best fit with the lowest number of parameters (Eq. (6.3)).

$$\begin{cases} \frac{d(me0(t))}{dt} = \gamma - \mu \cdot me0(t) - v_0 \\ \frac{d(me1(t))}{dt} = -\mu \cdot me1(t) + v_0 - v_1 \\ \frac{d(me2(t))}{dt} = -\mu \cdot me2(t) + v_1 - v_2 \\ \frac{d(me3(t))}{dt} = -\mu \cdot me3(t) + v_2 \end{cases} \tag{6.2}$$

$$\begin{cases} v_0 = k_0 \cdot dot \cdot me0(t) \\ v_1 = k_1 \cdot dot \cdot me1(t) \\ v_2 = k_2 \cdot dot \cdot me2(t) \end{cases} \tag{6.3}$$

Based on these results, it could be concluded that Dot1 is a distributive methyltransferase in vivo. This mechanism is supported by a crystal structure of yeast Dot1, which suggests that exchanging SAH for SAM requires active-site remodeling and release of the substrate lysine [19].

Through the model also rate constants for the three transitions were estimated (k_0,k_1,k_2 in Eq. (6.3)). Rate constants in these models represent the best estimations of the k_{cat}/K_m ratio for each methylation event. The kinetic constants in this study therefore represent apparent rate constants. More complicated enzyme kinetic equations can be considered but they had too many parameters to allow a unique fit to the available experimental data. The estimated rate constants of yeast Dot1 decreased as the number of methyl groups on the substrate increased [44]. One possible explanation for this trend is that the chance of properly orienting a free hydrogen on the terminal amino group of the lysine is reduced when fewer free hydrogen atoms are available, i.e., when one or two hydrogen atoms have already been replaced by methyl groups. Finally, a similar but semiquantitative strategy was applied to methyltransferase Set1 (SET-domain containing 1), which mono-, di-, and trimethylates histone H3K4. Simulations of the in vivo results strongly suggest a processive mechanism for this SET-domain methyltransferase [44].

A Distributive Mechanism of Methylation Provides an Explanation for Transhistone Cross-Talk Phenotypes

Ubiquitination of the C-terminal tail of histone H2B promotes H3K79 methylation [2]. The knowledge of the methylation mechanism of Dot1 has helped one to provide insight into this evolutionary conserved pathway of trans-histone cross-talk. Mutants of the H2B ubiquitination machinery show a striking loss of H3K79me3, which led to the suggestion that it might act by altering the processivity of Dot1 or specifically affecting the trimethylation activity of Dot1 [47,48]. By starting from the wild-type model, the methylation state of a strain without H2Bub was predicted after adjustment of the specific growth rate [44]. In accordance with the hypothesis that trimethylation is specifically affected by ubiquitination, first only the k_2 parameter, describing the rate of conversion of H3K79me2 to me3 was fitted on the basis of the data. As this did not significantly improve the data fit, and a good fit was only produced by changing all the parameters, the initial hypothesis could be rejected [44]. To validate the prediction that all transitions were compromised in the absence of H2Bub in the cell, we performed a genetic epistasis test: partial-loss-of-function alleles of *DOT1*, showing loss of H3K79me3 and gain of H3K79me1 (in agreement with the distributive mechanism) were combined with inactivation of the H2Bub machinery, which also partially compromises Dot1 activity. If H2Bub specifically promotes H3K79me3 formation, no additive effects would be expected. If

H2Bub promotes all transitions, it would further decrease the methylation in a background with a partial-loss-of-function allele of *DOT1*. The combination of mutations led to a loss of all methylation states, confirming the mechanism predicted by the mathematical model [43,44]. That H2Bub regulates all H3K79 methylation transitions is nicely in agreement with observations in fly and mammalian cells, in which H3K79me3 is present at very low levels and H2Bub promotes synthesis of H3K79me1 and H3K79me2 [2].

Histone Dilution as a Mechanism to Counteract the Distributive Build-Up of H3K79 Methylation

The steady-state model includes growth as a measure for histone dilution. A wave of dilution occurs during S phase when chromatin is duplicated by assembling a complement of newly synthesized, unmethylated histones. However, histone dilution can also occur outside S phase by so-called replication-independent (RI) histone turnover [49,50]. Likely the histone production and methylation rates are underestimated because of that. A prediction of the steady-state model is that in the absence of a demethylase, dilution of histones (be it replication-coupled or -independent) may be the main mechanism to counteract the slow build-up of H3K79me. To test the prediction that RI histone turnover negatively regulates H3K79 accumulation, we examined our dataset on histone inheritance, which can be considered as the inverse of histone turnover [51]. On a genome-wide scale, histone inheritance is not equal for each gene and a positive correlation was observed between high histone inheritance (i.e., low turnover) and H3K79me3 [51]. This observation is in agreement with the prediction of the model that histone turnover negatively regulates H3K79me3 in the cell.

A Single-Cell Model of H3K79 Methylation to Describe the Dynamics During the Cell Cycle

The steady-state model has been extremely helpful in understanding the kinetic mechanism of Dot1. To get more insight into the dynamics of H3K79 methylation throughout the cell cycle and specifically how the methylation pattern is reestablished following chromatin duplication, a different approach was needed. Instead of a time-averaged or population perspective this new model zoomed in on the cellular, more specifically the nuclear scale [44]. The previously obtained distributive kinetic model served as the starting point. However, for a single cell model the assumption of continuous histone production breaks down since it is known that histone synthesis is tightly connected with DNA duplication during S phase. To gain insight in the effect of periodic changes in histone production rate on the methylation state of a typical cell, two main phases were distinguished: a phase representative of the S phase and a phase representative of the G2-M-G1 phases combined

(Fig. 6.5). The S phase was assumed to take 20% of the total cell cycle time. Histone production was only taking place in the S phase, yet at a 5-fold higher rate to obtain the same overall histone production rate. Without an explicit variable for volume growth, a growth dilution is constantly acting on the histones (Dot1 concentration is a fixed parameter). Since according to the assumptions of the model methylation fractions are homogeneously distributed over the nuclear compartment, it follows that mitosis will not do more than just partition the original volume into two identical compartments. Since that does not alter the concentrations of the methylation states, cell division was not simulated. Since we further assumed that consecutive cell generations would pass on their methylation state, we repeated the single cell cycle simulation until a stable oscillatory pattern of methylation was reached, taking that as representative for the population (Fig. 6.5).

The single-cell model predicted that Dot1 is slow, and that the degree of H3K79 methylation is dynamic and is determined by the residence time of histones within the genome [44]. To validate this experimentally, we performed methylation analysis of biochemically purified old or aged histones from a cell. Histones that were synthesized at least four cell divisions prior to the purification showed substantially higher H3K79me3 and lower

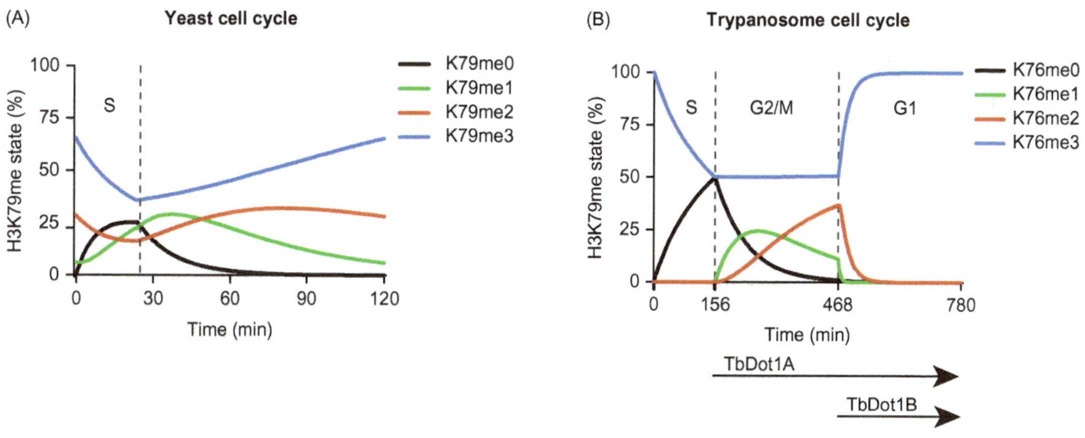

FIGURE 6.5 Single cell H3K79 methylation models.
(A) Simulation of H3K79 methylation in a single yeast cell starting at S phase. Methylation levels at the start of S phase are identical to the levels at the end of G1. (B) Simulation of H3K76 methylation in a single trypanosome, like in panel A but considering the sequential action of two different methyltransferases Dot1A and Dot1B in S, G2/M, and G1 phase. TbDot1A and -B expression could not be measured and were manually set to best fit the reported experimental data. *(A) modified from De Vos D, Frederiks F, Terweij M, van Welsem T, Verzijlbergen KF, Iachina E, et al. Progressive methylation of ageing histones by Dot1 functions as a timer. EMBO Rep 2011;12(9):956–62; and (B) modified from Stulemeijer IJE, De Vos D, van Harten K, Joshi OK, Blomberg O, van Welsem T, et al. Dot1 histone methyltransferases share a distributive mechanism but have highly diverged cata¬lytic properties. Sci Rep 2015;5:9824.*

H3K79me0/1/2 levels than histones of young average age. Histones of an average intermediate age showed an intermediate increase in H3K79me [44]. These results are in line with the idea that H3K79 methylation levels slowly accumulate over time. Of note, the accumulation progresses over multiple cell generations, indicating that the build-up of methylation is not finished at the end of a cell cycle just before a new S phase initiates. Slow accumulation of H3K79me on aging histones spanning multiple cell divisions has also been observed in human cells by independent pulse-chase proteomics methods [53]. Interestingly, these dynamics suggest what has been referred to as "scrambling" of epigenetic information and is incompatible with epigenetic memory models that assume exact duplication of the methylation patterns from mother to daughter cells [44,53]. Instead, the accumulation of H3K79me on aging histones or in aging cells could in principle act as a molecular timing mechanism. However, the physiological relevance of such a timer remains to be established.

Extending the Single-Cell Model With Demethylation

The modeling described above does not include demethylation. To date, there is no record of the existence of a H3K79 demethylase but demethylases have been found for other histone lysine methylation sites [27,54]. To examine the influence of a reverse reaction, a demethylase activity was included in the next modeling phase. Extending the steady state and single cell models in De Vos et al. [44] was straightforward; it suffices to mirror each methylation reaction with a demethylation reaction (Appendix: Eqs. (6.12) and (6.13)). The reaction rate was again assumed proportional to the enzyme (demethylase) concentration and to the substrate concentration. Simulations were done as before. In this model, the absence and presence of demethylases are both compatible with the same steady-state patterns. However, introducing a reverse process must be balanced by faster forward kinetics. An interesting consequence of this adjustment is a faster establishment of a steady state. Our previously described experiments in yeast showed that growth arrest or histone aging led to increased H3K79 methylation [44], which is inconsistent with fast establishment of an equilibrium and suggests that a H3K79 demethylase is absent or only a minor contributor. In contrast, growth-dependent methylation and accumulation of methylation on older histones did not happen in the case of the Set1 methyltransferase [44], which is known to operate in the presence of a demethylase activity that modulates the H3K4me1/3 levels [55–58].

Modeling H3K76 Methylation by Dot1 in African Trypanosomes

Having established the models to study the mechanism, regulation and cell cycle dynamics of H3K79 methylation in yeast, we applied this toolbox to examine the regulation of H3K76 methylation in African trypanosomes, the causative agents of sleeping sickness. In this unicellular parasite, two Dot1-like proteins, Dot1A and Dot1B are responsible for H3K76 methylation [9,10,13]. Together they set up a strictly controlled cell cycle pattern of H3K76me2 and H3K76me3, as nicely demonstrated by the Janzen lab using immunofluorescence [9,10,13]. This dynamic methylation pattern ensures proper cell cycle progression and modulates antigenic variation in African trypanosomes [9–13]. To understand the mechanistic basis for the observed cell cycle pattern, quantitative information on the Dot1A and Dot1B activities was obtained by expressing the proteins individually in yeast under the control of promoters of different strengths [52]. This approach was not possible in trypanosomes since misregulation of these enzymes is not tolerated [10,13]. We subjected Dot1A and Dot1B to a computational analysis using the same strategy as described for yeast Dot1. Simulations with the distributive model resulted in a better fit of the experimental data than the processive model [52]. Together with the observed characteristic changes in methylation this strongly suggested that Dot1A and Dot1B, like yeast Dot1, methylate with a distributive mechanism. Very elegant in vitro assays from the Janzen lab using recombinant enzymes and recombinant unmodified chromatin templates provided independent evidence for the distributive mechanism of Dot1A and Dot1B [9]. The two enzymes have very distinct estimated rate constants. For example, Dot1A displays a very low trimethylation rate (low *k2*). In contrast, Dot1B has a high rate of di- and trimethylation (high *k1* and *k2*), which led to the prediction that when expressed at sufficient levels, it can result in a binary switch: when a monomethylated state has been produced, Dot1B will very rapidly convert it into dimethylation and subsequently trimethylation [52]. This prediction could be experimentally validated by measuring the appearance of methylation states on newly synthesized histones using a genetic pulse-chase assay in yeast. Indeed, in the presence of Dot1B, newly synthesized histones rapidly accumulated the trimethylation state without the detectable appearance of mono- and dimethyl intermediates [52]. This assay also qualitatively agreed with the estimated rate constants for Dot1A.

A Two-Enzyme Single-Cell Model to Describe the Cell Cycle Control of H3K76 Methylation in Trypanosomes

Next we addressed the question whether dynamic Dot1 isoform expression patterns during the trypanosome cell cycle can explain the observed methylation dynamics. To that end we built a kinetic model that simulates the trypanosome cell cycle using previous models developed for yeast, but now taking into account a division of labor between the two distributive enzymes and the 13-hour trypanosome cell cycle profile, which is very different from the 90-minute cell cycle of yeast [10]. The trypanosome cell cycle was divided into three phases based on the observed fluctuations of the methylation states: an S phase (20%), G2/M phase (40%), and G1 phase (40%). The estimated rate constants of the steady-state models were used to parameterize a model that represents H3K76 methylation in trypanosomes (Fig. 6.5). However, importantly, here we could not assume that Dot1A and Dot1B activity are constant throughout the cell cycle. Instead we fitted a cell cycle-regulated activity profile based on fluctuations of the three methylation states observed by immunofluorescence data from the Janzen lab [10,13,52]. In this model, Dot1A and B are inactive during S phase when a complement of new histones is assembled. Dot1A becomes active just after S phase, leading to the sequential appearance of H3K76me1 and -me2 (Fig. 6.5). Dot1B is initially tightly repressed but activated at the start of G1, after which all H3K76 residues rapidly "switch" to the trimethyl state. The resulting two-enzyme cell cycle model provides a resource that can be used to gain more insight into how the cell cycle-regulated pattern is generated and how sensitive it is to perturbation. It also offers the possibility to predict the changes in methylation upon Dot1B overexpression or loss of Dot1A, two conditions that have been described to be lethal in trypanosomes [10,13]. The two conditions have in common that H3K76me1 is predicted to be virtually absent, whereas in wild-type cells it is predicted to peak in the G2/M phase [52]. It is tempting to speculate that H3K76me1 represents a biologically important signal in trypanosomes. Addressing this, prediction by experimental validation will require the development of tools to specifically control this methylation state in trypanosomes.

FUTURE DEVELOPMENTS AND APPLICATIONS OF THE MODELS

The approach of subsequent phases of genetic perturbation, quantitative measurements, model building, and experimental validation (Fig. 6.2) has provided very valuable insights into the regulation of H3K79 methylation by Dot1. One interesting result is that relatively straightforward and simple mathematical models and knowledge about the distributive mechanism very well explain the methylation defects seen in regulatory mutants. Although the methylation changes can be complex or seemingly specific for one methylation state, most phenotypes can be simply explained by altered overall activity of the distributive Dot1 enzyme. As such, the combined results and models provide a useful resource for future studies on regulation of Dot1. Indeed, they have been invaluable for interpreting H3K79 methylation levels seen in engineered H2Bub cross-talk strains [59]. Several assumptions were made to build the initial versions of the models. In the future, it will be important to refine the models. One aspect that deserves further attention is the nonhomogeneous H2Bub landscape: H2Bub is mainly found in transcribed regions and is absent in promoters, regulatory regions and silent chromatin [48,60]. It will be interesting to represent this dichotomy in the models to obtain rate constants for the activity of Dot1 on the two types of substrates. Interestingly, the general model is qualitatively in agreement with the methylation differences between the two chromatin compartments: low H2Bub goes together with low H3K79me3 and high H3K79me1 (like activity B in Fig. 6.4C), whereas high H2Bub correlates with high H3K79me3 and low H3K79me1 (like activity A in Fig. 6.4C) [48,59]. Not surprisingly, the in vivo situation includes additional layers of complexity, such as the recruitment of Dot1 to specific sites in the genome by its binding partners [2]. At such sites Dot1 may methylate by a pseudo-processive mechanism if efficient targeting would keep Dot1 close to its histone substrate so that it could methylate the same histone multiple times. Finally, the models described here can in principle also be applied to other histone methylation systems. Recently, the dynamics of several histone methylation marks has been measured in different biological contexts [36–41,53,61–63]. These studies can provide a good starting point for model development. The availability of CRISPR-Cas9 systems to repress and activate genes will further facilitate altering the levels of activity of methyltransferases in the cell system of interest. Modeling methylation systems at other sites will likely face several challenges, such as the control of a specific methylation event by multiple enzymes, the involvement of demethylases, the targeting or activation of methyltransferases by protein–protein interactions, and the influence of histone cross-talk. On the other hand, in vitro assays on defined peptide substrates may be available. Nevertheless, fundamental knowledge on the methylation mechanism, estimated rate constants, and simplified in vivo models can provide an important framework for understanding the regulation and dynamics of histone methylation in the cell.

Acknowledgments

We thank Iris Stulemeijer for fruitful discussions. FvL was supported by the Netherlands Organisation for Scientific Research (NWO-VICI-016.130.627) and the Dutch Cancer Society (KWF-NKI2014-7232). DDV was supported by the Belgian Science Policy Office by a MARS Inter University Attraction Poles project [IAP7/29 to DDV]. BMB was supported by Netherlands Organisation for Scientific Research (NWO-CSBR-853.00.110).

References

[1] Greer EL, Shi Y. Histone methylation: a dynamic mark in health, disease and inheritance. Nat Rev Genet 2012;13(5):343–57.

[2] Vlaming H, van Leeuwen F. The upstreams and downstreams of histone H3K79 methylation by Dot1L. Chromosoma 2016, Epub ahead of print.

[3] McLean CM, Karemaker ID, van Leeuwen F. The emerging roles of DOT1L in leukemia and normal development. Leukemia 2014;28:2131–8.

[4] Nguyen AT, Zhang Y. The diverse functions of Dot1 and H3K79 methylation. Genes Dev 2011;25(13):1345–58.

[5] Farooq Z, Banday S, Pandita TK, Altaf M. The many faces of histone H3K79 methylation. Mutat Res Rev Mutat Res 2016;768:46–52.

[6] van Leeuwen F, Frederiks F, Terweij M, De Vos D, Bakker BM. News about old histones—a role for histone age in controlling the epigenome. Cell Cycle 2012;11(1):11–12.

[7] Xu Y, Zhang M, Li W, Zhu X, Bao X, Qin B, et al. Transcriptional control of somatic cell reprogramming. Trends Cell Biol 2016;26:272–88.

[8] Brien GL, Valerio DG, Armstrong SA. Exploiting the epigenome to control cancer-promoting gene-expression programs. Cancer Cell 2016;29:464–76.

[9] Dindar G, Anger AM, Mehlhorn C, Hake SB, Janzen CJ. Structure-guided mutational analysis reveals the functional requirements for product specificity of DOT1 enzymes. Nat Commun 2014;5:5313.

[10] Gassen A, Brechtefeld D, Schandry N, Arteaga-Salas JM, Israel L, Imhof A, et al. DOT1A-dependent H3K76 methylation is required for replication regulation in *Trypanosoma brucei*. Nucl Acids Res 2012;40(20):10302–11.

[11] Frederiks F, van Welsem T, Oudgenoeg G, Heck AJ, Janzen CJ, van Leeuwen F. Heterologous expression reveals distinct enzymatic activities of two DOT1 histone methyltransferases of *Trypanosoma brucei*. J Cell Sci 2010;123(Pt 23):4019–23.

[12] Figueiredo L, Janzen C, Cross G. A histone methyltransferase modulates antigenic variation in African Trypanosomes. PLoS Biol 2008;6(7):e161.

[13] Janzen CJ, Hake SB, Lowell JE, Cross GA. Selective di- or trimethylation of histone H3 lysine 76 by two DOT1 homologs is important for cell cycle regulation in *Trypanosoma brucei*. Mol Cell 2006;23(4):497–507.

[14] White CL, Suto RK, Luger K. Structure of the yeast nucleosome core particle reveals fundamental changes in internucleosome interactions. EMBO J 2001;20(18):5201–18.

[15] Pettersen EF, Goddard TD, Huang CC, Couch GS, Greenblatt DM, Meng EC, et al. UCSF Chimera—a visualization system for exploratory research and analysis. J Comput Chem 2004;25(13):1605–12.

[16] van Leeuwen F, Gafken PR, Gottschling DE. Dot1p modulates silencing in yeast by methylation of the nucleosome core. Cell 2002;109(6):745–56.

[17] Ng HH, Feng Q, Wang H, Erdjument-Bromage H, Tempst P, Zhang Y, et al. Lysine methylation within the globular domain of histone H3 by Dot1 is important for telomeric silencing and Sir protein association. Genes Dev 2002;16(1222074915):1518–27.

[18] Feng Q, Wang H, Ng HH, Erdjument-Bromage H, Tempst P, Struhl K, et al. Methylation of H3-lysine 79 is mediated by a new family of HMTases without a SET domain. Curr Biol 2002;12:1052–8.

[19] Sawada K, Yang Z, Horton JR, Collins RE, Zhang X, Cheng X. Structure of the conserved core of the yeast Dot1p, a nucleosomal histone H3 lysine 79 methyltransferase. J Biol Chem 2004;279(41):43296–306.

[20] McGinty RK, Kim J, Chatterjee C, Roeder RG, Muir TW. Chemically ubiquitylated histone H2B stimulates hDot1L-mediated intranucleosomal methylation. Nature 2008;453(7196):812–16.

[21] Cheng X, Collins RE, Zhang X. Structural and sequence motifs of protein (histone) methylation enzymes. Annu Rev Biophys Biomol Struct 2005;34:267–94.

[22] Couture J-F, Trievel RC. Histone-modifying enzymes: encrypting an enigmatic epigenetic code. Curr Opin Struct Biol 2006;16:753–60.

[23] Del Rizzo PA, Trievel RC. Substrate and product specificities of SET domain methyltransferases. Epigenetics 2011;6:1059–67.

[24] Min J, Feng Q, Li Z, Zhang Y, Xu RM. Structure of the catalytic domain of human DOT1L, a non-SET domain nucleosomal histone methyltransferase. Cell 2003;112(5):711–23.

[25] Yu W, Chory EJ, Wernimont AK, Tempel W, Scopton A, Federation A, et al. Catalytic site remodelling of the DOT1L methyltransferase by selective inhibitors. Nat Commun 2012;3:1288.

[26] Smith BC, Denu JM. Chemical mechanisms of histone lysine and arginine modifications. Biochim Biophys Acta 2009;1789(1):45–57.

[27] Cheng X, Zhang X. Structural dynamics of protein lysine methylation and demethylation. Mutat Res 2007;618(1–2):102–15.

[28] Zhou L, Holt MT, Ohashi N, Zhao A, Muller MM, Wang B, et al. Evidence that ubiquitylated H2B corrals hDot1L on the nucleosomal surface to induce H3K79 methylation. Nat Commun 2016;7:10589.

[29] Holt MT, David Y, Pollock S, Tang Z, Jeon J, Kim J, et al. Identification of a functional hotspot on ubiquitin required for stimulation of methyltransferase activity on chromatin. Proc Natl Acad Sci USA 2015;112(33):10365–70.

[30] Fierz B, Muir TW. Chromatin as an expansive canvas for chemical biology. Nat Chem Biol 2012;8(5):417–27.

[31] Chatterjee C, Muir TW. Chemical approaches for studying histone modifications. J Biol Chem 2010;285(15):11045–50.

[32] Black JC, Van Rechem C, Whetstine JR. Histone lysine methylation dynamics: Establishment, regulation, and biological impact. Mol Cell 2012;48(4):491–507.

[33] Corpet A, Almouzni G. Making copies of chromatin: the challenge of nucleosomal organization and epigenetic information. Trends Cell Biol 2009;19(1):29–41.

[34] Probst AV, Dunleavy E, Almouzni G. Epigenetic inheritance during the cell cycle. Nat Rev Mol Cell Biol 2009;10(3):192–206.

[35] Alabert C, Groth A. Chromatin replication and epigenome maintenance. Nat Rev Mol Cell Biol 2012;13(3):153–67.

[36] Alvarez F, Muñoz F, Schilcher P, Imhof A, Almouzni G, Loyola A. Sequential establishment of marks on soluble histones H3 and H4. J Biol Chem 2011;286(20):17714–21.

[37] Scharf A, Imhof A. Every methyl counts—epigenetic calculus. FEBS Lett 2011;585(13):2001–7.

[38] Barth TK, Imhof A. Fast signals and slow marks: the dynamics of histone modifications. Trends Biochem Sci 2010;35(11):618–26.

[39] Scharf AN, Barth TK, Imhof A. Establishment of histone modifications after chromatin assembly. Nucl Acids Res 2009;37(15):5032–40.

[40] Zee BM, Levin RS, Dimaggio PA, Garcia BA. Global turnover of histone post-translational modifications and variants in human cells. Epigen Chrom 2010;3(1):22.

[41] Pesavento JJ, Yang H, Kelleher NL, Mizzen CA. Certain and progressive methylation of histone H4 at lysine 20 during the cell cycle. Mol Cell Biol 2008;28(1):468–86.

[42] Wang X, Chen CW, Armstrong SA. The role of DOT1L in the maintenance of leukemia gene expression. Curr Opin Genet Dev 2016;36:68–72.

[43] Frederiks F, Tzouros M, Oudgenoeg G, van Welsem T, Fornerod M, Krijgsveld J, et al. Nonprocessive methylation by Dot1 leads to functional redundancy of histone H3K79 methylation states. Nat Struct Mol Biol 2008;15(6):550–7.

[44] De Vos D, Frederiks F, Terweij M, van Welsem T, Verzijlbergen KF, Iachina E, et al. Progressive methylation of ageing histones by Dot1 functions as a timer. EMBO Rep 2011;12(9):956–62.

[45] Ghaemmaghami S, Huh WK, Bower K, Howson RW, Belle A, Dephoure N, et al. Global analysis of protein expression in yeast. Nature 2003;425(6959):737–41.

[46] Gavin AC, Bosche M, Krause R, Grandi P, Marzioch M, Bauer A, et al. Functional organization of the yeast proteome by systematic analysis of protein complexes. Nature 2002;415(6868):141–7.

[47] Shahbazian MD, Zhang K, Grunstein M. Histone H2B ubiquitylation controls processive methylation but not monomethylation by Dot1 and Set1. Mol Cell 2005;19(2):271–7.

[48] Schulze JM, Jackson J, Nakanishi S, Gardner JM, Hentrich T, Haug J, et al. Linking cell cycle to histone modifications: SBF and H2B monoubiquitination machinery and cell-cycle regulation of H3K79 dimethylation. Mol Cell 2009;35(5):626–41.

[49] Zentner GE, Henikoff S. Regulation of nucleosome dynamics by histone modifications. Nat Struct Mol Biol 2013;20(3):259–66.

[50] Venkatesh S, Workman JL. Histone exchange, chromatin structure and the regulation of transcription. Nat Rev Mol Cell Biol 2015.

[51] Radman-Livaja M, Verzijlbergen KF, Weiner A, van Welsem T, Friedman N, Rando OJ, et al. Patterns and mechanisms of ancestral histone protein inheritance in budding yeast. PLoS Biol 2011;9(6):e1001075.

[52] Stulemeijer IJE, De Vos D, van Harten K, Joshi OK, Blomberg O, van Welsem T, et al. Dot1 histone methyltransferases share a distributive mechanism but have highly diverged catalytic properties. Sci Rep 2015;5:9824.

[53] Sweet SM, Li M, Thomas PM, Durbin KR, Kelleher NL. Kinetics of re-establishing H3K79 methylation marks in global human chromatin. J Biol Chem 2010;285(43):32778–86.

[54] Kooistra SM, Helin K. Molecular mechanisms and potential functions of histone demethylases. Nat Rev Mol Cell Biol 2012;13(5):297–311.

[55] Osborne EA, Dudoit S, Rine J. The establishment of gene silencing at single-cell resolution. Nat Genet 2009;41(7):800–6.

[56] Tu S, Bulloch EM, Yang L, Ren C, Huang WC, Hsu PH, et al. Identification of histone demethylases in *Saccharomyces cerevisiae*. J Biol Chem 2007;282(19):14262–71.

[57] Huang F, Ramakrishnan S, Pokhrel S, Pflueger C, Parnell TJ, Kasten MM, et al. Interaction of the Jhd2 histone H3 Lys-4 demethylase with chromatin is controlled by histone H2A surfaces and restricted by H2B ubiquitination. J Biol Chem 2015;290(48):28760–77.

[58] Liang G, Klose RJ, Gardner KE, Zhang Y. Yeast Jhd2p is a histone H3 Lys4 trimethyl demethylase. Nat Struct Mol Biol 2007;14(3):243–5.

[59] Vlaming H, van Welsem T, de Graaf EL, Ontoso D, Altelaar AM, San-Segundo PA, et al. Flexibility in crosstalk between H2B ubiquitination and H3 methylation in vivo. EMBO Rep 2014;15:1077–84.

[60] Schulze JM, Hentrich T, Nakanishi S, Gupta A, Emberly E, Shilatifard A, et al. Splitting the task: Ubp8 and Ubp10 deubiquitinate different cellular pools of H2BK123. Genes Dev 2011;25(21):2242–7.

[61] Zee BM, Levin RS, Xu B, LeRoy G, Wingreen NS, Garcia BA. In vivo residue-specific histone methylation dynamics. J Biol Chem 2010;285(5):3341–50.

[62] Alabert C, Barth TK, Reveron-Gomez N, Sidoli S, Schmidt A, Jensen ON, et al. Two distinct modes for propagation of histone PTMs across the cell cycle. Genes Dev 2015;29(6):585–90.

[63] Scharf AN, Meier K, Seitz V, Kremmer E, Brehm A, Imhof A. Monomethylation of lysine 20 on histone H4 facilitates chromatin maturation. Mol Cell Biol 2008;29(1):57–67.

[64] Ashyraliyev M, Fomekong-Nanfack Y, Kaandorp JA, Blom JG. Systems biology: parameter estimation for biochemical models. Febs J 2009;276(4):886–902.

[65] Wolfram S. Mathematica. 6.0 ed. Champaign, IL: Wolfram Media, Inc.; 2007.

[66] Efron B, Tibshirani RJ. An introduction to the bootstrap. New York: Chapman and Hall/CRC Press; 1994.

GLOSSARY

Distributive mechanism a mechanism in which the enzyme releases from its substrate after one reaction step, requiring new binding for subsequent reactions.

Dot1/DOT1L a distributive histone lysine methyltransferase of histone H3 lysine 79, harboring a Class I domain and promoted by trans-histone cross-talk.

Histone methylation a posttranslational modification (PTM) of histone proteins that predominantly occurs on arginine and lysine residues; arginines can be mono- and dimethylated, lysines can be mono-, di-, and trimethylated.

Methyltransferase domain histone lysine methylation is carried out by proteins containing as the catalytic part a Su(var)3-9, Enhancer-of-zeste, Trithorax (SET) domain or a Class I methyltransferase domain, which is present in many other types of methyltransferases.

Processive mechanism a mechanism in which the enzyme remains bound to its substrate during several reaction steps, only releasing from its final product.

Trans-histone cross-talk one histone PTM controls another histone PTM on a different histone.

LIST OF ACRONYMS AND ABBREVIATIONS

Dot1 yeast Disruptor of Telomeric Silencing 1
DOT1L Dot1-like
H2Bub C-terminal H2B ubiquitination
H3K79 histone H3 lysine 79
H3K79me histone H3 lysine 79 methylation
$\boldsymbol{k_0/k_1/k_2}$ apparent rate constant for mono/di/trimethylation
***me*0** unmethylated
***me*1/2/3** mono/di/trimethylated
MT methyltransferase
ODE ordinary differential equation
SAM *S*-adenosyl-L-methionine
SAH *S*-adenosyl-L-homocysteine
SET Su(var)3-9, Enhancer-of-zeste, Trithorax
Set1 yeast SET-domain containing 1
$\boldsymbol{\mu}$ specific growth/dilution rate constant

APPENDIX: PARAMETER ESTIMATION AND FORMAL MATHEMATICAL DESCRIPTION OF THE MODELS

Parameter Estimation

Parameters are an essential part of most models and determining parameter values can be the primary reason to build a model in the first place. In many cases, no accurate parameter estimates are available for ordinary differential equation (ODE) models and so-called regression methods are used to obtain estimates [64]. It is important to realize that it can be impossible to determine a parameter simply because of the form of the mathematical expressions or because of data that are not providing sufficient information. Since such an unidentifiable parameter can compromise estimation of other parameters, it is preferably fixed. Parameter fixing can also be required if strong correlations between parameters exist. The question whether parameter values can be uniquely determined is studied in the mathematical field of system identification. Briefly said a good, i.e., information-rich, experiment, or experimental design elicits a specific system response through each underlying process (and its associated parameter(s)). If a good data series is available (sometimes after statistical outlier removal or data smoothing techniques), the challenge is still to find that optimal solution in parameter space starting from an initial guess of parameter estimates. The success of that search will critically depend on two factors: the fitting criterion and the search algorithm.

The fitting criterion or objective function value (also called cost or fitness function) is a measure for how well the simulated data correspond with the experimental data. The maximum likelihood estimator (MLE) is arguably the most widely used measure and is based on the assumption that the most likely parameter vector (or value) is the one that maximizes the probability or likelihood of the observed data. Importantly, if the assumption holds that the errors (differences between measured and predicted data) are independent and normally distributed, then the MLE estimate is the same as the parameter estimate that minimizes the sum of the squares. In other words in many cases the traditional method of minimizing the sum of the squared errors provides an acceptable criterion for data fitting.

Due to the often complex behavior of solutions of systems of ODEs as a function of parameters, the effect on the objective function (one can image this as a landscape with hills and valleys in the case of two parameters) can be very unpredictable. The risk is that the search method gets stuck in a local depression or minimum. The distinction is often made between local and global search methods. Methods of the first class (which is further divided into direct methods and gradient-based methods depending on the use of derivatives) can be proved to lead to a (local) minimum. They are typically fast, yet, cannot guarantee to find the global minimum, i.e., the lowest point in the landscape. Global methods on the other hand will not get stuck and over time find lower objective function values. However, the search process is very slow and in most cases not guaranteed to converge to the global minimum. Since in biological studies good initial estimates of parameters are the exception, a good strategy is to start with a global search to identify promising areas in the fitness landscape and then switch to a local search method for rapid convergence. That was also the strategy adopted in De Vos et al. [44] as we will further discuss below.

In De Vos et al. [44], data fitting is actually used not only for parameter estimation but also as a model selection technique as detailed in the main section. The fraction of the different H3K79 methylation states in 4 yeast strains expressing different amounts of a galactose-inducible *DOT1* gene (and growing generally at different rates) served as the experimental data used for determining parameters k_0, k_1, and k_2 of the Dot1 distributive model for a wild-type strain. The other parameters in the model were assumed to be known. Therefore, the differences in methylation states are assumed to reflect differences in the kinetic parameters that are fitted by the regression method and experimentally determined specific growth rates.

For the fitting criterion, the ordinary simple least squares measure was used with the Mathematica (6.0) NMinimize function for nonlinear global optimization [65]. We used the "Automatic" mode, which uses global search methods ("RandomSearch," "Simulated Annealing," "DifferentialEvolution") in combination with the direct local search method "Nelder-Mead." As discussed above, this is a good strategy to avoid getting stuck in a local minimum while still allowing fast convergence. Starting from a random set of parameter values (within a predetermined parameter space) the cost function value was calculated based on the methylation state of the model after convergence to its steady state value (typically 6000 minutes). Repeating that procedure yielded consistent parameter estimates (to a high precision). As a control the cost function value as a function of the kinetic parameters k_0, k_1, and k_2 was plotted to demonstrate that it is a sensitive and smooth (monotone) function of the parameter values in the neighborhood of the respective estimates. Such a relationship enables accurate parameter estimation.

As an extra test of the robustness of the fitting procedure random (Gaussian) noise was added to the original methylation data to a degree that exceeded the experimental noise on those data. Refitting consistently led to parameter estimates close to the reference estimates (from data without added noise). Moreover, this also provided a way to estimate the uncertainty of the parameter estimates. A modification of the classical bootstrapping approach (cf. for instance Efron [66]) was used. Instead of random resampling from a set of experimental data replicates, random Gaussian noise (with sigma = ± 2.5% methylation) was added to the reference dataset, before running the fitting routine on the "resampled" data. The standard deviation of the refitted parameters then served as estimate of the parameter error. A series of randomized datasets were generated and fitting results were used for statistical analysis. The average of the parameter estimates and the cost function were plotted as a function of the number of runs until their values sufficiently converged. The standard deviation corresponding to that set of runs was then used as a rough estimate of the parameters' standard error. Since (artificial) experimental errors are in this approach strictly mapped to the errors in the three fitted parameters the corresponding error estimates are potentially overestimates. Indeed, unlike in reality in the model other parameters (and processes) do not contribute to variation in the methylation states.

Formal Mathematical Description of the Models

Diverse ways exist to construct a mathematical model that describes a biochemical process. As in many other cases, for the studies described here systems of ODEs were used. ODEs can accurately represent time-dependent changes in well-mixed (homogenous) fixed-volume compartments such as approximately found in cells or organelles. All systems described below express how the four states of the histone H3K79, represented by the (concentration) variables *me*0, *me*1, *me*2, and *me*3, change over time. Two main classes of models can be distinguished with the first aimed at representing the steady-state methylation over a population of cells. The second class represents the temporal changes in methylation state in a typical single cell.

1. Kinetic models for steady state methylation

   Initially, a set of ODE models was defined for fitting to in vitro data from nuclear extracts incubated with different amounts of recombinant Dot1 [43]. This did not produce satisfactory results (see main text) and therefore the models were adapted for in vivo data. More specifically, the originally fixed concentration of histone substrate was replaced with a constant histone production rate (γ) and first-order dilution/degradation rates (with rate constant μ). In this way, a steady-state total histone concentration H_{tot} is reached over time with:

$$\gamma = \mu * (me0 + me1 + me2 + me3) = m * H_{tot} \tag{6.4}$$

   a. Processive mechanism

$$\begin{cases} \dfrac{d(me0(t))}{dt} = \gamma - \mu \cdot me0(t) - v_0 - v_{0'} - v_{0''} \\ \dfrac{d(me1(t))}{dt} = -\mu \cdot me1(t) + v_0 \\ \dfrac{d(me2(t))}{dt} = -\mu \cdot me2(t) + v_{0'} \\ \dfrac{d(me3(t))}{dt} = -\mu \cdot me3(t) + v_{0''} \end{cases} \tag{6.5}$$

   The ν_0 terms represent different conversion steps starting from nonmethylated H3K79 and leading to different methylation end products. The following kinetic expressions were used for model comparison:

Same conversion rate constants k for all steps (dot is the parameter for Dot1 concentration):

$$\begin{cases} v_0 = k \cdot dot \cdot me0(t) \\ v_{0'} = k \cdot dot \cdot me0(t). \\ v_{0''} = k \cdot dot \cdot me0(t) \end{cases} \tag{6.6}$$

Different rate conversion constants k_0, $k_{0'}$, and $k_{0''}$:

$$\begin{cases} v_0 = k_0 \cdot dot \cdot me0(t) \\ v_{0'} = k_{0'} \cdot dot \cdot me0(t) \; . \\ v_{0''} = k_{0''} \cdot dot \cdot me0(t) \end{cases} \tag{6.7}$$

b. Distributive mechanism

$$\begin{cases} \dfrac{d(me0(t))}{dt} = \gamma - \mu \cdot me0(t) - v_0 \\ \dfrac{d(me1(t))}{dt} = -\mu \cdot me1(t) + v_0 - v_1 \\ \dfrac{d(me2(t))}{dt} = -\mu \cdot me2(t) + v_1 - v_2 \\ \dfrac{d(me3(t))}{dt} = -\mu \cdot me3(t) + v_2 \end{cases} \tag{6.2}$$

For a distributive mechanism, $me0$, $me1$, and $me2$ are substrates for methylation as represented by the terms v_0, v_1, and v_2. Below the main tested rate expressions for those conversion steps are presented.

One rate constant k for all simple mass-actions steps:

$$\text{A}\begin{cases} v_0 = k \cdot dot \cdot me0(t) \\ v_1 = k \cdot dot \cdot me1(t) \\ v_2 = k \cdot dot \cdot me2(t) \end{cases} \tag{6.8}$$

Different rate constants k_0, k_1, and k_2 for all mass-actions steps:

$$\text{B}\begin{cases} v_0 = k_0 \cdot dot \cdot me0(t) \\ v_1 = k_1 \cdot dot \cdot me1(t) \\ v_2 = k_2 \cdot dot \cdot me2(t) \end{cases} \tag{6.3}$$

This model can be simulated and analyzed interactively via the JWS online model database on http://jjj.biochem.sun.ac.za/models/devos/.

Michaelis–Menten kinetics with the same K_m and different catalytic constants $k_{cat,0}$, $k_{cat,1}$, $k_{cat,2}$:

$$C\begin{cases} v_0 = k_{cat,0} \cdot dot \cdot \dfrac{\dfrac{me0(t)}{K_m}}{1+\dfrac{me0(t)}{K_m}} \\ v_1 = k_{cat,1} \cdot dot \cdot \dfrac{\dfrac{me1(t)}{K_m}}{1+\dfrac{me0(t)}{K_m}} \\ v_2 = k_{cat,2} \cdot dot \cdot \dfrac{\dfrac{me2(t)}{K_m}}{1+\dfrac{me0(t)}{K_m}} \end{cases} \tag{6.9}$$

Michaelis–Menten kinetics with the same k_{cat} and different constants K_{m0}, K_{m1}, and K_{m2} constants:

$$D\begin{cases} v_0 = k_{cat} \cdot dot \cdot \dfrac{\dfrac{me0(t)}{K_{m_0}}}{1+\dfrac{me0(t)}{K_{m_0}}} \\ v_1 = k_{cat} \cdot dot \cdot \dfrac{\dfrac{me1(t)}{K_{m_1}}}{1+\dfrac{me1(t)}{K_{m_1}}}, \\ v_2 = k_{cat} \cdot dot \cdot \dfrac{\dfrac{me2(t)}{K_{m_2}}}{1+\dfrac{me2(t)}{K_{m_2}}} \end{cases} \tag{6.10}$$

Michaelis–Menten kinetics with product inhibition:

$$E\begin{cases} v_0 = k_{cat,0} \cdot dot \cdot \dfrac{\dfrac{me0(t)}{K_{m_0}}}{1+\dfrac{me0(t)}{K_{m_0}}+\dfrac{me1(t)}{K_{m_1}}+\dfrac{me2(t)}{K_{m_2}}+\dfrac{me3(t)}{K_{m_3}}} \\ v_1 = k_{cat,1} \cdot dot \cdot \dfrac{\dfrac{me1(t)}{K_{m_1}}}{1+\dfrac{me0(t)}{K_{m_0}}+\dfrac{me1(t)}{K_{m_1}}+\dfrac{me2(t)}{K_{m_2}}+\dfrac{me3(t)}{K_{m_3}}}, \\ v_2 = k_{cat,2} \cdot dot \cdot \dfrac{\dfrac{me2(t)}{K_{m_2}}}{1+\dfrac{me0(t)}{K_{m_0}}+\dfrac{me1(t)}{K_{m_1}}+\dfrac{me2(t)}{K_{m_2}}+\dfrac{me3(t)}{K_{m_3}}} \end{cases} \tag{6.11}$$

c. Reversible reactions (distributive)

In the case of reversible kinetics, terms were added to the system of Eq. (6.2) that represent demethylation of *me*1, *me*2, and *me*3 (ν_1', ν_2', and ν_3', respectively; Eq. (6.12)). Demethylase activity was integrated into the apparent rate constant $k_{r,app}$ (Eq. (6.13)).

$$\begin{cases} \frac{d(me0(t))}{dt} = \gamma - \mu \cdot me0(t) - v_0 + v_{1'} \\ \frac{d(me1(t))}{dt} = -\mu \cdot me1(t) + v_0 - v_1 - v_{1'} + v_{2'} \\ \frac{d(me2(t))}{dt} = -\mu \cdot me2(t) + v_1 - v_2 - v_{2'} + v_{3'} \\ \frac{d(me3(t))}{dt} = -\mu \cdot me3(t) + v_2 - v_{3'} \end{cases} \tag{6.12}$$

$$\begin{cases} v_{1'} = k_r \cdot dem \cdot me1(t) = k_{r,app} \cdot me1(t) \\ v_{2'} = k_r \cdot dem \cdot me2(t) = k_{r,app} \cdot me2(t) \\ v_{3'} = k_r \cdot dem \cdot me3(t) = k_{r,app} \cdot me3(t) \end{cases} \tag{6.13}$$

2. Single-cell model for temporal methylation during the cell cycle

The system of Eq. (6.2), with the kinetic expressions of Eq. (6.3), also forms the basis of the single cell cycle models for yeast and trypanosomes. As explained in the main text, two and three cell cycle phases are defined for them, respectively, with histone synthesis restricted to the first phase (S). For the yeast model, Dot1 activity is assumed constant during the cell cycle. For the trypanosome model(s), different expression profiles for the A and B isoforms of Dot1 over the three phases are proposed. Detailed information, including the corresponding parameter sets, can be found in Refs. [44,52].

IV

SECTION

Reconciling Randomness and Precision: Bistable Epigenetic Memory and Switching: Stochastic Models

The previous section dealt with kinetic ordinary differential equation (ODE) models, describing changes in behavior of many molecules. The models described in Chapters 5 and 6 are "deterministic" in nature: This means that with a given model and given inputs, a simulation will always give exactly the same results. This is appropriate if we are interested in global properties or, to return to our bird analogy: We are interested in the average size of the flock at a particular time or place. But ultimately, epigenetic regulation works at the level of individual molecular events regulating individual genes. In the case of the birds of the beach, our big wave might send some seagulls flapping into the distance, whereas others merely take a few sedate steps up the beach. The next time a wave comes in, these same birds might flap away.

Stochastic simulations are performed using probabilities. Imagine we are watching a single bird on the beach and can influence its behavior by rolling a dice. When the wave comes in, we throw a six, and the bird stays put. If we throw a four or lower, the bird moves. At the next time point, we evaluate again whether the bird moves. In this way, the random behavior of the bird is captured by a simple stochastic model. If we imagine doing this operation 1000 times, then we will approximate the average behavior of the whole flock of birds (if they all move with the same probability and do not influence each other). So, the stochastic model will give essentially the same result as the equivalent deterministic model if repeated a sufficient number of times. So, why do we need stochastic models? If the fluctuations in behavior of individual molecules have consequences for the system, then we need to take them into account. Imagine if our bird, when it moves, can influence three other birds and cause them to move at the same time? Then, our population would be very influenced by the random decisions of a single member, and its behavior would no longer be captured by the deterministic model described above.

The models described in the following two chapters give beautiful examples of the use of stochastic modeling to describe bistable epigenetic systems, and how they switch stochastically between states. A bistable system has two extreme states, but intermediate states are unstable, rather like a flipping a light switch. Bistability is a fundamental property of many epigenetic systems and is essential for epigenetic memory, because each of the stable states reinforce themselves, and is robust enough to withstand and recover from the disruptions of replication and cell division. However, each state can also switch with a certain probability to the opposite state, allowing room for regulated change. Chapter 7 (Modeling Bistable Chromatin States) gives an excellent insight into the general properties of stochastic models of bistable systems. The authors explain how they can be applied to many different epigenetic systems to understand what makes them stable, and how they can be induced to switch. Chapter 8 (Quantitative Environmentally Triggered Switching Between Stable Epigenetic States) describes the application of models of this kind to quantitative experimental data from a superb model for epigenetic memory, namely the quantitative memory of the length of winter in the plant *Arabidopsis thaliana*.

CHAPTER 7

Modeling Bistable Chromatin States

Ian B. Dodd[1] and Kim Sneppen[2]
[1]*University of Adelaide, Adelaide, SA, Australia*
[2]*University of Copenhagen, Copenhagen, Denmark*

CONTENTS

CELL MEMORY BY CHROMATIN-BASED EPIGENETICS

Cells need to be responsive to changes in their environment but they also need to maintain memory of their identity independent of their environment. A cell's genomic sequence represents an essential long-term "species memory"; a record of successful solutions to life that stretch back over 4 billion years. Shorter term "cell memory" is used to generate multiple cell types, allowing cells of the same species that share the same environment to exist in distinct, persistent states. This cell differentiation provides environmental preadaptation and sexual reproduction for unicellular organisms [1] and enables the often vast number of cell specializations required in multicellular organisms.

Cell memory can be stored as programed genetic changes, such as the DNA rearrangements in the acquisition of immunity in bacterial CRISPR systems, in yeast mating-type switching, or in the generation of antibody diversity in vertebrates. However, particularly in multicellular organisms, most cell memory is achieved through epigenetic mechanisms that do not involve changes to the genome sequence but nevertheless allow cells under the same conditions to exist in alternative gene expression states that are stable over time and heritable through cell replication [2,3]. Thus, random events or specific transient environmental signals can cause a long-lived and heritable change in gene expression state—a change of cell identity.

Positive feedback in circuits of diffusible regulators can produce stable alternative regulatory states that are inherited by the random distribution of the regulators to both daughter cells during cell division. Such circuits operate in bacteria and eukaryotes and have been extensively modeled [4] and synthesized [5,6].

Epigenetics and Systems Biology. DOI: http://dx.doi.org/10.1016/B978-0-12-803075-2.00007-6

The other major class of epigenetic mechanism, which can be termed chromatin-based epigenetics, is relatively poorly understood. DNA in cells is subject to chemical modifications and bound by a host of different proteins and RNA molecules, producing a heterogeneous and dynamic complex termed chromatin [7]. The basic idea of chromatin epigenetics is that the same region of DNA can exist in alternative chromatin states that affect the expression of the underlying genes, and that these states are stable and can persist through DNA replication.

The complexity of gene regulation makes it difficult to identify chromatin-based epigenetics in eukaryotes. First, alternative chromatin states of the same region of DNA may be caused by genetics rather than epigenetics, due to unknown sequence differences that can affect the behavior of the region even when located far away. Second, it is often difficult to exclude the possibility that chromatin states are a downstream effect of alternative states of diffusible regulators. However, a unique feature of chromatin-based epigenetics is that it acts *in cis*, it is a specific state of a specific DNA molecule. Thus, unlike epigenetics by diffusible regulators, chromatin epigenetics allows identical DNA regions to exist stably and heritably in different states within the same cell or nucleus (Fig. 7.1). Observation of this "*cis*-acting" memory, combined with demonstration that the same DNA molecule is capable of existing in more than one state, excluding a genetic cause, provide the "smoking gun" for chromatin-based epigenetics [8–11].

The basic premise underlying our work is that chromatin epigenetics is due to dynamic bistability generated by localized positive feedback within chromatin.

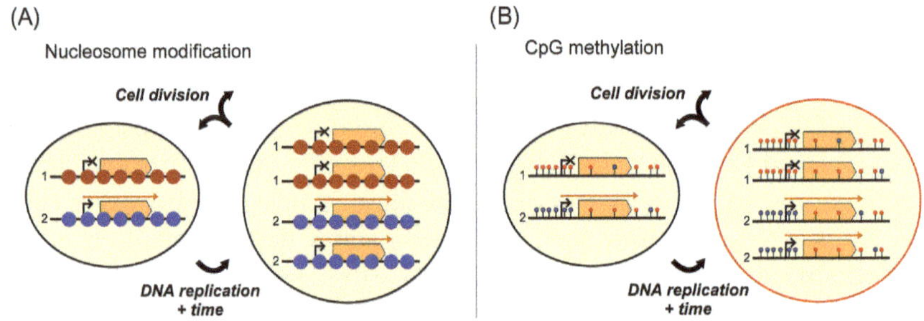

FIGURE 7.1 Chromatin-based epigenetics allows identical DNA sequences to be in different regulatory states in the same cell.

Even when DNA molecules 1 and 2 are identical (i.e., alleles of the same genes) chromatin-based epigenetics allows them to exist in alternative chromatin states that can affect gene expression. The left panel shows occupation of each allele by alternatively modified nucleosomes; the right panel shows alternative methylation states of CpG dinucleotides. This "*cis*-independent" property is not possible with epigenetic systems that operate through positive feedback in circuits of diffusible regulators. Chromatin-based epigenetics is not limited to such cases—DNAs 1 and 2 could be different loci.

We have focused our modeling on systems where the evidence for chromatin-based epigenetics is most compelling, examining nucleosome-mediated epigenetics in yeasts and DNA methylation-mediated epigenetics in mammals.

MODELING APPROACH AND PHILOSOPHY

Benefits of Mathematical Modeling

Humans make models to make sense of the world—to organize information and provide predictability. In the biological sciences, models are commonly expressed using words or diagrams, which often suffer from a lack of precision. Mathematics is the most precise language we have for describing the processes we think are happening and for comparing this with what we can observe. Even attempting to make a mathematical model helps to clarify ideas, define assumptions and identify missing information. A working mathematical description is a potent tool, simplifying large amounts of information, providing precise predictions to aid further experimental tests, and allowing rapid exploration of the range of properties of the system to help understand design–function relationships. Importantly, a mathematical model also aids generalization of insights gained from one system to other systems.

However, the rigour of a mathematical model is no guarantee that it is correct. Models will always be based on incomplete and often incorrect knowledge. They should be judged by their usefulness—their ability to explain current knowledge and to guide the attainment of new knowledge.

Our Modeling Approach—Simplified Stochastic Simulations

There are many kinds of mathematical models (many examples are provided in this book). Our approach has been to use computer-based stochastic simulations, which are a relatively "mathematics-light" way to add or remove possible processes and which naturally provide noise. The battle between a parameter-averse physicist ("We don't need to put that in") and a parameter-addicted biologist ("But we know it's not that simple") results in a compromise where complex processes are represented in simplified or idealized ways. Our hope is that this simplification exposes rather than obscures the fundamentals of the system. Details are important, of course, but fundamentals are more important.

NUCLEOSOME-MEDIATED EPIGENETICS

Eukaryotic DNA is packaged by histone proteins into nucleosomes. An octamer of histone proteins (two each of histones H2A, H2B, H3, and H4) wraps ~150 bp of DNA almost twice around itself. Together with linker H1 histones, nucleosomes are spaced roughly once per 200 bp and affect various DNA-based processes, including transcription. Nucleosomes come in many "flavors" due to variant histone proteins and various chemical modifications to the histones, such as acetylations and methylations of specific amino acids, that are placed or removed by a large number of specific enzymes (writers). These nucleosome modifications affect gene expression by strengthening or weakening the histone–DNA interaction and thus altering the accessibility of the DNA to transcription factors, and by creating or abolishing recognition sites for transcription-regulating proteins that bind to the nucleosomes (readers).

Such properties do not by themselves provide for alternative nucleosome states that are stable and heritable. However, long-standing evidence, supported by more recent data, indicated that nucleosomes, or at least the H3–H4 component of nucleosomes, are partitioned to each daughter strand during DNA replication [12]. Thus, the nucleosome modification state can persist through replication, though diluted by the addition of new nucleosomes. Almost two decades ago, this nucleosome inheritance was combined with the idea that linking a modification reading capability to a modification writing capability might produce positive feedback, where an enzyme could bind to a particular nucleosome modification and create that same modification on nearby nucleosomes [13,14]. This positive feedback by recruitment means that when DNA replication across a patch of nucleosomes in a particular modification state produces daughter DNAs containing a mixture of modified and newly inserted nucleosomes, the enzymes recruited by the modification on the old nucleosomes could potentially act to modify the new nucleosomes to reestablish the modified patch on both DNAs (Fig. 7.2). This positive feedback also allows the possibility of spreading of a modification state into neighboring chromatin, a feature apparent in some systems.

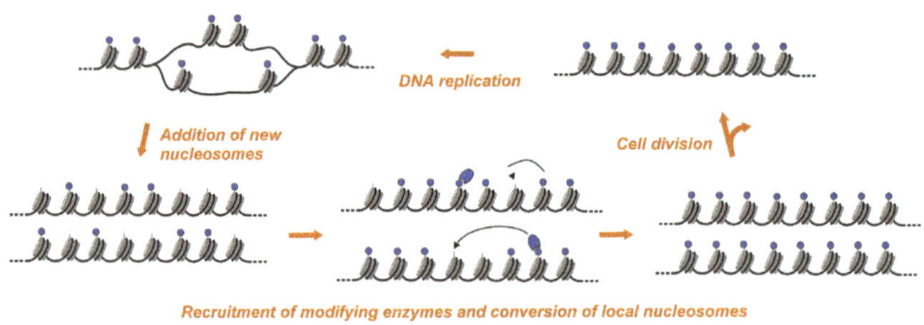

FIGURE 7.2 Model for inheritance of nucleosome modification states.

Fundamental Lessons From Modeling Nucleosome-Mediated Bistability in Fission Yeast

To test and explore this idea for nucleosome-based epigenetics more rigorously, we applied this mechanism in a stochastic model inspired by a well-defined experimental epigenetic system in the fission yeast *Schizosaccharomyces pombe* [15].

Wild-type *S. pombe* contains inactive copies of its alternative mating-type genes within a 20 kb "silenced" region of chromatin. However, deletion of part of this region and insertion of a *ura3* reporter gene, weakens the initiation of silencing and causes the region to alternate, apparently randomly, between periods of activity and periods of inactivity that can persist for thousands of cell divisions. *Cis*-memory was shown by the ability of two copies of the region within the same cell to exist stably and heritably in different states [8,9]. A nucleosome modification-based mechanism was implicated by the lack of an alternative mechanism (there is no DNA methylation in *pombe*), the requirement for the histone H3 Lys9 (H3K9) methylase Clr4 and other histone modifying proteins for silencing, and the correlation between silencing and the presence across the region of methylated histone H3K9 (H3K9me) (for reference, see Ref. [15]).

The model specifies an isolated system of 60 nucleosomes (~12 kb of DNA). Each nucleosome can be interconverted between three modification states: A, U, or M (Fig. 7.3A). M is the modification state associated with silencing (H3K9me in this system), A is the state associated with gene activity (perhaps acetylated H3K9), and U is a neutral, unmodified nucleosome state. Nucleosome interconversion can be by nonfeedback processes ("noise") which occur at a constant rate per nucleosome, or by feedback reactions that are dependent on the state of other nucleosomes in the system. The simulation is simple; each step involves choosing a random "target" nucleosome, choosing whether the reaction applied to it would be feedback or nonfeedback, and then attempting the reaction. A nonfeedback reaction is a simple change of the target's state one step along the A–U–M pathway (straight arrows in Fig. 7.3A). The feedback reaction attempt involves choice of a random second "recruiter" nucleosome from any of the remaining 59 nucleosomes—if this is A, then the target's state is moved one step toward the A state; if this is M, then the target's state is moved one step toward the M state. (These recruited reactions are indicated by curved arrows in Fig. 7.3A.) U nucleosomes do not act as recruiters. These rules for feedback reactions mean that many reaction attempts fail due to unproductive target-recruiter combinations. Although this is clearly a gross simplification and idealization of the complex biochemistry taking place at this locus inside cells, we hope that it is a useful approximation.

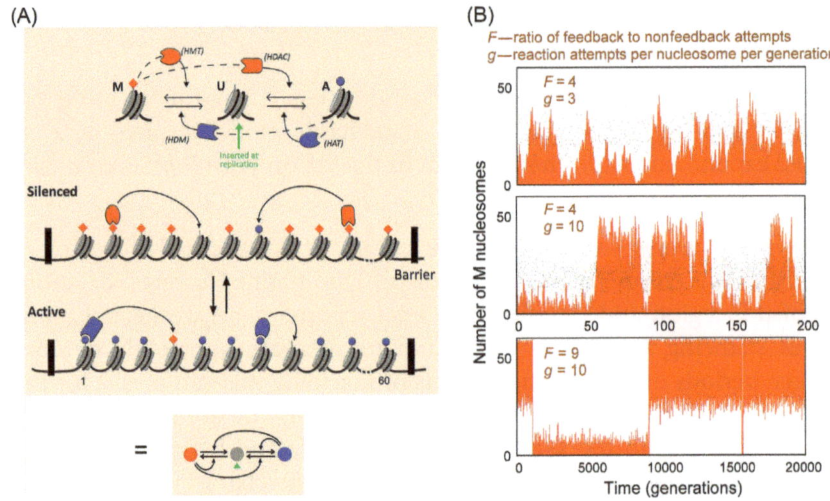

FIGURE 7.3 Model for bistable nucleosome modification states in *S. pombe*.
(A) The upper panel shows the model scheme for recruited and nonrecruited interconversions between three nucleosome types (M, U, A) [15]. The reactions are iterated within an isolated system of 60 nucleosomes (middle panel). Note that the recruited enzymes can act on nucleosomes anywhere in the system. DNA replication is simulated by conversion of each nucleosome to the U type with $P = 0.5$. The bottom panel shows a symbolic representation of the reaction scheme, used throughout this chapter. Straight arrows represent nonrecruited interconversions; curved arrows represent recruited conversions, with the arrow starting from the recruiting nucleosome type and ending at the reaction catalyzed. Recruitment reactions are nonlocal (global) by default. The nucleosome type inserted after replication is indicated by the green triangle. (B) Results of stochastic simulations with different ratios of feedback reaction attempts to nonfeedback reaction attempts (F), and different average numbers of reaction attempts per nucleosome between each DNA replication (g; generation). See text for discussion.

The simulations showed that the system is capable of strong bistable behavior, existing in either a high-M system state, where most nucleosomes are M, or a high-A system state, where most nucleosomes are A. The modeling revealed three key properties that we believe are fundamental for achieving bistable chromatin states: Positive feedback, cooperativity, and nonlocal interactions [4].

Positive Feedback

Unsurprisingly, the positive feedback reactions were necessary for bistability in this model, but we found it remarkable that strong bistability could be maintained even with a relatively high level of nonfeedback interconversions (Fig. 7.3B). It seems likely that all stable epigenetic states will require some active process for correcting the inevitable errors that occur in cellular systems.

A higher level of positive feedback was needed to cope with the disruption caused by DNA replication and thus to provide faithful heritability of the states (Fig. 7.3B). We simulated replication by randomly removing nucleosomes (with 50% probability) and replacing them with U nucleosomes to create a daughter DNA. However, we found that the system could recover from this disruption if there were at least 10 attempted reactions per nucleosome between replication events, equating to one reaction attempt per nucleosome every 10 min over the 100-min generation time for the *pombe* cells. To achieve the ~1000 cell generation stabilities for each state observed in the pombe system, required that only 80% of these reaction attempts were feedback reactions ($F = 4$).

Another type of positive feedback, where binding of the enzyme to the appropriately modified nucleosome allosterically activates its enzyme activity, has been observed in vitro [16–18]. We note that such feedback cannot replace feedback by recruitment. Allosteric activation has the benefit of reducing errors through unrecruited conversions, but it also potentially has the disadvantage of increasing errors due to recruited conversions by nucleosomes from outside the system.

Cooperativity

We saw bistability only when the positive feedback involved cooperativity between nucleosomes. Cooperativity is a way of providing ultrasensitivity, which is well understood to be critical for bistability in systems of diffusible regulators [19,20]. Ultrasensitivity indicates a sigmoidal stimulus–response curve. At low or high concentrations of the feedback regulator, the stimulus–response curve is flat and fluctuations in regulator concentration have small effects on its feedback activity, allowing the low or high concentration states to be stable. Between these regimes, the stimulus–response curve is steep, so that the feedback drives the system toward the high or low regulator concentrations.

The *pombe* model gives bistability without any explicit cooperativity in the feedback, as only a single recruiting nucleosome is needed for each recruitment reaction. However, the two steps required to convert between the opposing A and M states mean that in most cases two recruitment reactions are needed for these conversions, providing an implicit cooperativity. We found that modified reaction schemes where only one of the A↔M interconversion reactions was subject to recruitment feedback gave bistability only if explicit cooperativity was added (i.e., two recruiting nucleosomes were required).

We expect that cooperativity in the positive feedback is an essential property of all systems of chromatin epigenetics that are based on bistability.

Nonlocal Interactions

The standard model imposed no restrictions on the relative locations of target and mediator nucleosomes ("global" conversions). However, experimentalists are attracted to the idea of neighbor-only nucleosome interactions because this would give a simple stepwise spreading of nucleosome modifications that can be easily blocked by barriers on the DNA. We found that restricting recruitment feedback to nearest neighbors ("local" conversions) essentially eliminated true bistability, producing a system dominated by slowly advancing or retreating boundaries between patches of A or M nucleosomes (Fig. 7.4). Nonlocal interactions provide the necessary whole-system "awareness" that allows even a small excess of one nucleosome type to use cooperativity to drive it to system dominance. We found that sufficient nonlocality for strong bistability could be provided by an interaction probability that decayed as a power law with $d^{-1.0}$ or $d^{-1.5}$, where d is distance along the DNA.

Positive feedback, with at least some of this feedback utilizing ultrasensitivity and nonlocal interactions, are common features of our models and those of others [21–25] that successfully produce bistable behavior. While maintaining these key properties, a variety of models that capture specific features of different systems are able to provide heritable bistability.

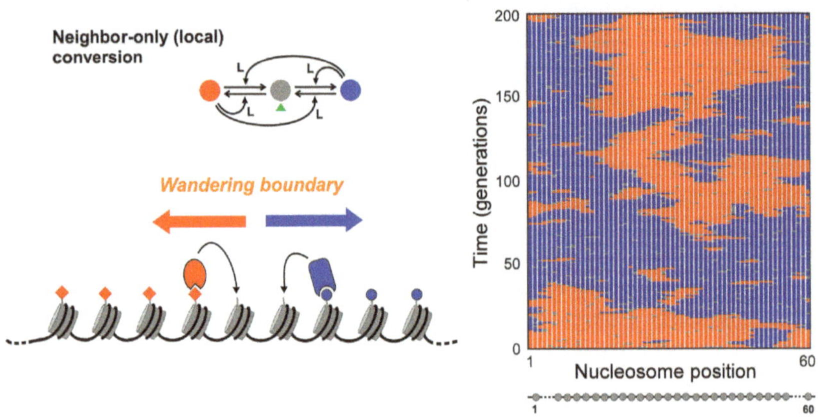

FIGURE 7.4 Neighbor-restricted feedback does not generate bistability.
Simulation using the *pombe* model of Fig. 7.3 but with all recruitment reactions acting only on nearest neighbor nucleosomes (indicated by L = local in the reaction scheme). The "battle" between different nucleosome types is restricted to boundaries of patches of each type, and when the reactions are balanced (as in the simulation) the boundary moves randomly back and forth. Long-range interactions are needed for each nucleosome to "sense" the state of the whole system and to allow a majority of one nucleosome type to exert its dominance and drive out the opposing type.

Different Reaction Schemes

Our *pombe* model is very simple, but it is possible to simplify it even further (Fig. 7.5A). This "bottom-down" approach makes the system more mathematically tractable [26] and has been applied by us and others to the silent information regulator (Sir) system that controls epigenetic silencing of the *Saccharomyces cerevisiae* mating-type loci and telomeric genes [22,27]. The Sir models specify only two different nucleosome states (acetylated, A, and unacetylated, U) and only one recruitment reaction (Sir2 is recruited by U nucleosomes and catalyzes the A→U reaction on other nucleosomes). Explicit Sir2 cooperativity is needed for such single-step models. Asymmetric reactions and the asymmetric disruption caused by replication (A-type nucleosomes are inserted) also make it more difficult to tune the system to achieve the balance needed for bistability.

Of course, it is relatively easy to devise *more* complex schemes than our *pombe* model (Fig. 7.5). In general, more complex systems provide more options for producing heritable bistability.

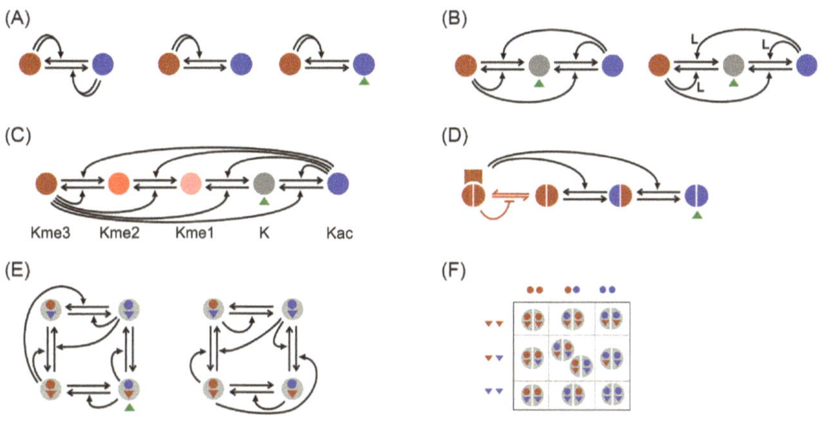

FIGURE 7.5 Many reaction schemes can generate bistability.
A huge variety of reaction schemes involving modification-specific recruitment of read–write enzymes is possible, many of which can generate bistable states. Multiple nucleosome types can be generated by multiple modification states at a single amino acid position (A–F), the two copies of each amino acid position per nucleosome (D and F), modifications at multiple amino acid positions (E and F), as well as protein-bound states (D). Schemes can vary with respect to which nucleosome type(s) are inserted after DNA replication and which type(s) are associated with gene silencing or activity. Modification reactions can vary with respect to their rates and whether they involve recruitment. Recruited reactions can in theory have different distance constraints (e.g., global vs local, B) and be intrinsically cooperative (A) or noncooperative (B–E).

Modifications at a particular histone residue can generate more than three nucleosome types, for example a lysine can be acetylated, unmodified, or contain one, two or three methyl groups (Fig. 7.5C). Furthermore, nucleosomes generally carry two copies of any histone residue and can thus be heterotypically modified (Fig. 7.5D). Readers that can differentiate these states can give them distinct recruitment properties.

Modification states of *different* histone residues can also be read in combination (forming a "histone code"; [28]), multiplying the potential number of individual nucleosome states. In a systematic search, we found that hundreds of different reaction schemes were able to give heritable bistability when each of the four different nucleosome states generated by a pair of two-state modifications (e.g., H3K4/H3K4me × H3K27/H3K27me) was allowed to recruit specific writers (Fig. 7.5E) [29]. Ten distinct nucleosome types can be generated by two-state modifications at two different amino acids in both half-nucleosomes (Fig. 7.5F).

In our *pombe* model, stable complexes between nucleosomes and other proteins, such as Swi6 (HP1), were not specified. However, addition of modification-dependent stably bound entities adds extra nucleosome states and epigenetic capabilities. Specifying a Sir complex-bound to fully unacetylated nucleosomes with distance-dependent Sir–SIR cooperative interactions (Fig. 7.5D), and with binding inhibiting and being inhibited by Dot1 methylation of H3K79, allowed simulation of many features of this epigenetic silencing system [30].

Thus, modeling shows that bistability can be generated by a variety of reaction schemes.

Temporal Regulation—Tilting the System

In order to be biologically useful, epigenetic systems are regulated over time. In our standard *pombe* model the processes in the system are held constant. However, modeling shows that such systems can be manipulated temporally by changing the strengths of the reactions.

A change in the strengths of one or more of the reactions in one direction (e.g., toward M) relative to the other direction (e.g., toward A) can shift the system out of the bistable regime into M-only or A-only monostable regimes [15,26]. For example, tuning the strength of the unrecruited U→A reaction within a *pombe* model system that is biased toward the M state is able to control the state of the system and is able to act as a transient signal to switch the system from one stable state to the other [31]. Even when bistability is weak or absent, the positive feedback can confer an ultrasensitive response, making the system highly sensitive to changes in the reaction activity [31].

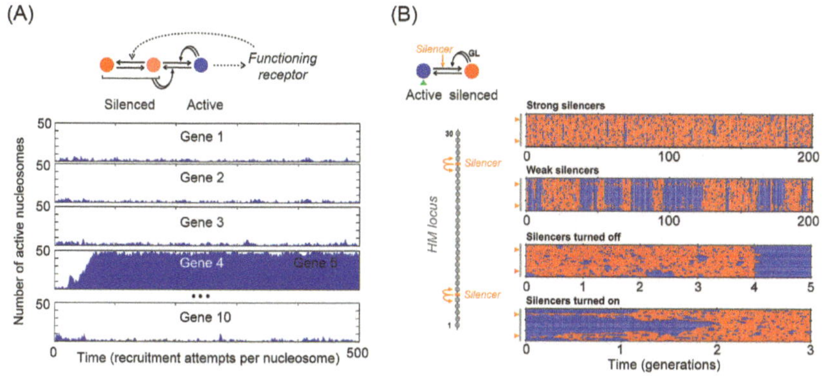

FIGURE 7.6 Controlling bistability in time and space.
(A) A model of odorant receptor (OR) choice (32). All OR genes begin in the S (silenced) state. At first, the S→A recruitment reaction is stronger than the A→S recruitment reaction with parameters such that each gene will rarely transition to the active A state. Expression of an OR protein from an active gene feeds back to inhibit the S→A reactions at all OR genes by fostering formation of a resistant nucleosome type. This puts the system into a regime that fixes the active gene in the A state and the other genes in the S state (32). (B) Model showing the ability of localized silencer elements to control the state of a chromatin region (27). The general A→S recruitment reaction required two recruiting S nucleosomes, one from any position (global) and one as the neighbor of the target A nucleosome (local). Each silencer was able to cause the A→S reaction on its immediate neighbor nucleosomes. The state of the 30 nucleosome region could be controlled by changing the probability of the silencer reaction.

The combination of memory and regulation provided by nucleosome modification feedback is illustrated by a model of odorant receptor (OR) choice [32]. Mammals have >1000 OR genes but each olfactory neuron must express just one of these for the life of the neuron. In the model (Fig. 7.6A), where the neuron begins life with all OR genes in the S (silenced) state but with the S→A recruitment reaction stronger than the A→S recruitment reaction, and the genes will slowly but randomly make a transition to the active A state. The model proposes an extra feedback, where the first gene to become active expresses its OR protein, which in addition to its normal function causes an inhibition of the S→A reactions at all OR genes by fostering formation of a resistant nucleosome type. This alters the balance of the S→A and A→S reactions, allowing the active gene to remain in the A state but preventing the other genes from leaving the S state [32]. An alternative model for OR choice involving histone modifications but without positive feedback has been proposed [33]. The basic idea that the first active OR gene inhibits activation of the other genes is the same; however, this model does not provide any explanation for the stability of the active state.

In the above models, the regulation of reaction rates is general, applied uniformly over the whole system. However, biological control is expected to involve targeting of enzymes to specific DNA sites. Modeling of the action of silencer elements in budding yeast, probably the best understood natural case of targeted epigenetic regulation, indicates that such localized reactions can nevertheless have system-wide effects (Fig. 7.6B). Silencers are DNA elements that recruit the histone deacetylase Sir2. Each silenced mating-type locus (*HM* locus) contains two silencers, which when operating normally keep the region continuously in an inactive state. Inactivation of the silencers results in the continuous activity of the region. The underlying bistability in the system is revealed when the silencers are partially inactivated, which can cause the region to switch between active and inactive states that persist for tens of generations [34]. To model this system, we used a 2-nucleosome state model (Silenced or Active) with the A→S reaction subject to positive feedback by recruitment of Sir2 by S nucleosomes [27]. Each silencer was modeled as a "special nucleosome" that was fixed in the S state and recruited Sir2 with high efficiency. Even when the silencer-mediated A→S reaction was limited to its immediate neighboring nucleosomes, the activity of the silencer could control the behavior of the whole system: Strong silencers gave a continuous high-S state, weak silencers gave A/S heritable bistability, and absent silencers gave a continuous high-A state (Fig. 7.6B).

A combination of general and localized temporal regulation is employed in the more complex variant of our *pombe* model used by Angel et al. to describe epigenetic regulation of the *Arabidopsis* FLC locus [23].

Spatial Regulation—Containing Positive Feedback

Epigenetic regulation must also be controlled spatially, that is, directed and confined to the proper genomic regions. Models based on nucleosome modification have not yet fully addressed this issue, though a variety of mechanisms have been explored.

Recruitment of histone modifying enzymes by sequence-specific DNA binding proteins, such as for the budding yeast silencers or *Drosophila* Polycomb response elements [35] seems to be the primary mechanism for targeting epigenetic regulation spatially. However, the positive feedback that underlies the bistability in these systems means that the nucleosome modification state of the targeted region could inappropriately influence the modification state of other chromatin regions. There would seem to be two kinds of solutions: (1) inhibit interactions between the targeted region and other regions and (2) allow the interactions but make the modification state of other regions resistant to the effects of the interactions.

Inhibiting Interactions

Interactions between chromatin regions may occur in two basic ways: Via a one-dimensional linear spreading along the chromatin, or via contact through three-dimensional space. Inhibition of three-dimensional interactions, such as by enhancer-blocking insulator elements or loop domains is poorly understood and the magnitude of the contact inhibition is unclear [36].

One-dimensional spreading is thought to be able to be blocked by DNA elements that are unable to carry or propagate the relevant modification (e.g., nucleosome-free DNA or DNA bound by modification-resistant nucleosomes). However, such simple barriers cannot block the long-range interactions that we expect to operate in these bistable systems. One type of model in which a simple barrier can inhibit spreading yet which retains the necessary long-range interactions, is if the recruitment reaction for the modification step that creates the final active or silenced nucleosome state is local, that is, only able to modify a nearest neighbor of the recruiting nucleosome, whereas long-range cooperativity is implemented in other recruitment reactions [27]. Recent data on the time to switch from the active to the silenced state of first one and then another of two separated promoters in the wild-type mating-type region of *pombe* [37] are reproduced by a model where only the deacetylation reaction is mediated by long-range recruitment, whereas all other recruitment reactions only act between nearest neighbors (Fig. 7.5B, right). Noticeably the robustness of this asymmetric model against the strength of nonrecruited events relies on having recruitment acting on conversions in both directions, and is increased if one extends the model by separating the active from the silenced state by more intermediate states.

However, blockage of one-dimensional influence by these kinds of systems is fragile, since the long-range interactions can still exert influence that can aid the growth of rare modification events in neighboring regions.

Encoding Resistance to Outside Influences

A straightforward way in which different chromatin regions could exist stably in different modification states despite interactions between them is if they contain different DNA elements that continuously target different modification enzymes.

For example, the Sir silencer elements in budding yeast act continuously to promote silencing modifications in their vicinity. This means that the chromatin regions near these elements are exposed to different rates of the modification reactions than the rest of the genome: The silenced regions are in a monostable regime of the Sir system, while outside DNA exists either in a monostable active regime or in the active state of a bistable regime. However, in our models of the Sir system, we found that this localization of silencing-promoting reactions did not by itself provide sufficient contrast to prevent silencing propagating into neighboring chromatin [27,30].

This contrast between the behavior of differently targeted chromatin regions can be increased in models where the supply of the reader–writer feedback enzymes is limited. Their recruitment to certain regions "titrates" their availability at other regions and the targeted regions become the most favorable locations for these enzymes. However, none of these models has properly addressed the issue of competition between different similarly targeted regions of the genome, such as between the 2 *HM* loci and 32 telomeric sites silenced by Sir in budding yeast. Titration can easily limit spreading at one region, but one would expect large fluctuations in the sizes of the modified regions when multiple similar regions compete for the same enzymes.

Another mechanism to aid discrimination between genomic regions is if there is continuous targeting of opposing modification enzymes, for example silencing-fostering modification enzymes to some regions and activity-fostering modification enzymes to other regions [27]. Thus, some regions may be so strongly located in monostable silenced or active regimes that they resist the influence of neighboring regions. For regions that are required to be bistable, the levels of continuous targeting of modification enzymes would depend on the constitutive state of the neighboring DNA. For example, a region always surrounded by active DNA may need a constant bias toward silencing to balance the outside influence and allow it to be bistable.

As chromatin-modifying activities can be associated with elongating RNA polymerase [38], the transcription process itself, with its clearly defined start and end points, provides a way to localize modification reactions to a region of chromatin. Potential positive feedback between these modifications and the initiation of transcription adds potential for generating bistability [30,39] (see in particular Chapter 8: Quantitative Environmentally Triggered Switching Between Stable Epigenetic States).

We expect that different combinations of these mechanisms, and potentially new mechanisms, will be needed to explain the full patterning of epigenetic nucleosome modifications across whole genomes.

DNA METHYLATION-MEDIATED EPIGENETICS

CpG Methylation

Covalent chemical modifications of DNA occur in all kingdoms of life, with a variety of roles. An epigenetic function in a variety of animal and plant systems has been attributed to methylation of cytosine residues in DNA to create 5meC. The most studied of these is 5meC modification at 5′-CG-3′ dinucleotide sequences, or CpG sites, termed CpG methylation.

One of the clearest cases for an epigenetic role of CpG methylation, at least in animals, is genomic imprinting in mammals, where one of the two copies of a gene in a diploid cell (the maternally- or paternally derived copy) can be active whereas the other is inactive [10]. This *cis*-restricted activity, together with the heritability of the choice of active/inactive allele within a clone of cells or an entire animal, and the fact that the same DNA can be active in some cases and inactive in others (in different clones or animals), argue strongly for a chromatin-based epigenetic mechanism. All cases of genomic imprinting seem to be associated with differential CpG methylation, where the same DNA region on the active allele is either highly or lowly methylated (a high or low proportion of CpGs are methylated), while it is in the opposite state on the inactive allele. If this differential CpG methylation is removed by inactivation of DNA methylation enzymes, the differential expression of the two alleles is lost [10]. However, it is not clear whether CpG methylation alone is sufficient for epigenetic memory or must work together with nucleosome-based epigenetics.

The Classical Model

This field has been founded on an elegant model for the inheritance of CpG methylation proposed over 40 years ago [40,41]. Because a CpG site on double-stranded DNA contains a C on each strand, replication of a fully methylated site (both C's methylated) generates hemimethylated CpGs, containing one original methylated C and a newly inserted unmethylated C. If hemimethylated CpGs were to be reliably recognized by "maintenance" enzymes that methylate the unmethylated C before the next replication event, then fully methylated CpGs can be reliably inherited. Unmethylated CpGs are naturally maintained upon DNA replication and their reliable inheritance thus relies on their avoiding de novo enzymatic methylation. This "classical" model was supported by the discovery of DNA methyl transferase 1 (DNMT1), a DNA methylation enzyme that has a strong preference for methylation of hemimethylated CpGs over unmethylated CpGs (see Ref. [42]).

Inadequacy of the Classical Model and Other Independent-CpG Models

Stable memory by the classical model relies on the absence of errors (failed maintenance methylation, or mistaken methylation of an unmethylated CpG), as once the state of a CpG is changed there is no force to restore the original state. Errors seem unavoidable, since enzymes that can catalyze each step of the unmethylated (u)↔hemimethylated (h)↔fully methylated (m) pathway for a CpG are now known. In addition to 'writer' enzymes that add methylation (DNMT1, DNMT3A, DNMT3B, [42]), enzymes are known that remove methylation (TET enzymes, APOBEC [43]). Given finite rates for these reactions (potentially different for different CpGs), each CpG has a single equilibrium state, dividing its time between the u, h, and m forms. Thus a single CpG, or even a group of CpG sites each acting independently, that is set into either the u or m state will steadily revert to the same equilibrium mixture, erasing memory [44–48].

Models where individual CpGs act independently also provide no explanation for the observed correlation of methylation states between nearby CpGs. CpGs are highly clustered in vertebrate genomes, forming small high density patches set within large regions of very low density [49,50]. CpG clusters that are sufficiently large and dense are termed CpG islands and are often associated with promoters. Cluster methylation is strikingly bimodal—either most CpGs in a cluster are methylated or most are unmethylated, with intermediate methylation levels rare [50,51]. To explain this bimodal behavior, independent-CpG models must invoke external factors that act over whole clusters to effect a strong bias toward either high or low methylation.

The independent-CpG models have even more difficulty explaining observations of dynamic bistability of CpG methylation. Lorincz et al. [52] made precise genomic insertions of a promoter-reporter gene DNA fragment pretreated such that none, some or all of its CpGs were methylated. The unmethylated DNA gave genetically active clones (stable for many cell generations), whereas fully methylated DNA gave genetically inactive clones. The partially methylated DNA gave either active clones that had lost the premethylations, or inactive clones that had gained new methylations. This experiment shows imperfect inheritance of single CpGs, and bistable behavior of a CpG cluster that cannot be due to constitutive external factors. One could, of course, invoke external factors that are different for the active and inactive clones, but this would be to jettison the idea of CpG methylation having a true epigenetic role.

Modeling Bistability Through CpG Collaboration

We have been exploring models where CpGs do not act independently but rather interact with each other in a manner similar to that proposed for nucleosomes. We find that this CpG collaboration allows for stable memory in the presence of errors, produces bimodality and bistability naturally through positive feedback, and retains a central role for CpG methylation in memory.

Spatially Independent Models

The first demonstration that a collaborative model could produce bistability was by Sontag et al. [45], who showed that in a system without demethylation reactions, bistability could be achieved if the rate of the u→h reaction was dependent on cooperativity between h sites. Cooperativity was implemented in a manner that ignores spatial relationships between CpGs, by using the square of the overall fraction of h sites in the whole system after replication.

Our CpG methylation models are similar to the *pombe* nucleosome model in having three possible states of each CpG, arranged in a linear reaction

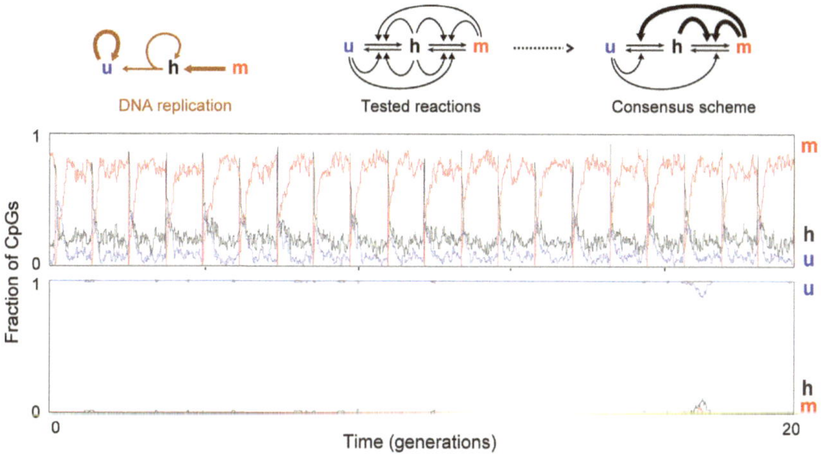

FIGURE 7.7 Collaboration between CpGs can generate heritable bistability.
DNA replication gives a strong push toward the u state. The tested reaction scheme shows the different reactions for which over 10^8 combinations of rates were scanned for their ability to generate heritable bistability [48]. The CpG modification reaction scheme symbolism is the same as for nucleosome-based systems. The consensus scheme shows those reactions that favor bistability and their favored relative strengths. The simulation shows a 20 generation time course of the fractions of u, h, and m sites in an 80 CpG system controlled by a consensus-type reaction scheme. The upper panel was started in a high-*m* state and the lower panel in a low-*m* state. Each state was stable for at least 100 generations.

sequence (Fig. 7.7). A major difference is DNA replication, which gives a bias toward the unmethylated state, converting all m's to h's and half the h's to u's (we assume that there is no strand bias for the mC in an h site). We allowed all interconversion steps to be catalyzed. We note that our representation of the m→h and h→u demethylation reactions as single steps oversimplifies the multistep reactions carried out by the demethylases ([43]; effectively, our h and u sites may contain one or two Cs that carry other modifications). "Reader" protein domains that can recognize each of the modification states of CpGs are also known, such as methyl-binding domains and Zn-CXXC domains [53], which make a variety of recruitment reactions theoretically possible.

To explore which reactions are needed for bistability of CpG clusters, we sampled a range of rates for the four unrecruited conversions and eight recruited conversions (Fig. 7.7), testing which combinations of reactions produced bistability in a system of 80 CpGs by simulation [48]. As in our nucleosome models, each reaction attempt involves choosing a random CpG site as a target, choosing a random reaction according to fixed probabilities, and, if appropriate, choosing random recruiting CpGs from among all CpGs in the system. The reaction is performed if there is a match between target, reaction and recruiter. Note that spatial relationships between CpGs are ignored, also that there is no explicit cooperativity as in the Sontag et al. model [45]. We used an average of 100 reaction attempts per CpG between each DNA replication event.

A large number, but a small fraction, of our sampled reaction schemes gave bistability over at least 100 cell generations (Fig. 7.7). The "consensus" reaction scheme (Fig. 7.7) contained the four recruitment reactions where u or m sites stimulate conversions toward their state, as in the pombe nucleosome model except that the reactions in the m direction were stronger in order to counteract the replication bias toward u. In addition, h sites could be involved in recruitment, in particular collaborating to foster the h→m reaction, providing maintenance methylation to cope with replication. Reaction schemes with these five recruitment reactions were capable of >1000 generation stabilities [48].

One of the attractions of collaboration is that it removes the need for symmetrical DNA modifications in inheritance. A sequence modified on only one strand will produce a modified and an unmodified daughter sequence, but as long as there is a cluster of such sequences (and they are not all oriented in the same way), then a reader–writer can be recruited to nearby modified sequences to restore the modification on the unmodified sequences, akin to the mechanism proposed for nucleosomes. Thus, any DNA modification could be used to create chromatin-based epigenetic memory.

Models Exploiting Spatial Heterogeneity of CpGs

A distinctive feature of CpG methylation, compared with nucleosome modification, is that the arrangement of CpGs can be very different at different genomic locations. It is likely that interactions between CpGs, which we propose are critical in controlling methylation, are affected by the DNA distance between them. Thus, the methylation behavior of CpGs should depend on their spatial relationship to other CpG sites.

It was known that CpGs in low CpG density regions generally show a high level of methylation (~70%), while CpGs in CpG islands tend to be bimodally methylated: Either poorly methylated or highly methylated [51]. To look at systematically, we used existing genome-wide methylation data from human cells to examine the methylation levels of reasonably isolated CpG clusters, that is, clusters in which the distances (in base pairs) between CpGs in the cluster are all substantially smaller than the distances from the edge of the cluster to neighboring CpGs [50]. We found a strong dependence of methylation on the number of CpGs in such clusters (termed cluster "size"). Small clusters of 1–5 CpGs tend to be highly methylated, intermediate-sized

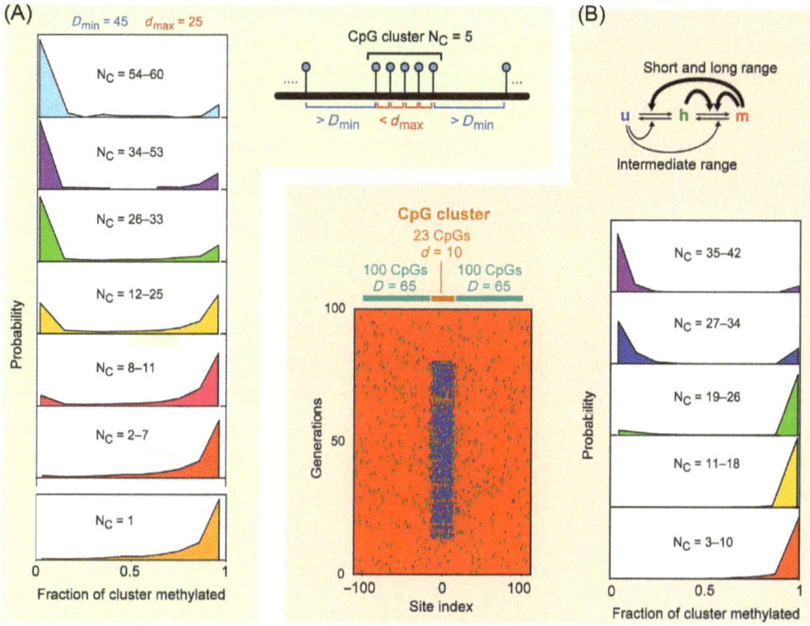

FIGURE 7.8 Distance-dependent CpG collaboration can reproduce the effect of CpG cluster size on methylation.
(A) The schematic shows the definition of a CpG cluster based on two distance parameters: A maximal distance between CpGs within the same cluster and a minimal distance between the cluster and outside CpGs. The histograms give the average fractional methylation (fraction of CpGs methylated, averaged over many DNA fragments) for clusters defined by $D_{min} = 45$ and $d_{max} = 25$ of different sizes (N_C), observed in four human methylomes [50]. Small CpG clusters tend to be highly methylated, large CpG clusters tend to be unmethylated, whereas intermediate size clusters show bimodal methylation (note that in almost all cases these are different clusters). (B) The reaction scheme shown was used in stochastic simulations in which CpG clusters of different sizes were set within a larger region of low-density CpGs. Special functions were used to define how the methylation and demethylation recruitment reactions act over DNA distances. The simulation illustrates that a CpG cluster can display heritable bistability while surrounded by constitutively methylated low-density CpG regions (note that the DNA scale is compressed for the low-density regions). The histograms show that this model was also able to reproduce the effect of cluster size on methylation propensity.

clusters (~10–30 CpGs) show either high or low methylation levels, whereas large clusters (>40 CpGs) tend to have low methylation levels (Fig. 7.8A). There was also a systematic density effect, with higher density clusters being more likely to be unmethylated, a feature observed previously [54]. Such systematic effects are strongly indicative of CpG interactions and suggest a spatial asymmetry in methylation-promoting and demethylation-promoting interactions.

Further analysis supported quite different spatial interactions for methylated and unmethylated CpGs. We found that the methylation levels of CpGs *around* intermediate-sized clusters is elevated for those clusters that are highly methylated, but not for those clusters that are unmethylated [50]. This suggested that methylation-promoting interactions can act over distance, whereas demethylation-promoting interactions are more spatially restricted.

We showed that by imposing certain spatial constraints on the consensus recruitment reactions in our previous model the model could reproduce the ability of a CpG cluster to exist in either a high or low methylation state (bistability) when set within a larger region of low-density CpGs that remain highly methylated (Fig. 7.8B). This model was also able to reproduce the dependence of cluster methylation probability on cluster size (Fig. 7.8B) but did not reproduce the effect of cluster methylation on CpGs surrounding the cluster. Though the spatial constraints used were admittedly somewhat arbitrary, this shows that much of the genome-wide variation in CpG methylation *could* in theory be explained by relatively simple, universal collaboration between CpGs, without the need to invoke a plethora of specific external factors.

Although some experimental observations are consistent with recruitment in methylation and demethylation reactions (see Refs. [48,50]), clear evidence for this feedback is not available. One complication is that feedback in methylation may be complex, involving other chromatin marks, such as nucleosome modifications (e.g., as for CHG methylation in plants [55]).

We have explored alternative mechanisms to positive feedback by recruitment. It has been proposed that binding of proteins to unmethylated CpG clusters may inhibit methylation enzymes and therefore help maintain low methylation levels [56]. By specifying a distinct inhibitor-bound state for u sites, we showed that his type of feedback (with cooperativity and some spatial activity of the inhibitors) was able to replace the recruitment demethylation reactions and allow a CpG cluster to be bistable when surrounded by highly methylated, low CpG density regions [57]. However, this model did not reproduce the effects of cluster size or density on methylation.

An alternative model, which replaces direct recruitment of demethylases with a feedback between methylation status, nucleosome occupancy and methylation sensitivity, is able to reproduce cluster size and density effects, as well as an effect of a highly methylated cluster on outside CpGs [58]. Here, the spatial range of the demethylation feedback is determined by the length of DNA occupied by the nucleosome. The model is based on evidence that gene silencing is associated with nucleosome occupation of methylated CpG islands, while promoter activation is associated with eviction of these nucleosomes and CpG demethylation [59]. Cooperative feedback in the model is provided by indirect interaction [60] between unmethylated CpGs in resisting nucleosome occupation, and preferential targeting of DNA methylation to nucleosomal DNA [61].

OUTLOOK

The field of chromatin epigenetics, which for a long time has been largely dominated by identification of yet more molecular players, has started to move to a new phase of examining fundamental mechanisms and how they generate epigenetic behavior. We hope that modeling will continue to play a part in this progress.

The biologist is excited by the prospect of more "realistic" models that deepen the connection with experiments: Such as more complete models of genome-wide nucleosome states and of the relationship between CpG spatial patterns and methylation states; and more detailed models informed by, and informing, new experimental approaches. These include highly controlled in vivo targeting of specific modification enzymes [62–64] or insertion of synthetic CpG-containing DNA into defined genomic regions [54], as well as in vitro experiments that use purified reader–writer enzymes and reconstructed chromatin templates to test positive feedback by recruitment and explore its cooperativity and nonlocality [17,18].

The physicist is also excited by the prospect of developing simplifying models that highlight fundamentals: Such as distilling the complex extant knowledge of the Polycomb–trithorax system down to its epigenetic core; and exploring how interactions between nucleosome positioning and CpG methylation could reduce the number of processes required to explain how spatial patterning of CpG sites underpins their methylation potential. From a broader perspective, the combined experimental and theoretical understanding of the dynamics of epigenetics provides a playing ground that can teach us about diversity in life and how this is implemented in terms of a variety of dominance and cooperation games [4,65–67].

References

[1] Veening JW, Smits WK, Kuipers OP. Bistability, epigenetics, and bet-hedging in bacteria. Annu Rev Microbiol 2008;62:193–210.

[2] Ptashne M. On the use of the word "epigenetic." Curr Biol 2007;17(7):R233–6.

[3] Felsenfeld G. A brief history of epigenetics. Cold Spring Harb Perspect Biol 2014;6(1):a018200

[4] Sneppen K. Models of life: dynamics and regulation in biological systems. Cambridge, UK: Cambridge University Press; 2014.

[5] Elowitz MB, Leibler S. A synthetic oscillatory network of transcriptional regulators. Nature 2000;403(6767):335–8.

[6] Ajo-Franklin CM, Drubin DA, Eskin JA, et al. Rational design of memory in eukaryotic cells. Genes Dev 2007;21(18):2271–6.

[7] Felsenfeld G, Groudine M. Controlling the double helix. Nature 2003;421(6921):448–53.

[8] Grewal SI, Klar AJ. Chromosomal inheritance of epigenetic states in fission yeast during mitosis and meiosis. Cell 1996;86(1):95–101.

[9] Thon G, Friis T. Epigenetic inheritance of transcriptional silencing and switching competence in fission yeast. Genetics 1997;145(3):685–96.

[10] Ferguson-Smith AC. Genomic imprinting: the emergence of an epigenetic paradigm. Nat Rev Genet 2011;12:565–75.

[11] Berry S, Hartley M, Olsson TS, Dean C, Howard M. Local chromatin environment of a Polycomb target gene instructs its own epigenetic inheritance. eLife 2015;4.

[12] Alabert C, Groth A. Chromatin replication and epigenome maintenance. Nat Rev Mol Cell Biol 2012;13(3):153–67.

[13] Grunstein M. Yeast heterochromatin: regulation of its assembly and inheritance by histones. Cell 1998;93(3):325–8.

[14] Turner BM. Histone acetylation as an epigenetic determinant of long-term transcriptional competence. Cell Mol Life Sci 1998;54(1):21–31.

[15] Dodd IB, Micheelsen MA, Sneppen K, Thon G. Theoretical analysis of epigenetic cell memory by nucleosome modification. Cell 2007;129(4):813.

[16] Margueron R, Justin N, Ohno K, et al. Role of the polycomb protein EED in the propagation of repressive histone marks. Nature 2009;461(7265):762–7.

[17] Al-Sady B, Madhani HD, Narlikar GJ. Division of labor between the chromodomains of HP1 and Suv39 methylase enables coordination of heterochromatin spread. Mol Cell 2013;51(1):80–91.

[18] Muller MM, Fierz B, Bittova L, Liszczak G, Muir TW. A two-state activation mechanism controls the histone methyltransferase Suv39h1. Nat Chem Biol 2016;12(3):188–93.

[19] Ferrell Jr. JE. Self-perpetuating states in signal transduction: positive feedback, double-negative feedback and bistability. Curr Opin Cell Biol 2002;14(2):140–8.

[20] Sneppen K, Krishna S, Semsey S. Simplified models of biological networks. Annu Rev Biophys 2010;39:43–59.

[21] Mukhopadhyay S, Nagaraj VH, Sengupta AM. Locus dependence in epigenetic chromatin silencing. Biosystems 2010:49–54.

[22] Sedighi M, Sengupta AM. Epigenetic chromatin silencing: bistability and front propagation. Phys Biol 2007;4(4):246–55.

[23] Angel A, Song J, Dean C, Howard M. A Polycomb-based switch underlying quantitative epigenetic memory. Nature 2011;476(7358):105–8.

[24] Binder H, Steiner L, Przybilla J, Rohlf T, Prohaska S, Galle J. Transcriptional regulation by histone modifications: towards a theory of chromatin re-organization during stem cell differentiation. Phys Biol 2013;10(2):026006.

[25] Anink-Groenen LC, Maarleveld TR, Verschure PJ, Bruggeman FJ. Mechanistic stochastic model of histone modification pattern formation. Epigenetics Chromatin 2014;7(1):30.

[26] Micheelsen MA, Mitarai N, Sneppen K, Dodd IB. Theory for the stability and regulation of epigenetic landscapes. Phys Biol 2010;7(2):026010.

[27] Dodd IB, Sneppen K. Barriers and silencers: a theoretical toolkit for control and containment of nucleosome-based epigenetic states. J Mol Biol 2011;414(4):624–37.

[28] Rando OJ. Combinatorial complexity in chromatin structure and function: revisiting the histone code. Curr Opin Genet Dev 2012;22(2):148–55.

[29] Sneppen K, Dodd IB. A simple histone code opens many paths to epigenetics. PLoS Comput Biol 2012;8(8):e1002643.

[30] Sneppen K, Dodd IB. Cooperative stabilization of the SIR complex provides robust epigenetic memory in a model of SIR silencing in *Saccharomyces cerevisiae*. Epigenetics 2015;10(4):293–302.

[31] Sneppen K, Micheelsen MA, Dodd IB. Ultrasensitive gene regulation by positive feedback loops in nucleosome modification. Mol Syst Biol 2008;4:182.

[32] Alsing AK, Sneppen K. Differentiation of developing olfactory neurons analysed in terms of coupled epigenetic landscapes. Nucleic Acids Res 2013;41(9):4755–64.

[33] Tan L, Zong C, Xie XS. Rare event of histone demethylation can initiate singular gene expression of olfactory receptors. Proc Natl Acad Sci U S A 2013;110(52):21148–52.

[34] Xu EY, Zawadzki KA, Broach JR. Single-cell observations reveal intermediate transcriptional silencing states. Mol Cell 2006;23(2):219–29.

[35] Steffen PA, Ringrose L. What are memories made of? How Polycomb and Trithorax proteins mediate epigenetic memory. Nat Rev Mol Cell Biol 2014;15(5):340–56.

[36] Krivega I, Dean A. Enhancer and promoter interactions-long distance calls. Curr Opin Genet Dev 2012;22(2):79–85.

[37] Obersriebnig MJ, Pallesen EM, Sneppen K, Trusina A, Thon G. Nucleation and spreading of a heterochromatic domain in fission yeast. Nat Commun 2016;7:11518.

[38] Tanny JC. Chromatin modification by the RNA Polymerase II elongation complex. Transcription 2014;5(5):e988093.

[39] Kitada T, Kuryan BG, Tran NN, et al. Mechanism for epigenetic variegation of gene expression at yeast telomeric heterochromatin. Genes Dev 2012;26(21):2443–55.

[40] Holliday R, Pugh JE. DNA modification mechanisms and gene activity during development. Science 1975;187(4173):226–32.

[41] Riggs AD. X inactivation, differentiation, and DNA methylation. Cytogenet Genome Res 1975;14(1):9–25.

[42] Jones PA, Liang G. Rethinking how DNA methylation patterns are maintained. Nat Rev Genet 2009;10(11):805–11.

[43] Wu H, Zhang Y. Reversing DNA methylation: mechanisms, genomics, and biological functions. Cell 2014;156(1–2):45–68.

[44] Pfeifer GP, Steigerwald SD, Hansen RS, Gartler SM, Riggs AD. Polymerase chain reaction-aided genomic sequencing of an X chromosome-linked CpG island: methylation patterns suggest clonal inheritance, CpG site autonomy, and an explanation of activity state stability. Proc Natl Acad Sci U S A 1990;87(21):8252–6.

[45] Sontag LB, Lorincz MC, Georg Luebeck E. Dynamics, stability and inheritance of somatic DNA methylation imprints. J Theor Biol 2006;242(4):890–9.

[46] Genereux DP, Miner BE, Bergstrom CT, Laird CD. A population-epigenetic model to infer site-specific methylation rates from double-stranded DNA methylation patterns. Proc Natl Acad Sci U S A 2005;102(16):5802–7.

[47] Jeltsch A, Jurkowska RZ. New concepts in DNA methylation. Trends Biochem Sci 2014;39(7):310–18.

[48] Haerter JO, Lövkvist C, Dodd IB, Sneppen K. Collaboration between CpG sites is needed for stable somatic inheritance of DNA methylation states. Nucleic Acids Res 2014;42(4):2235–44.

[49] Gardiner-Garden M, Frommer M. CpG islands in vertebrate genomes. J Mol Biol 1987;196(2):261–82.

[50] Lovkvist C, Dodd IB, Sneppen K, Haerter JO. DNA methylation in human epigenomes depends on local topology of CpG sites. Nucleic Acids Res 2016;44(11):5123–32.

[51] Zhang Y, Rohde C, Tierling S, et al. DNA methylation analysis of chromosome 21 gene promoters at single base pair and single allele resolution. PLoS Genet 2009;5(3):e1000438.

[52] Lorincz MC, Schübeler D, Hutchinson SR, Dickerson DR, Groudine M. DNA methylation density influences the stability of an epigenetic imprint and Dnmt3a/b-independent de novo methylation. Mol Cell Biol 2002;22(21):7572–80.

[53] Du Q, Luu PL, Stirzaker C, Clark SJ. Methyl-CpG-binding domain proteins: readers of the epigenome. Epigenomics 2015;7(6):1051–73.

[54] Lienert F, Wirbelauer C, Som I, Dean A, Mohn F, Schubeler D. Identification of genetic elements that autonomously determine DNA methylation states. Nat Genet 2011;43(11):1091–7.

[55] Johnson LM, Bostick M, Zhang XY, et al. The SRA methyl-cytosine-binding domain links DNA and histone methylation. Curr Biol 2007;17(4):379–84.

[56] Thomson JP, Skene PJ, Selfridge J, et al. CpG islands influence chromatin structure via the CpG-binding protein Cfp1. Nature 2010;464(7291):1082–6.

[57] Sormani G, Haerter JO, Lovkvist C, Sneppen K. Stabilization of epigenetic states of CpG islands by local cooperation. Mol Biosyst 2016;12(7):2142–6.

[58] Sneppen K, Dodd IB. Nucleosome dynamics and maintenance of epigenetic states of CpG islands. Phys Rev E Stat Nonlin Soft Matter Phys 2016;93(6):062417.

[59] Lin JC, Jeong S, Liang G, et al. Role of nucleosomal occupancy in the epigenetic silencing of the MLH1 CpG island. Cancer Cell 2007;12(5):432–44.

[60] Polach KJ, Widom J. A model for the cooperative binding of eukaryotic regulatory proteins to nucleosomal target sites. J Mol Biol 1996;258(5):800–12.

[61] Chodavarapu RK, Feng S, Bernatavichute YV, et al. Relationship between nucleosome positioning and DNA methylation. Nature 2010;466(7304):388–92.

[62] Ragunathan K, Jih G, Moazed D. Epigenetics. Epigenetic inheritance uncoupled from sequence-specific recruitment. Science 2015;348(6230):1258699.

[63] Keung AJ, Bashor CJ, Kiriakov S, Collins JJ, Khalil AS. Using targeted chromatin regulators to engineer combinatorial and spatial transcriptional regulation. Cell 2014;158(1):110–20.

[64] Bintu L, Yong J, Antebi YE, et al. Dynamics of epigenetic regulation at the single-cell level. Science 2016;351(6274):720–4.

[65] Bornholdt S, Jensen MH, Sneppen K. Emergence and decline of scientific paradigms. Phys Rev Lett 2011;106(5):058701.

[66] Sneppen K, Mitarai N. Multistability with a metastable mixed state. Phys Rev Lett 2012;109(10):100602.

[67] Mitarai N, Mathiesen J, Sneppen K. Emergence of diversity in a model ecosystem. Phys Rev E Stat Nonlin Soft Matter Phys 2012;86(1 Pt 1):011929.

GLOSSARY

De novo methylation methylation of an unmethylated CpG site; the u→h reaction

CHG methylation methylation of C when part of sequence CHG, where H = not G.

LIST OF ACRONYMS AND ABBREVIATIONS

5meC 5-methyl cytosine
APOBEC apolipoprotein B mRNA editing enzyme, catalytic polypeptide-like
CRISPR clustered regularly interspaced short palindromic repeats
DNMT DNA methyl transferase
kb kilobasepair
Sir silent information regulator
TET ten eleven translocase

CHAPTER 8

Quantitative Environmentally Triggered Switching Between Stable Epigenetic States

Rea Antoniou-Kourounioti and Martin Howard
Computational & Systems Biology, John Innes Centre, Norwich Research Park, Norwich, United Kingdom

CONTENTS

INTRODUCTION

In recent years, the field of epigenetics has advanced very rapidly, with the components regulating epigenetic dynamics increasingly well characterized at a molecular level. Furthermore, bioinformatic analyses of genome wide data sets have shed light on large-scale epigenetic behavior. However, on their own, these approaches are frequently insufficient to unpick the underlying mechanisms that control epigenetic responses. For example, results from epigenomic studies typically reveal correlations at a whole genome level between, say, histone modifications and expression states [1–3]. But such correlative results may not shed much light on the underlying mechanisms at work or often even on the causality of the system. For example, it is still debated whether histone modifications are a cause or consequence of gene expression [4,5]. As a result, we believe that our understanding of the fundamental mechanisms of epigenetics has lagged well behind our ability to collect large-scale epigenomic data. To overcome these difficulties, we believe it is necessary to delve more deeply into the dynamics of individual epigenetic systems, thoroughly dissect their modes of regulation and then work back outwards toward a genomic scale, with the knowledge that the foundations of that expansion are well justified at individual loci.

Clearly, even dissecting epigenetic dynamics at a single locus is a formidable task, bearing in mind the complexity of many of the elements involved, including histone modifications and modifiers, nucleosome dynamics and remodelers, larger scale chromatin structure, and the transcriptional machinery itself. For this reason, we have adopted an interdisciplinary approach, fusing focused experiments with mathematical modeling in an effort to more rapidly elucidate the underlying principles. Our modeling philosophy has been strictly minimal, incorporating only those processes that are absolutely

Epigenetics and Systems Biology. DOI: http://dx.doi.org/10.1016/B978-0-12-803075-2.00008-8

necessary, and attempting to use the model predictively to accelerate our understanding of the mechanisms involved.

For our experimental system, we use the model plant *Arabidopsis thaliana* and focus on the floral repressor *Flowering Locus C* (*FLC*). We now briefly introduce the key features of *FLC*, which is a vital component of flowering time regulation. *FLC*'s genetically determined "starting" expression levels early in the plant's development determine the plant's subsequent life cycle, separating winter annual and summer annual plants. Winter annuals germinate in the autumn but remain vegetative during the winter to flower in the more favorable conditions of spring, whereas summer annuals both germinate and flower during the summer. This separation is therefore physiologically defined through the need, or not, for vernalization, the exposure to a prolonged period of cold leading to accelerated flowering. Vernalization acts through *FLC*, where *FLC* in turn functions by repressing the floral activator *Flowering Locus T* (*FT*). High levels of *FLC* expression thereby delay flowering [6]. Vernalization represses *FLC* and thus allows the expression of *FT*, which in turn enables flowering.

Other pathways also regulate *FLC* expression, the most prominent being the autonomous pathway, which acts to constitutively repress *FLC* expression. Another factor that regulates *FLC* is *FRIGIDA* (*FRI*). In contrast to the autonomous pathway, FRI activates *FLC*. Importantly, if FRI is dominant over the autonomous pathway, leading to high constitutive levels of *FLC* expression, vernalization is then necessary to reduce *FLC* expression to trigger flowering. Indeed, the vernalization requirement in natural ecotypes is usually determined by the presence or absence of an active *FRI* allele [7]. This means that *FRI* is associated with a vernalization requirement, despite the fact that environmental signals are not thought to be an input into FRI's regulation of flowering. In this chapter, we will focus on the cold-dependent vernalization pathway, though we emphasize that other inputs into the floral transition are also important. For example, *FT* is itself regulated by many other signals, including photoperiod [8] (Fig. 8.1A).

During vernalization, *FLC* expression is quantitatively downregulated by low temperatures (Fig. 8.1B and C). In other words, *FLC* levels at the end of the cold are quantitatively determined by the duration of the period of cold experienced by the plant (Fig. 8.1B and C). A set of noncoding antisense transcripts (*COOLAIR*) is upregulated during the early phase of the cold, and this has been linked to *FLC* shutdown [9,10]. Crucially, the reduced *FLC* levels at the end of cold exposure are maintained post-cold in the absence of the originating environmental signals, thereby helping to determine flowering time. These lowered *FLC* levels are inherited through multiple DNA replications, and therefore this process is necessarily an example of quantitative mitotically heritable epigenetic memory. Histone modifications have

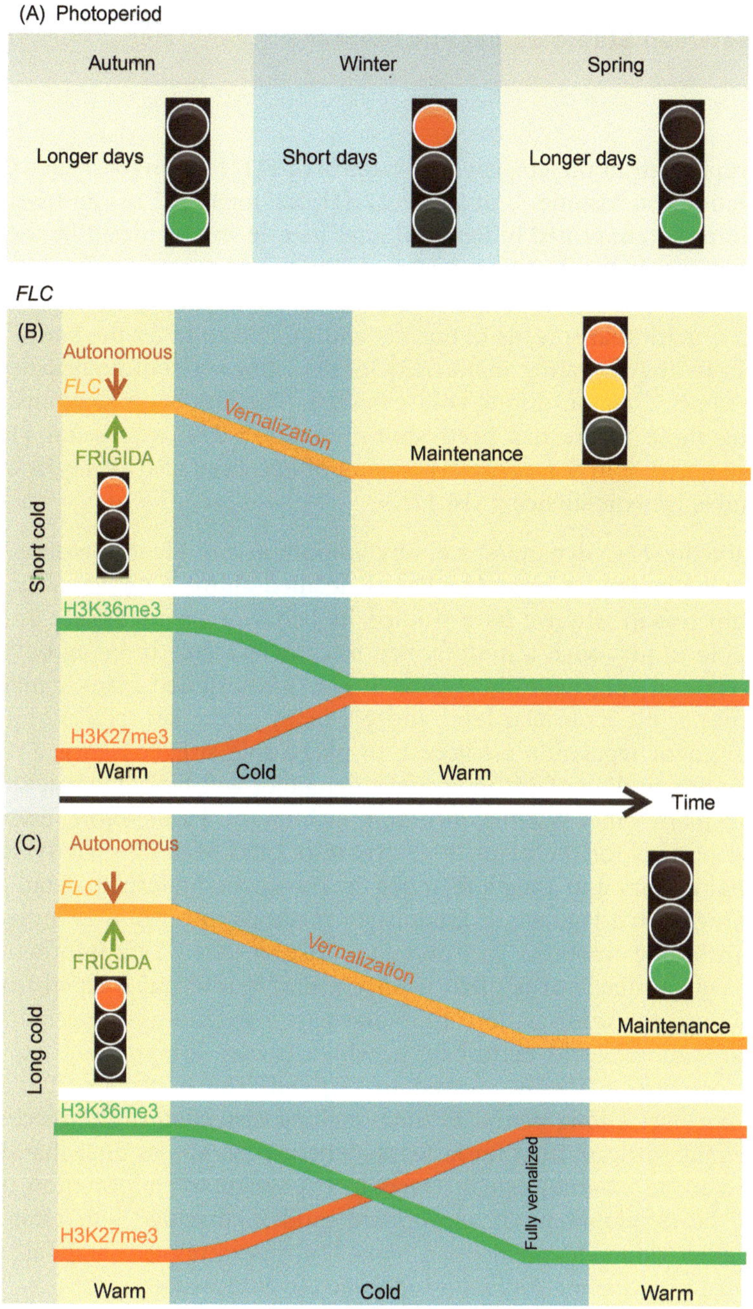

FIGURE 8.1

Flowering regulation.

Traffic lights indicate activating or repressive signals from pathways shown. (A) Photoperiod can promote flowering during autumn and spring when day-lengths are longer. (B and C) *FLC* shutdown and quantitative memory of winter cold. *FLC* expression levels are determined by the autonomous and *FRIGIDA* pathways before cold. In winter annual accessions, levels of *FLC* are high and flowering is prevented. During the cold, overall *FLC* expression is downregulated gradually from this starting level. This downregulation is coupled to changes in histone marks at *FLC* chromatin. The active mark H3K36me3 is reduced in the nucleation region while the silencing mark H3K27me3 is increased in the same region. After cold exposure, *FLC* expression is maintained at the reduced level reached at the end of cold, at either (B) intermediate or (C) fully repressed levels after shorter or saturating durations of cold, respectively.

been implicated in the epigenetic dynamics of *FLC* [11], in particular the tri-methylation, on histone 3, of lysine 27 (H3K27me3) [12,13] and lysine 36 (H3K36me3) (confirmed in Ref. [14] after its role was predicted by the work discussed in this chapter) (Fig. 8.1B and C). During the cold, a local increase in the silencing mark, H3K27me3, can be observed in a region at the 5′ end of *FLC*, a mark that spreads out at the end of cold to cover the entire locus. Conversely, the activating mark, H3K36me3, shows a coordinated decrease in the same 5′ region during cold exposure. The loss of methyltransferases affecting these marks has been shown to affect *FLC* expression [14,15]. Genetic variation in the noncoding region of *FLC* itself also affects its expression and epigenetic silencing [16,17].

Undoubtedly *FLC* dynamics are very important for plants because of its important function in reproduction, as detailed above. However, this is not the main reason why we have studied its behavior so intensively. In order to be able to play such a multifaceted role, for example in being both constitutively and epigenetically regulated, *FLC* has acquired many remarkable properties at the molecular level. Indeed, many of the key factors of genetic and epigenetic regulation are known to play a role at *FLC*. Hence, a mechanistic understanding of *FLC* epigenetic dynamics will be immediately applicable to many other systems, including in humans. Furthermore, because it is so important for the plant, *FLC* dynamics have been extensively studied over many years and therefore much is known in molecular detail about the pathways that regulate it. Established methods to investigate these pathways, including resources and mutants, exist so that where information is not already available, it is often relatively simple to acquire experimentally. In addition, for vernalization, the epigenetic regulation is triggered by an external stimulus, cold perturbation, which is easy to apply and to which the plants have naturally evolved to respond. The dynamics of the process occur over a long period of time, making dissection of the mechanism more straightforward than for faster epigenetic processes or ones that do not respond to an external signal. Overall, this wealth of information, which allows the development of appropriate models, together with established methods to test model predictions, make *FLC* an exceptional candidate for an integrated experimental/modeling approach to epigenetics.

In this chapter, the key feature of *FLC* that we wish to understand is its quantitative memory of the duration of cold that the plant has experienced. This is a highly nontrivial feature, especially considering that while the plants are growing, they must maintain a particular, quantitative level of *FLC* expression through many cell divisions, long after the cold signal that originally set this level has disappeared. We will therefore describe the mathematical modeling work that was used to decipher the underlying mechanisms of this process, as well as the complementary experimental work used to advance and validate the model.

MEMORY OF THE COLD IS DIGITAL AND IS STORED LOCALLY IN THE CHROMATIN

Efforts to investigate epigenetic dynamics at *FLC* using a combined experimental/modeling approach began with Angel et al. [13]. Two features of the process were especially important for the mathematical model to capture: Epigenetically stable, graded levels of *FLC* expression at a whole plant level, and an ability to switch a stable epigenetic state in response to an external stimulus (here, cold). A "core" model, not including the switching capability was used to answer the first question, with extensions to a more complete model to handle switching. This work was inspired by previous work on cell-autonomous, bistable (ON/OFF) epigenetic dynamics in yeast [18]. If a similar bistable mechanism were used at *FLC*, the population average graded reduction of *FLC* could potentially be explained by a change in the fraction of transcriptionally ON versus OFF *FLC* loci.

Memory

A key idea behind the model at a molecular level is that chromatin modifications can be self-sustained by feedback that adds more of the same modification to the locus. It has been shown that including only a single type of chromatin mark (either active or repressive), together with unmodified histones, is sufficient to obtain bistable ON/OFF epigenetic states. However, in that case, cooperativity has to be added to the feedback in the system explicitly to satisfy the requirement for the nonlinearity needed for robust bistability [18–20]. An attractive alternative is to include a second chromatin mark that antagonizes the first, leading to histones in three states: activating, unmodified, and silencing [13,18]. A particular mark was then assumed to be able to recruit factors that catalyze the addition of the same mark and that stimulate the removal of the opposing mark (Fig. 8.2). In this case, the nonlinearity needed for bistability is implicitly included in the positive feedback two-step process needed to transition from one histone mark to the opposing one.

A further assumption of the model necessary to ensure robust epigenetic stability was long-range interactions. One could imagine that fluctuations caused by random enzymatic events could add a small number of modifications that are in the opposing state to the majority of modifications at a locus. These opposing modifications will find it more difficult to spread if opposed by all the other modifications at the locus as compared to only those of their neighbors. Accordingly, it was assumed that each histone mark could effect a change in any other mark at the locus with equal probability. Effectively, therefore, the majority mark at the locus collectively functions to stamp out the presence of any minority marks. Indeed it has been shown that silent loci have compact nucleosome packaging [21] that could facilitate

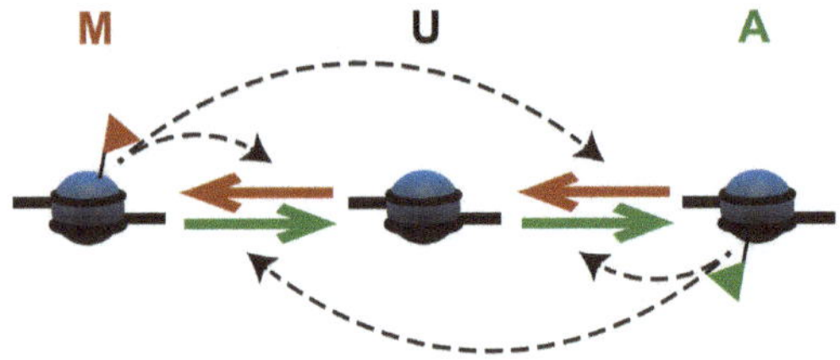

FIGURE 8.2

Antagonism of A and M chromatin marks.

The M mark promotes the addition of more M marks at the locus and the removal of A marks. Similarly, the A mark promotes the addition of more A marks and the removal of M marks. The positive feedback (dashed arrows) shown affects transitions between marks (solid arrows) and is necessary for stable epigenetic states without introducing additional assumptions of cooperativity.

such long-range interactions, and where the compact chromatin could in turn stimulate polycomb repressive complex 2 (PRC2) [22]. The authors of [13] also suggested rapid diffusion of recruited factors along noncoding RNAs as a mechanism for long-range interactions.

With these assumptions, bistable, digital ON/OFF states with loci predominantly covered with either active modifications (ON) or silencing modifications (OFF) are then the natural outcome of model dynamics. In the context of *FLC*, the silencing mark was already known to be H3K27me3, the mark added by PRC2. The model also predicted the existence of an activating mark that antagonizes H3K27me3, recruiting factors that remove H3K27me3, and adding further active marks. This modeling thus directly predicted the existence of H3K27me3 demethylases, which were at the time not identified in plants (but were soon after discovered; *REF6* [23] and *ELF6* [24]). Furthermore, the removal of the activating mark in the model was promoted by the presence of H3K27me3.

The model [13] was implemented using a stochastic simulation that tracked the state of each histone at an *FLC* locus using the above three state model, with silencing M, unmarked U, and activating A marks. The histones could transition from modified to unmodified, losing their mark, or acquire either mark if initially unmodified. For the silencing M modification, these events correspond to the loss or gain of H3K27 trimethylation by the action of the appropriate enzymes. These changes could occur via the stimulated transitions discussed above, or through noisy addition/removal. Furthermore, nucleosome swap out was incorporated through random loss of pairs of histone marks, simulating nucleosome turnover. DNA replication was also included in the model as the synchronized loss of, on average, half the nucleosomes at a locus, thereby accounting for the random insertion of parental nucleosomes into the two-daughter DNA strands, with the remaining spaces being filled by unmarked nucleosomes.

The key prediction of this model was digital ON/OFF expression of *FLC* at individual loci. Individual *FLC* loci switch off during the cold, with the number of silent loci predicted to be quantitatively related to the duration of cold. This prediction was confirmed by a reporter fused to *FLC* and imaged in the roots post-cold. Three different reporters were tested, GUS [13], Venus and mCherry [25]. It was first shown that the reporter gene was expressed in all root cells before the cold and, as predicted, an increasing number of silenced cells appeared with longer cold treatment. Additionally, automated image analysis of the intensity of the Venus reporter protein showed clear bimodal expression for intermediate lengths of cold, confirming the digital state of the loci [25].

Overall, this modeling work has provided new insight into epigenetic regulation and its consequences at different scales (Fig. 8.3). A whole plant average is important in determining flowering as seen from the correlation of *FLC* average transcript levels and flowering time. This is consistent with a digital ON/OFF response at individual loci because the downstream gene that is regulated by *FLC*, *FT*, encodes for mobile products [26,27] and therefore averages out the digital *FLC* response. The techniques used to quantify *FLC* mRNA and chromatin modifications also measure whole plant averages and show a graded response (Fig. 8.3—whole plant level). The modeling work suggested, however, that looking at a single cell level might give a very different picture (Fig. 8.3—cell level), as was subsequently found experimentally. Moreover, the model also provided an explanation for the stability of the epigenetic memory: strong positive feedback can easily fill in any missing histone modifications, even in the extreme case of DNA replication where histone modification levels are on average halved. The model also elucidated the dynamics of *FLC* chromatin (Fig. 8.3—gene level), highlighted the key players involved (e.g., H3K27me3), and predicted additional factors and interactions that were tested in subsequent work (see "Model Validation" section).

Pushing this line of investigation further, it was then demonstrated that two copies of the *FLC* gene within a single cell can be regulated independently [25]. Plants were generated with a fluorescent reporter fused to *FLC*, using two different reporters giving rise to a plant line with yellow-labeled FLC (from *FLC:Venus*) and another line with a red-labeled FLC (from *FLC:mCherry*). These plant lines were crossed together to give a plant line whose cells had one copy of the *FLC* gene making a product that was yellow-labeled and another that was red-labeled. These plants were cold treated and it was shown that one of the *FLC* copies could be stably in an OFF state whereas the other was stably ON and vice versa, as well as both copies being

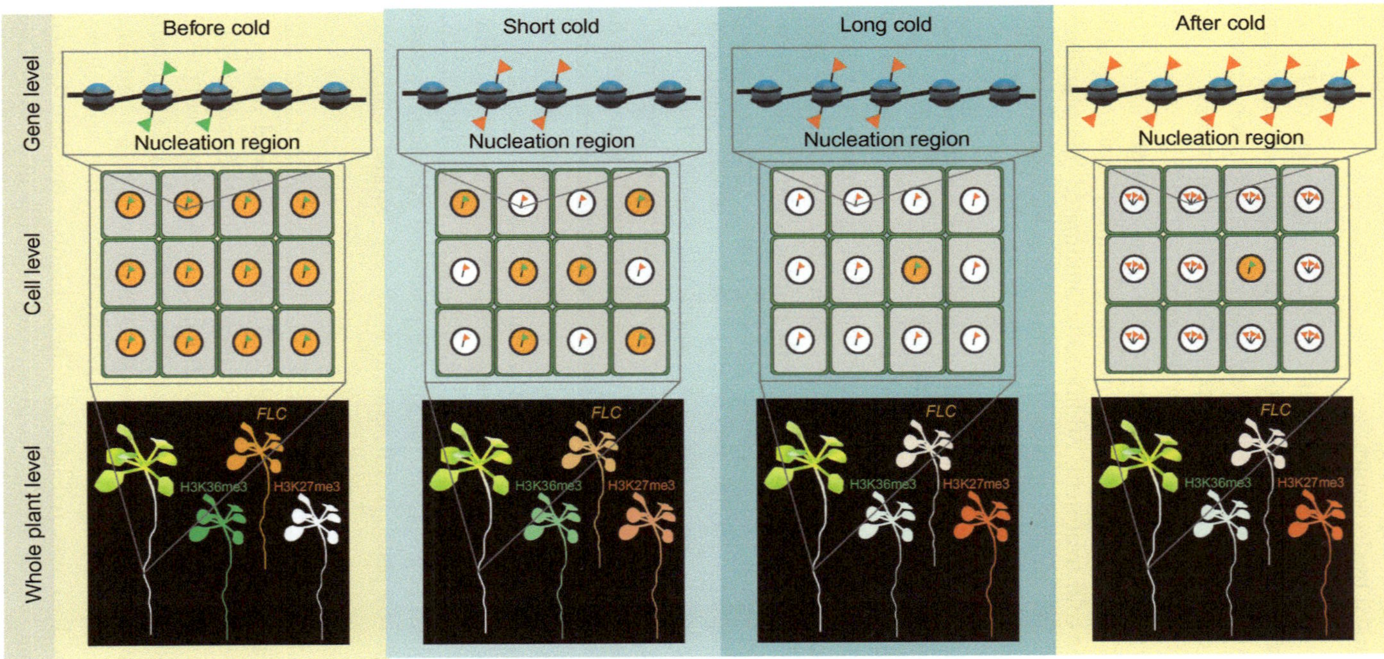

FIGURE 8.3

Quantitative silencing at different scales.

Gene level: The state of the *FLC* locus is determined, at least in part, by histone modifications. In the model, the ON state is associated with A modifications while the OFF state is covered with M modifications. Cold registration happens with the forced addition of M in the nucleation region. Full epigenetic switching happens post-cold with the spreading of M across the whole locus. Cell level: ON/OFF loci can be observed using reporters fused to *FLC*. Loci switch from ON to OFF independently and stochastically in the cold and this state is maintained post-cold. Whole plant level: Measurements such as qPCR and ChIP give whole plant average levels of expression or histone modification, respectively, and therefore cannot distinguish between continuously varying (analog) versus ON/OFF (digital) regulation of a gene.

able to be ON or both OFF. This provided strong evidence that the epigenetically stable transcriptional state of the gene is not determined by a diffusible "trans" factor, which would couple the states of the two copies, but rather the memory of the cold is stored locally at the chromatin (in cis). The latter mechanism naturally allows for two copies of *FLC* to be in different epigenetic states inside the same cell. This finding resolved a long-standing and deep question in the field about where epigenetic memory was actually stored and provides strong support for the notion that histone modifications are indeed causal for epigenetic memory.

Switching

The second aim of the model was to investigate switching between stable epigenetic states. Each state of the above model, either predominantly A's or M's, is stable through many cell cycles, as is the experimentally observed *FLC* expression state [25]. Nevertheless, an ability to switch from one state to the other is clearly absolutely fundamental to vernalization, since the number of loci that have switched out of the A state into the M state must progressively increase with the duration of cold. To allow switching of a locus, bias must be introduced by external factors. Data obtained from Chromatin Immunoprecipitation (ChIP) experiments suggested a mechanism for how this might occur, involving the plant homeodomain (PHD) proteins VIN3 and VRN5 [28], which are part of the PHD-PRC2 complex needed for vernalization and for H3K27me3 accumulation in a small subregion of the gene [13,29].

The accumulation of H3K27me3 in a region of approximately 5 histones at the 5′ end of the *FLC* gene (termed the nucleation region) and the presence of VRN5 at the *FLC* locus post-cold suggested two features that were used in the model to stimulate switching. In the model, nucleation could be stochastically triggered at a locus in the cold, associated with an increased probability of addition of the M mark in the nucleation region, even allowing direct transitions of A to M. The probability of nucleation occurring depended on the accumulation of another PHD protein, VIN3, which is known to be cold-induced and disappears rapidly after the cold. These rules introduced a strong bias toward the M state in the nucleation region during the cold. The second feature was a VRN5 dependent bias toward the M state throughout the locus that was only applied post-cold, when VRN5 persists at the locus.

These additional features were added to the model: nucleation by the incorporation of steps in the simulation specific to nucleation region histones and M state bias by a post-cold increase in the parameter for the stimulated addition of silencing modifications. The model was then simulated using these additional rules, and it was shown that switching from a state where the locus is predominantly covered by A marks to one where it is predominantly covered by M's can be triggered by cold-induced nucleation, as described above. In essence, the forced nucleation of M marks in the nucleation region is able to induce a flip of the entire locus into the M state, with the spreading transition mostly occurring immediately after plants are returned to the warm. ChIP data of *FLC* H3K27me3 in the nucleation region and elsewhere (gene body region) were used to fit the model. The parameters were constrained by the qualitative and quantitative features of the system, in particular the high stability of the epigenetic states while still retaining an ability to be controllably switched from one state to the other depending on the external perturbations. The model was not sensitive to changes in most of the parameters. This conclusion has the important biological implication that the memory should therefore be robust to many perturbations that could occur through mutational or environmental changes. However, a notable exception was the size of the nucleation region which needed to be strongly correlated with the size of the locus whose state was to be flipped by its action. This finding suggests that the physical sizes of both the gene and nucleation region might be under selection to allow epigenetic stability through positive feedback (which is stronger in larger regions) while retaining the ability to switch (by scaling the nucleation region size with the overall size of the locus).

Overall, we can conclude that a combined modeling/experimental approach has cast significant light on the switching of epigenetic states through the process of nucleation. Specifically, it was shown that antagonistic active and silent marks at *FLC* can lead to stable ON/OFF states and that cold-induced nucleation can allow switching of a stable ON locus to a stable OFF locus after the cold. These ON/OFF states have been observed experimentally, with the proportion of loci in each state changing quantitatively with the duration of the cold. However, these investigations opened up further questions about the nature of nucleation that we address in the next section.

COLD REGISTRATION IS DIGITAL

In the previous section, we established that the memory of cold is stored in a digital all-or-nothing fashion at each gene copy. An obvious next question is how the cold is itself sensed and registered. The registration of the cold is associated with the appearance of the nucleation peak, i.e., an accumulation of H3K27me3 in the nucleation region. However, once again a digital versus analog question arises: the peak could either appear digitally at each locus in an all-or-nothing fashion at a random time in the cold or else the height of all individual peaks could increase continuously over time during the cold in an analog manner. In the model, a digital peak could then be converted into long-term digital memory with high probability after the cold, with high levels of the M modification across the locus. However, for a gradually accumulating (analog) peak, the probability of switching the body of the *FLC* locus to a high M state would be determined by the height of the M peak at each locus. Unfortunately, this height cannot be determined experimentally at single loci with the techniques currently available, since ChIP gives a population average measurement. A more indirect approach was therefore necessary to decipher the underlying mechanism, again employing mathematical modeling [30].

In the analog case, all individual loci have increasing levels of a signal (nucleation region H3K27me3, Fig. 8.4), and this signal needs to overcome a threshold in order to switch the digital memory state. In the digital case, a one-step-increase of the signal past the threshold will straightforwardly lead to switching. Of course, there will be noise in the system giving a distribution of starting levels in a population (Fig. 8.4A). During the time that the external stimulus is applied, the population levels of the signal can increase at the same rate in both the analog and digital cases. In the analog case, the signal is increasing more or less equally at all loci, whereas in the digital case an increasing number of loci switch to a nucleated state while the other loci remain unchanged (Fig. 8.4B). Therefore, if the stimulus is removed after only a short time, for the same average levels, a small number of cells will have subsequently switched in the digital case, but fewer or none will have reached the necessary threshold to switch in the analog case. Conversely, if the cold stimulus is applied for a sufficiently long time, the two cases will become indistinguishable again (Fig. 8.4C).

The prediction is therefore that after short periods of cold *FLC* will not be downregulated efficiently if the nucleation peak is analog. To quantify the expected efficiency of the different cases and compare to experimental results, Angel et al. [30] developed three models based on the model described previously (see previous section, "Memory of the Cold is Digital and is Stored locally in the Chromatin" and Ref. [13]). In the first, the nucleation peak rises continuously in each cell in the cold. This is simulated by a gradual increase in the probability of adding an M mark to a histone in the nucleation region. In the second model, a full nucleation peak appears in one step in the cold, independently and stochastically at each locus, and this transition is irreversible in the cold. The final model is also digital, but now the digital nucleation peak appears and disappears, with the fraction of time that it is present increasing with the duration of cold exposure.

Angel et al. [30] fitted all three models to ChIP and *FLC* expression data obtained by quantitative polymerase chain reaction (qPCR) from plants that had been treated with varying lengths of cold and found that the second model was superior to the other two, with the analog model giving the worst fit. The analog model severely underestimated the effects of vernalization

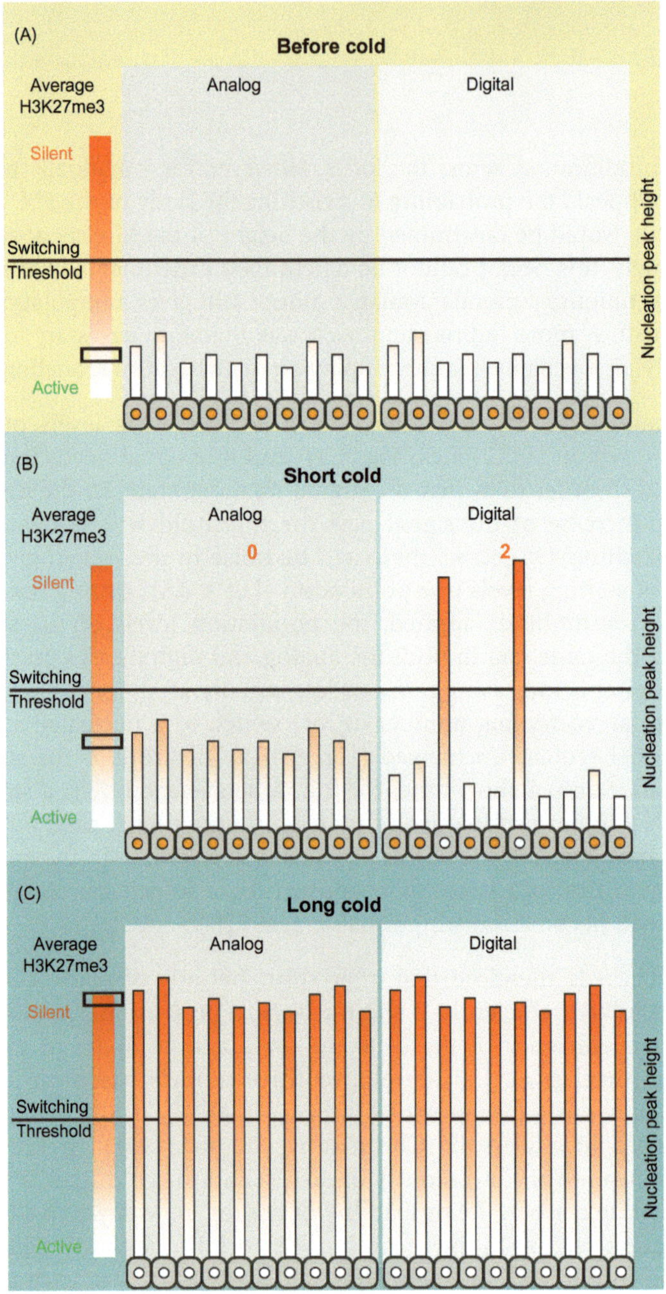

FIGURE 8.4

Digital versus analog cold registration.

Average H3K27me3 mark levels in the nucleation region are indicated (left) together with levels at individual loci, represented by the height of the nucleation peak at each locus (right). A switching threshold is shown that must be crossed to allow conversion of the peak to a fully stable memory state with M modifications coating the entire locus. (A) Before the plant has experienced any cold, all loci are in the active state with low nucleation region H3K27me3. The analog and digital cases are the same at this point. (B) After a short period of cold, no loci have switched in the analog case while, for the same average H3K27me3 levels, a few loci have nucleated and will be epigenetically switched in the digital case. (C) After a saturating period of cold, the two cases again look identical.

for short cold exposures, as expected from Fig. 8.4. The system could in theory compensate for the slow switching early on by a faster analog nucleation peak rise at that time, but further constraints apply in the form of the H3K27me3 ChIP profiles which in fact suggest that the population-level peak rise is slower rather than faster at that time. Other compensatory mechanisms also encounter limitations in other aspects of the mechanism. For example, modifying the threshold to allow enhanced switching in the analog model would risk noisy fluctuations giving rise to prematurely silenced loci. These and other constraints led to the optimal fit for each model that was used in a further experimental comparison, as we now describe.

The experimental protocol was designed to maximize the differences between the predicted results of the three models and thereby distinguish between analog and digital registration of cold at *FLC* by comparing the three predictions with the experimental results. Plants were treated with multiple short periods of cold, interrupted by a few warm days and their expression levels were compared to those of plants which had experienced a continuous but equal length of time in the cold. Various interruption protocols were tested experimentally and the same protocols were simulated using the three models to predict the expected *FLC* levels in each case. The experiments found that the plants buffered interruptions to cold even better than was predicted by any of the three models. However, the digital model with irreversible nucleation emerged as being the most similar to the experimental observation of efficient vernalization in interrupted cold.

Consequently, it was concluded that cold registration, like epigenetic memory, is digital, to allow efficient conversion of shorter-duration cold temperature signals to stable *FLC* downregulation. In field conditions, where temperatures can fluctuate significantly, the use of digital registration is therefore likely to be a key mechanistic feature of vernalization.

MODEL VALIDATION

One of the predictions of the model [13] was the presence of an opposing histone mark ("A") at *FLC* that antagonizes the M mark (H3K27me3) (Fig. 8.2). Similarly, the M mark was expected to antagonize the A mark. The A mark was therefore expected to show a mirror image profile at *FLC* compared to H3K27me3 (Fig. 8.5A—predicted). Yang et al. [14] tested the profiles of four known activating chromatin marks over *FLC*, namely H3K36me3, H3K36me2, H3K4me3, and H3K4me2. None of these marks had the mirror image profile expected, but H3K36me3 did show an anticorrelation with H3K27me3 throughout vernalization at *FLC*, and in particular at the nucleation region (Fig. 8.5A—observed).

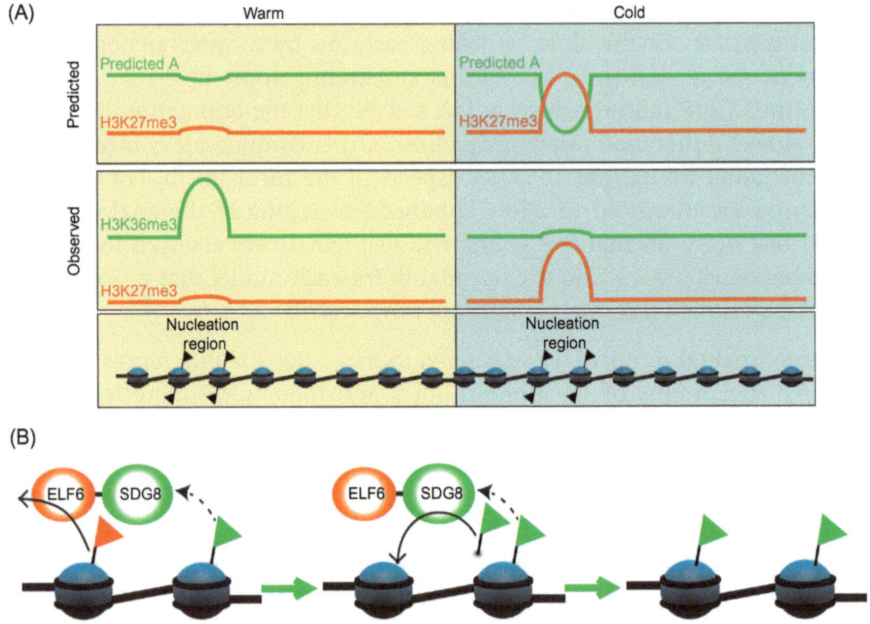

FIGURE 8.5

Discovery of predicted A mark and coupling of proteins involved in the feedback between A and M.

(A) Predicted ChIP profile along the *FLC* locus for the A mark is shown and compared to a schematic of the experimental H3K36me3 profile. (B) A physical coupling was discovered between ELF6, the H3K27me3 demethylase, and SDG8, the H3K36me3 methyltransferase, as predicted by the model. Dashed arrows represent recruitment reactions, whereas solid arrows represent mark removal/addition reactions.

To examine whether this anticorrelation was functional, Yang et al. [14] investigated *sdg8* mutant plants lacking an H3K36me3 methyltransferase, where the addition of H3K36me3 is impaired. This mutation showed increased H3K27me3 levels both at *FLC* and genome-wide. *FLC* expression was also silenced as would be expected from high H3K27me3 levels. Furthermore, the two modifications, H3K27me3 and H3K36me3, could not be detected at the same histone. These results suggested that the predicted antagonism does exist between these two marks and such antagonism is necessary for bistability, at least in the nucleation region.

Another prediction of the model [13] was that the A mark would recruit factors that remove the M mark as well as add more of itself, and similarly though in reverse for the M mark. Such a coupling was confirmed for the factors acting at *FLC* to remove H3K27me3 (ELF6) and add H3K36me3 (SDG8)

[31] (Fig. 8.5B). Both factors promote the active state and therefore mutants of either gene have low *FLC* expression and early flowering phenotypes. This physical coupling was detected in vivo by purifying tagged SDG8 and analyzing the proteins that were bound to it [31]. The ELF6 protein was detected, as well as proteins involved in transcription. To confirm that the two proteins function in the same genetic pathway, plants defective in both SDG8 and ELF6 (double mutants) were compared to the single mutants of each. If the function of the two proteins was independent, it was expected that the double mutant phenotype would be stronger than either of the single mutants. However, in this case the double mutant showed the same flowering as the *sdg8* single mutant, indicating that the two genes work in the same pathway. This finding was confirmed by measurements of *FLC* expression as well as H3K27me3 and H3K36me3 ChIP at *FLC*. The localization of the proteins at *FLC* was also investigated using ChIP against the SDG8 and ELF6 proteins in the same study, where it was shown that both proteins are needed for the localization of either one to the *FLC* chromatin.

Finally, vernalization dynamics in the *elf6* mutant were investigated and it was shown that ELF6 functions in setting the *FLC* expression levels in the warm. However, despite this difference in starting levels the downregulation of *FLC* was not perturbed in the absence of this protein, implying that ELF6 does not affect nucleation. This conclusion in turn suggests that nucleation is imposed externally to the otherwise bistable dynamics of the nucleation region, in accordance with how nucleation was actually imposed in the modeling.

Overall, the models have demonstrated significant ability to help uncover mechanistic aspects of epigenetic regulation at *FLC*. Specific examples include the digital nature of the cold temperature registration and quantitative epigenetic memory at *FLC*, as well as the physical coupling of activation and derepression embodied in the properties of ELF6 and SDG8. These were all predictions of the model that have been validated by experimental work. Hence, we believe that we have been successful in our aim to use predictive modeling to accelerate our understanding of mechanism.

OUTLOOK

The combination of modeling and experimental work described in this chapter has led to the decryption of key epigenetic mechanisms relating to quantitative memory and switching. The lessons learned at *FLC* will, we hope, be widely applied to other epigenetic systems, as there are extensive parallels at the genetic and molecular levels between different organisms [32]. However, many open questions remain, relating both to epigenetic regulation in general, and to *FLC* and flowering regulation in particular. We will conclude by outlining some of these remaining issues.

In terms of flowering, an interesting further question relates to thermosensing. As we have discussed, it has already been shown that the cold induces digital nucleation of silencing marks at *FLC*, but it remains unknown how the cold is actually sensed. One of the upstream regulators of *FLC*, *VIN3*, is cold-induced and during cold exposure also appears to hold a memory of the duration of cold, involving histone modifications [33]. Antisense transcription from the *FLC* locus is also cold-induced and appears to play a role in the locus transcriptional shutdown [10]. All these features have so far been investigated under rather unrealistic temperature profiles (e.g., constant 4°C), which are not representative of what plants experience in reality. In the interrupted cold experiments [30], a step was taken toward more complicated vernalization treatments. There, it was shown that plants are well equipped to filter fluctuations and extract the core signal of cold from a varying temperature profile. Indeed, digital rather than analog nucleation is a central part of this robustness. However, an essential next step is to investigate the vernalization process in real-world field conditions and examine whether the mechanistic framework that we have built up here can satisfactorily handle more realistic, noisier temperature inputs.

For *FLC* epigenetic regulation, a question that remains unanswered from this work is the identity of the A mark in the gene body. Another chromatin mark could take the role of the A mark in this region or a separate mechanism could be responsible for keeping the M mark out of the body region before the cold. The long-debated question of whether heritable information can be stored locally in the chromatin was recently answered in the positive, using the *FLC* system [25]. Though this does not prove that the histone modifications are themselves the memory, it does nevertheless strongly support their involvement. In the future, it will be interesting to investigate what defines the heritable transcriptional state of the locus, both at *FLC* and elsewhere: is it primarily the marks, or a combination of those marks with locus-associated proteins [34]?

The questions answered in the studies described here, as well as the remaining open questions, are far from trivial and in many cases it has been difficult to make progress with experiments alone. Mathematical modeling is rightly entering the epigenetics field and, when used judiciously, can provide insight that might take years to gain without it. So far, within epigenetics, this is the only system where experimentalists and modelers have worked together so closely for an extended period of time. We hope that the clear achievements of this collaboration will inspire others to undertake similar interdisciplinary collaborations and thereby accelerate the dissection of complex underlying mechanisms.

References

[1] Li B, Carey M, Workman JL. The role of chromatin during transcription. Cell 2007;128(4):707–19.

[2] Zhang X, Clarenz O, Cokus S, Bernatavichute YV, Pellegrini M, Goodrich J, et al. Whole-genome analysis of histone H3 lysine 27 trimethylation in *Arabidopsis*. PLoS Biol 2007;5(5):e129.

[3] Ha M, Ng DWK, Li W-H, Chen ZJ. Coordinated histone modifications are associated with gene expression variation within and between species. Genome Res 2011;21(4):590–8.

[4] Ptashne M. On the use of the word "epigenetic." Curr Biol 2007;17(7):R233–6.

[5] Henikoff S, Shilatifard A. Histone modification: cause or cog? Trends Genet 2011;27(10): 389–96.

[6] Andrés F, Coupland G. The genetic basis of flowering responses to seasonal cues. Nat Rev Genet 2012;13:627–39.

[7] Johanson U, West J, Lister C, Michaels S, Amasino R, Dean C. Molecular analysis of FRIGIDA, a major determinant of natural variation in *Arabidopsis* flowering time. Science 2000;290(5490):344–7.

[8] Song YH, Shim JS, Kinmonth-Schultz HA, Imaizumi T. Photoperiodic flowering: time measurement mechanisms in leaves. Annu Rev Plant Biol 2015;66(1):441–64.

[9] Swiezewski S, Liu F, Magusin A, Dean C. Cold-induced silencing by long antisense transcripts of an *Arabidopsis* Polycomb target. Nature 2009;462(7274):799–802.

[10] Csorba T, Questa JI, Sun Q, Dean C. Antisense *COOLAIR* mediates the coordinated switching of chromatin states at *FLC* during vernalization. Proc Natl Acad Sci USA 2014;111(45):16160–5.

[11] Bastow R, Mylne JS, Lister C, Lippman Z, Martienssen RA, Dean C. Vernalization requires epigenetic silencing of *FLC* by histone methylation. Nature 2004;427(6970):164–7.

[12] Schubert D, Primavesi L, Bishopp A, Roberts G, Doonan J, Jenuwein T, et al. Silencing by plant Polycomb-group genes requires dispersed trimethylation of histone H3 at lysine 27. EMBO J 2006;25:4638–49.

[13] Angel A, Song J, Dean C, Howard M. A Polycomb-based switch underlying quantitative epigenetic memory. Nature 2011;476(7358):105–8.

[14] Yang H, Howard M, Dean C. Antagonistic roles for H3K36me3 and H3K27me3 in the cold-induced epigenetic switch at Arabidopsis *FLC*. Curr Biol 2014;24(15):1793–7.

[15] Wood CC, Robertson M, Tanner G, Peacock WJ, Dennis ES, Helliwell CA. The *Arabidopsis thaliana* vernalization response requires a polycomb-like protein complex that also includes VERNALIZATION INSENSITIVE 3. Proc Natl Acad Sci U S A 2006;103(39):14631–6.

[16] Coustham V, Li P, Strange A, Lister C, Song J, Dean C. Quantitative modulation of polycomb silencing underlies natural variation in vernalization. Science 2012;337(6094):584–7.

[17] Li P, Filiault D, Box MS, Kerdaffrec E, van Oosterhout C, Wilczek AM, et al. Multiple *FLC* haplotypes defined by independent cis-regulatory variation underpin life history diversity in Arabidopsis thaliana. Genes Dev 2014;28:1635–40.

[18] Dodd IB, Micheelsen MA, Sneppen K, Thon G. Theoretical analysis of epigenetic cell memory by nucleosome modification. Cell 2007;129(4):813–22.

[19] David-Rus D, Mukhopadhyay S, Lebowitz JL, Sengupta AM. Inheritance of epigenetic chromatin silencing. J Theor Biol 2009;258(1):112–20.

[20] Ferrell JE. Self-perpetuating states in signal transduction: positive feedback, double-negative feedback and bistability. Curr Opin Cell Biol 2002;14(2):140–8.

[21] Boettiger AN, Bintu B, Moffitt JR, Wang S, Beliveau BJ, Fudenberg G, et al. Super-resolution imaging reveals distinct chromatin folding for different epigenetic states. Nature 2016;529(7586):418–22.

[22] Yuan W, Wu T, Fu H, Dai C, Wu H, Liu N, et al. Dense chromatin activates polycomb repressive complex 2 to regulate H3 lysine 27 methylation. Science 2012;337(6097):971–5.

[23] Lu F, Cui X, Zhang S, Jenuwein T, Cao X. Arabidopsis REF6 is a histone H3 lysine 27 demethylase. Nat Genet 2011;43(7):715–19.

[24] Crevillén P, Yang H, Cui X, Greeff C, Trick M, Qiu Q, et al. Epigenetic reprogramming that prevents transgenerational inheritance of the vernalized state. Nature 2014;515(7528): 587–90.

[25] Berry S, Hartley M, Olsson TSG, Dean C, Howard M. Local chromatin environment of a Polycomb target gene instructs its own epigenetic inheritance. eLife 2015;4:e07205.

[26] Jaeger KE, Wigge PA. FT protein acts as a long-range signal in Arabidopsis. Curr Biol 2007;17(12):1050–4.

[27] Corbesier L, Vincent C, Jang S, Fornara F, Fan Q, Searle I, et al. FT protein movement contributes to long-distance signaling in floral induction of *Arabidopsis*. Science 2007;316(5827):1030–3.

[28] De Lucia F, Crevillen P, Jones AME, Greb T, Dean C. A PHD-polycomb repressive complex 2 triggers the epigenetic silencing of *FLC* during vernalization. Proc Natl Acad Sci USA 2008;105(44):16831–6.

[29] Finnegan EJ, Dennis ES. Vernalization-induced trimethylation of histone H3 lysine 27 at *FLC* is not maintained in mitotically quiescent cells. Curr Biol 2007;17(22):1978–83.

[30] Angel A, Song J, Yang H, Questa JI, Dean C, Howard M. Vernalizing cold is registered digitally at *FLC*. Proc Natl Acad Sci U S A 2015;112(13):4146–51.

[31] Yang H, Howard M, Dean C. Physical coupling of activation and de-repression activities to maintain an active transcriptional state at *FLC*. Proc Natl Acad Sci USA 2016;113:9369–9374.

[32] Hennig L, Derkacheva M. Diversity of Polycomb group complexes in plants: same rules, different players? Trends Genet 2009;25(9):414–23.

[33] Finnegan EJ, Bond DM, Buzas DM, Goodrich J, Helliwell CA, Tamada Y, et al. Polycomb proteins regulate the quantitative induction of VERNALIZATION INSENSITIVE 3 in response to low temperatures. Plant J: Cell Mol Biol 2011;65:382–91.

[34] Steffen PA, Ringrose L. What are memories made of? How Polycomb and Trithorax proteins mediate epigenetic memory. Nat Rev Mol Cell Biol 2014;15(5):340–56.

GLOSSARY

Analog in this chapter, a system is said to be analog if it has the ability to take continuously varying values.

Bistable a system with two stable states. Perturbations can temporarily take the system away from one such state, but the system will spontaneously return to one of the two.

Chromatin DNA in eukaryotic cells is wrapped around protein complexes called nucleosomes, which are formed of several histone proteins (histone H2A, H2B, H3, and H4), whose function is to pack the DNA. The combination of DNA, histones and other DNA-bound proteins is collectively called chromatin. Histone proteins have "tails" protruding from the core nucleosome that can carry "histone modifications."

Chromatin immunoprecipitation (ChIP) ChIP can be used to quantify the binding of proteins to particular DNA regions in vivo. The binding is usually first stabilized by crosslinking treatments that form covalent bonds between the DNA and proteins that interact with it. The DNA is then fragmented to give sequences of appropriate length. The protein of interest can be fused to a tag such as the tandem-affinity purification tag (TAP-tag), which was developed to allow easy and efficient purification. Using antibodies that bind to this tag or to the protein itself, the crosslinked protein and DNA are extracted from the total chromatin in a process called immunoprecipitation. Following the enrichment of the chromatin of interest, the crosslinking is reversed and the DNA is purified. Finally, the amount of DNA of a particular, known, sequence can be quantified by qPCR to determine the relative binding of the protein to that region, or the DNA can be sequenced for genome-wide analysis. The same method can also be applied to find sites with particular chromatin modifications. In the case of H3K27me3, an antibody is available that binds to this mark specifically so a fusion protein is not necessary. The regions of DNA bound to histones with this modification can thus be identified and the levels of this modification relative to the total H3 histone (immunoprecipitated using a generic H3 antibody) can be quantified.

Digital in this chapter digital is used to describe the property of a system to be in discrete states, as opposed to taking continuously varying values. In particular, for *FLC*, two such epigenetically stable transcriptional states are possible (ON/OFF). Therefore this term is used interchangeably with bistable.

Histone modification post-translational covalent modifications of histone tails, which are thought to be involved in transcriptional regulation and epigenetic memory.

Polycomb repressive complex 2 (PRC2) protein complex initially discovered in *Drosophila* that is conserved in many organisms and is involved in gene silencing through the histone modification H3K27me3.

Quantitative polymerase chain reaction (qPCR) sequence-specific DNA quantification method. Cycles of DNA replication are coupled to measurement of the intensity of a DNA-binding dye (or alternatively fluorescent probes). The specific primers used will determine the sequence that is amplified. The number of cycles it takes to reach a critical level of fluorescence is used to estimate the concentration of the sample. This method can also be applied to RNA by adding a reverse transcription step before the quantification, to produce DNA from the sample RNA. Internal controls are very important for this method.

Vernalization acceleration of flowering by a long period of cold.

LIST OF ACRONYMS AND ABBREVIATIONS

ChIP chromatin immunoprecipitation
ELF6 *Early Flowering 6*
FLC *Flowering Locus C*
FRI *FRIGIDA*
FT *Flowering Locus T*
PHD plant homeodomain
PRC2 polycomb repressive complex 2
qPCR quantitative polymerase chain reaction
REF6 *Relative of Early Flowering 6*
SDG8 *SET Domain Group 8*

SECTION V

The Third and Fourth Dimensions: Chromosomal Long Range Interactions: Polymer Models

Epigenetic regulation occurs within a tightly packed and folded genome, in which distant loci may be brought close together in three-dimensional (3D) space. Furthermore, the chromatin template is forever changing, due to chromatin modifications, chromosome movements, and the disruptions of the cell cycle. However, many of these contacts between distant regions are not random: they reform after cell division, many are specific to different cell types and disruption of essential contacts can lead to disease. The last decade or so has seen an explosion in the development of techniques that are able to measure chromosome conformation at ever increasing resolution, and a concomitant surge in the application of biophysical polymer models to the problem of deconvolving these data sets. Understanding the principles governing 3D genome organization within a predictive theoretical framework will be essential to understanding the disruptions that occur to 3D structure during gene expression changes, differentiation and cellular stress, and the genome rearrangements associated with specific diseases.

Chapter 9 (On the Nature of Chromatin 3D Organization: Lessons From Modeling) and Chapter 10 (From Chromosome Conformation Capture to Polymer Physics and Back) discuss the use of models based on the principles of polymer physics to interpret data generated by "chromosome conformation capture" techniques, which give a snapshot of chromosome contacts in fixed cells. Chapter 11 (A Combination Approach Based on Live-Cell Imaging and Computational Modeling), and Chapter 12 (Capturing Chromosome Structural Properties From Their Spatial and Temporal Fluctuations) focus on dynamic chromosome behavior in living cells, and the combination of experimental measurements in real time with models that capture these dynamics.

CHAPTER 9

On the Nature of Chromatin 3D Organization: Lessons From Modeling

Mario Nicodemi[1,2] and Ana Pombo[3]

[1]Dipartimento di Fisica "E. Pancini", Universita' di Napoli "FedericoII", and INFN, Complesso Universitario di Monte S. Angelo, Napoli, Italy [2]Berlin Institute of Health, MDC-Berlin, Germany [3]Berlin Institute for Medical Systems Biology, Max Delbrueck Center for Molecular Medicine, Berlin-Buch, Germany

CONTENTS

INTRODUCTION

New technologies, such as Hi-C methods, can measure the frequency of physical contacts between DNA segment pairs genome-wide and are revealing that the genome of mammalian cells has a complex 3D architecture, comprising extensive long-range, functional interactions [1–4]. Chromosomes are folded into a sequence of 0.5–1.0 Mb long domains, known as topological associated domains (TADs) [5,6], comparatively conserved between mice and humans. TADs, in turn, are thought to be one level of a hierarchy of higher order domains (metaTADs), extending up to chromosomal scales [7]. Such chromatin interactions have key functional roles, as for instance they control gene activity through the formation of physical loops between regulatory regions and target remote genes, disruptions being linked to human conditions [1–4]. Yet, the basic molecular mechanisms whereby chromosome 3D organization is orchestrated in health and disease remain mysterious. To make sense of genome-wide contact data and to expose the principles shaping chromosome 3D structure, models from polymer physics have been introduced (see recent reviews in Refs. [8–12]).

Here we briefly review the basic models of polymer physics introduced to explain the folding principles of chromatin and the computational approaches developed to visualize and reconstruct from high-throughput contact data, such as Hi-C data, chromosome 3D conformations, and its link to epigenetics.

POLYMER MODELS OF CHROMATIN

The gold-standard model of a free polymer in polymer physics is the self-avoiding-walk (SAW) chain of beads (Fig. 9.1). It represents a string-of-beads

Epigenetics and Systems Biology. DOI: http://dx.doi.org/10.1016/B978-0-12-803075-2.00009-X

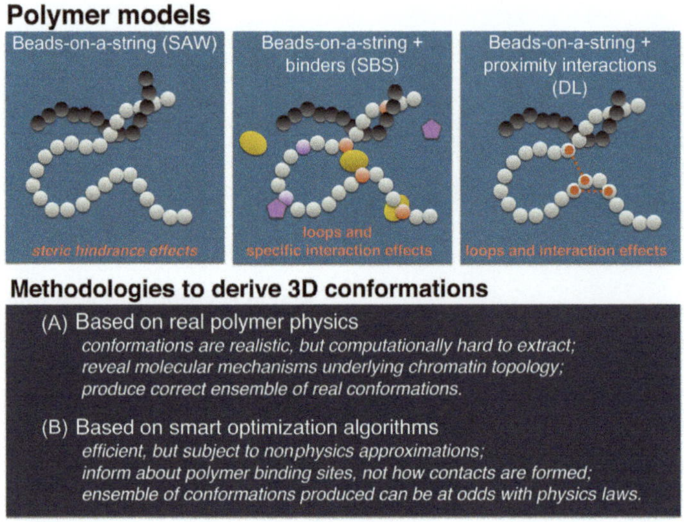

FIGURE 9.1 ***Top:*** **The self-avoiding-walk (SAW) model of polymers takes into account steric hindrance effects (*top-left*).**
The strings & binders switch (SBS) model also describes the interaction of the SAW chain with binding, and looping, molecular factors (*top-center*). By integrating out the binders, the Dynamic Loop model (DL, *top-right*) is obtained, where polymer beads have a direct interaction. *Bottom:* Polymer models can be used to make sense of Hi-C and imaging data. (A) Their conformations can be investigated under the rigorous constraints of the laws of polymer physics, which can be computationally difficult; or (B) they can be employed under efficient optimization algorithms, which may be subject to nonphysical approximations.

experiencing only self-avoidance effects, which folds spontaneously in a random conformation. A chromosome model based on a SAW bead chain under a Brownian dynamics has been originally proposed to describe the formation of chromosome territories in mammalian genomes by a purely kinetic effect [13]. In this model, the segregation of chromosomes into discrete territories observed in the nucleus of mammalian cells is modeled as resulting from topological barriers observed when highly condensed unentangled ring polymers are allowed to unfold, without invoking protein-mediated interactions between DNA sequences.

The fractal globule (FG) model of chromosome conformation [1] also focuses on a string of ideal beads but explores a transient, and highly unstable [14], state of the polymer encountered, for instance, when it expands from an unknotted compact initial conformation, well before approaching equilibrium. The FG model, however, does not describe well chromatin contact matrices or their variation between chromosomes, cell differentiation, or cell cycle states [15,16].

A polymer model made of a freely jointed chain of cylinders has been considered for Saccharomyces cerevisiae chromosomes [17]. Interestingly, it reproduces genome-wide DNA contact frequencies in that system and recapitulates the position and morphology of the nucleolus, as well as subchromosomal contact features. The model was also able to describe nuclear reorganization events that follow a reduction in ribosomal DNA transcription. It returned a view where large-scale *Saccharomyces cerevisiae* nuclear architecture can be largely understood from the generic properties of crowded noninteracting polymers rather than of specific DNA-binding factors. A recent study of high resolution genome interactions in Schizosaccharomyces pombe using synchronized cells, complemented with a string-of-beads polymer model investigated by Monte Carlo computer simulations and incorporating biological constraints, such as centromere and telomere localization, nucleolus excluded volume and constraints on chromosome interactions derived by the genome conformation capture experimental techniques, supported a more complex view where groups of genes with high and low transcript levels have preferred positions within the nucleus, which is spatially divided into functional subnuclear domains that correlate with gene activity [18].

Those models have captured the important role of steric hindrance on chromatin motion. However, they do not explain the observed behavior of chromosome architecture in mammals, such as the dependency of locus contact probabilities on genomic distance and their mean 3D spatial distances. Importantly, these models also fail to explain the formation of TADs with preferential boundaries at specific genomic coordinates: as they neglect specific interactions, and all beads are equal, they predict uniform contact matrices, completely lacking topological domains.

Polymer Models of TADs

The first model capable of explaining all available Hi-C and fluorescence in-situ hybridization (FISH) data from basic polymer physics, including the formation of TADs, has been the strings and binders switch (SBS) model [15,16] (reviewed in Refs. [11,12], Fig. 9.1). The SBS model explores the possibility that chromosome 3D conformations and their specific, functional contacts, cannot be established only by randomness but require interactions of chromatin mediated through other nuclear components, such as DNA-binding molecular factors. In the SBS model, chromatin is schematically represented as a SAW chain of beads that contains binding sites for diffusing molecules (positioned across the polymer at specific coordinates). The binders can bridge the polymer binding sites and hence loop the chain. From polymer physics, it is known that the stable conformational classes of the system correspond to its thermodynamic phases. By changing the values of its control parameters, such as concentration and affinity of binding molecules, or the distribution of binding sites, the system thermodynamic phase is changed and, correspondingly, its related allowed conformations. The ensemble of modeled polymers represents the multitude of polymer conformations that can emerge in each thermodynamic phase, reminiscent of the multitude of spatial distances that is observed experimentally for contacting loci across a population of cells.

In the simplest cases, two main conformational classes exist, which resemble the simple view of eu- and heterochromatin: an open state, analogous to the SAW random folded state, and a compact state corresponding to a conformation where the polymer binders have fully looped the polymer by occupying the binding sites, except for steric hindrance effects. The SBS model envisages also a third state, the theta-point state, which is encountered at the transition point between the open and closed states. Analysis of FISH data, such as the distance momenta ratio of different genomic regions, appears to support the idea that such a state is indeed seen in a fraction of loci in real cells [16]. Future work will clarify in more detail the existence of theta-point states in chromatin folding in living nuclei.

The SBS model was able to explain, in particular, how specific architectural patterns can be established by randomly diffusing binding molecules, how different stable conformations can be produced and reliably regulated by well known cell strategies, such as protein upregulation or epigenetic chromatin modification, by exploiting fundamental thermodynamic mechanisms.

Within the SBS model, the formation of TADs, as well as the looping out of loci, can be rationalized by specialization of polymer binding sites and binding molecules. For instance, when two binding domains are present for two distinct binding factors, i.e., a polymer with blocks of different binding sites (technically, a block copolymer), clearly two separate TADs are formed. The combinatorial action of a few types of factors could explain the variety of existing TADs. The emerging picture from the SBS model is that chromatin is a complex mixture of differently folded regions controlled by basic mechanisms of polymer physics, and by locus-specific regulation of chromatin binding and modifications. A locus can be folded in different conformations across a population of cells, as much as different loci can be folded in different conformations inside a single cell. An SBS model informed with biological details has already been used to predict the folding of specific loci [19,20], showing that Hi-C pairwise contact data can be reproduced with 95% accuracy.

In the same class of models that take into account chromatin interactions, we have the Dynamic Loop (DL) model [21], a beads-on-a-string chain where the beads have a binding probability to form a loop when they make physical contact (Fig. 9.1). The DL model corresponds to a version of the SBS model where only the net effects of the binders is considered, i.e., a net short-range effective attraction between the chain beads. In fact, the broad predictions of the DL and SBS models are similar. Both models predict (1) the existence of a folded conformation made of loops with a broad length distribution, i.e., a broad variability in lengths genome-wide, as in interphase chromatin; (2) the assembly of chromosome territories, by entropy-driven segregation of looped polymers; (3) the existence of subchromosomal domains whose compaction level depends on chromatin contact probability; and (4) cell-to-cell variations in chromatin folding topologies.

The DL model has also been employed to explain, for instance, the effect of knock-down of CTCF and Rad21 (known transcription factors involved in chromatin folding) on chromosome 3D structure, which results in increased chromatin compaction. Consistent with experimental observations, the DL model finds that the increased compaction can be explained by selectively decreasing the number of short-range loops, leaving long-range looping unchanged [22]. The observation that depletion of single chromatin binders has intermediate effects on chromatin folding suggests that chromosome architecture is controlled by a fine balance between short-range and long-range contacts, allowing for dynamic switching mechanisms regulated by soluble chromatin activities between, for instance, open and more compact chromatin states.

More recent models based on polymer chains that interact with diffusing proteins have been developed for example to show that spontaneous binding at equilibrium (even in the absence of explicit interaction between proteins) triggers local DNA compaction and protein aggregation in specific shapes [23]: for instance, small bivalent proteins align into rows [as in binding of the bacterial histone-like nucleoid-structuring protein (H-NS)]; larger proteins form quasispherical aggregates (as in nanoparticle binding), whereas cylinders with eight binding sites (representing nucleosomal cores) assemble into irregularly folded clusters (like those seen in nucleosomal strings). Those structures are supposed to arise by an entropic bridging-induced attraction that minimizes bending and looping penalties in the template. Analogously, structures similar to transcription factories and TAD forming intrachromosomal contacts can be found [23].

The view where different TADs correspond to the different segments of a block copolymer model, each associated to a given epigenomic state (e.g., a given set of histone markers), has been further discussed in a model to describe the folding patterns observed in Hi-C experiments in *Drosophila* [24]. The model suggests that TADs can be formed in the previously measured epigenomic domains and proposes that TAD–TAD interactions correspond to multistable, transient states, continuously changing even when their related epigenomic profiles remain unaltered.

Importantly, other molecular mechanisms are likely to contribute to the formation of chromatin folding domains, beyond binding factors. Modeling of thermally fluctuating chromatin fibers as beaded chains with bending and torsional resistance demonstrated that chromatin supercoiling can establish chromatin domains that resemble TADs characterized by otherwise random contacts by proximity within these domains [25]. Chromatin domains found in bacterial genomes could also be generated by polymer modeling of plectonemes (looped sequences of nucleic acids twisted together), which alone can enhance local random contacts within each domain [26].

One major division in the above list of models is whether they consider that the contacts within TADs are random (established by constraints mostly established at TAD boundaries) or result from specialized chromatin contacts that represent specific chromatin states (active, silent, Polycomb-repressed, enhancer, etc.). Both mechanisms can be present in real cells.

More recently, an interesting model named the Extrusion Loop Model (ELM) has been proposed focused, in particular, on the role of CTCF binding sites and cohesin mediated interactions to explain folding [4,27]. The ELM is made of a chain of beads all interacting with all, where the beads corresponding to pairs of forward and reverse CTCF binding sites can be strongly bound by an additional molecular factor (which is thought to be cohesion rings) that produce intense pairwise contacts and loops. The ELM is a special case of the SBS or DL models and, consistent with their concepts, it has been shown to explain well the folding of loci dominated by CTCF factors at scales in the 10–500-kb range [4,27].

Polymer Models of Mitotic Chromosomes

Chromatin folding at mitosis is an interesting biological situation as most, but not all, chromatin binding sites dissociate from chromatin, and chromatin condensation occurs. Polymer models specific to mitotic chromosomes have also been developed. They often fall in the same categories listed above. In particular, a polymer chain with binding sites arranged in a bottle-brush like structure was shown to recapitulate the decay of the average contact probability with genomic separation in mammalian mitotic chromosomes [28]. Cheng et al. used computer polymer models to understand the folding of mitotic chromosomes in budding yeast [29]. A chromosome is modeled as a bead chain under a Brownian dynamic trajectory. Some of the beads, corresponding to measured condesin sites (a molecular factor involved in chromatin binding), can bridge and loop the chain (as by action of condensing). They showed that in the absence of condensin the resulting open chain conformation describes well the observed open state of the yeast chromosomes. Upon condensin introduction, as expected, chromosomes became more compacted, recapitulating the known features of yeast mitotic chromosomes.

MODELS TO RECONSTRUCT 3D-CONFORMATION FROM CONTACT DATA

The genome-wide data on chromatin contacts derived by chromosome conformation capture based technologies, such as Hi-C and TCC, concern pairwise contact matrices. To infer the corresponding full 3D chromosomal structures, they must be complemented with models. This has been done by use of polymer models in different ways. Polymer physics models as those discussed in the above section naturally provide realistic 3D conformations, yet, they can be computationally extremely expensive. A different approach (see references below) was based on the idea to find the optimal relative positions of the beads of a polymer chain to best describe the observed contact frequencies, under some given constraint, without considering the real dynamics of polymer physics. In this way different visualization tools have been produced, offering the option to add layers of additional information, such as epigenetic maps or transcription levels.

Toward producing a 3D spatial understanding of folding of specific genomic regions, a method has been developed to represent a genomic domain as a set of points having spatial restraints (or springs) between them, optimized in a way that their equilibrium distances become proportional to the observed frequency of interaction to generate the average, or *consensus*, three-dimensional conformations [30]. This technique has been applied, for instance, to the α-globin gene domain, finding the presence, at the megabase scale, of globules enriched for transcribed genes [30]. Different methods to reconstruct 3D conformations have exploited a metric that converts interaction probabilities into nuclear distances. One of the first models of that type was used to generate a map at kilobase resolution of the haploid genome of Saccharomyces cerevisiae, considering intra- and interchromosomal interactions [31]: The resulting map suggested that chromosomes cluster via centromeres at one pole of the nucleus, with one chromosome (chr. XII) extending toward the nucleolus. More recently, computational tools have been developed, such as TADbit [32], that allow the automatic determination of 3D structures of chromosomes and their epigenetic domains integrating different experimental data.

Along similar lines, efficient algorithms, capable at accommodating sparse and noisy contact map data, have also been introduced [33]. Bayesian methods have been considered to find the positions of the beads of a "beads-on-a-string" model best representing Hi-C population average data [34]. Applied to a high resolution Hi-C dataset, it showed that local genomic regions typically have homogeneous 3D chromosomal structures, and chromatin domains link to genomic and epigenetic features: for instance, gene rich, accessible and early replicated genomic regions tend to be more elongated and have higher structural variations than gene poor, inaccessible and late replicated genomic regions [34].

One major limitation of models that reconstruct average 3D conformations by optimization of a cost function is that they often fail to summarize an ensemble of biologically different states of chromatin folding, for example the distributions of distances between pairs of genomic loci measured by imaging. Efforts to extract a population of 3D structures of chromosome from TCC data, rather than the average contacts, have been made through producing a coarse-grained structural representation of chromosomes in beads, subjected to restraints according to contact frequencies. This work supports the view that each chromosome can be partitioned into "blocks" of consecutive regions with similar contact profiles, and capable of reproducing the approximate positioning of chromosomal territories in the nucleus [35].

The above strategies are effective at exploiting polymer chains to reconstruct conformations; however, they do not ensure that the obtained 3D structures are consistent with the constraints of polymer physics (e.g., the distribution of conformations derived is not compatible with the basic laws of thermodynamics), as their 3D structures result from smart optimization algorithms rather than by solving the real equation of motion of polymers. That further step has been attempted in a different type of approach. A polymer chain under an approximated confinement potential has been considered [36] to estimate size and structure of chromosomal territories and confined domains from Hi-C data in yeast by the mean time for two monomers of the chain to encounter from the calculations. A different method based on a beads-on-a-string model of chromatin was developed to extract the ensemble of chromatin conformations corresponding to TADs in population-based 5C data on the *Xist* locus and to identify the key interactions that stabilize TAD boundaries [37]. In this model, each bead interacts via a spring-like potential with other beads located within a short distance range. The method combines an expectation-maximization algorithm with equilibrium thermodynamics. By Monte Carlo simulations, the polymer chain is taken to thermodynamic equilibrium and its predicted contact matrix compared to experimental 5C contact frequencies ("expectation step"). In a second ("maximization") step, the interaction potentials are optimized to derive a new ensemble closer to the 5C data. And the cycle is iterated to converge. The optimized final ensemble of configurations was then used to predict the distances between genomic loci and their distribution, which were tested against single-cell-based FISH assays producing successful matches [37]. Although this approach can identify candidate sequences that may participate in chromatin contacts, it does not clarify the mechanisms underlying the folding of *Xist*.

CONCLUSIONS

Polymer models of chromatin are becoming a key tool to make sense of high-throughout architectural data and their corresponding chromosome 3D conformations. As we briefly reviewed here, they are returning a picture of the mechanisms shaping the complex structures that mark chromatin folding across genomic scales, ranging from loops, to TADs and higher order domains such as metaTADs. And they are providing the tools to visualize chromosome conformations, as emerging from contact frequency data, and to integrate them with additional, key datasets, such as epigenetics and transcriptional data, to derive a comprehensive view of chromatin. The ambition for the coming years is to start developing novel approaches, combining computational and experimental methods, for a deeper understanding of chromatin folding and its links to gene regulation, and for implementing diagnostic and therapeutic tools of diseases linked to chromatin misfolding, such as congenital diseases [38] and cancers [39].

References

[1] Lieberman-Aiden E, van Berkum NL, Williams L, Imakaev M, Ragoczy T, Telling A, et al. Comprehensive mapping of long-range interactions reveals folding principles of the human genome. Science 2009;326:289–293.

[2] Bickmore WA, van Steensel B. Genome architecture: domain organisation of interphase chromosomes. Cell 2013;152:1270–1284.

[3] Hughes JR, et al. Analysis of hundreds of *cis*-regulatory landscapes at high resolution in a single, high-throughput experiment. Nat Genet 2014;46:205–212.

[4] Dekker J, Mirny L. The 3D genome as moderator of chromosomal communication. Cell 2016;164:1110–21.

[5] Dixon JR, Selvaraj S, Yue F, Kim A, Li Y, Shen Y, et al. Topological domains in mammalian genomes identified by analysis of chromatin interactions. Nature 2012;485:376–380.

[6] Nora EP, Lajoie BR, Schulz EG, Giorgetti L, Okamoto I, Servant N, et al. Spatial partitioning of the regulatory landscape of the X-inactivation centre. Nature 2012;485:381–385.

[7] Fraser J, Ferrai C, Chiariello AM, Schueler M, Rito T, Laudanno G, et al. Hierarchical folding and reorganisation of chromosomes are linked to transcriptional changes during cellular differentiation. Mol Syst Biol 2015;11:852.

[8] Langowski J. Polymer chain models of DNA and chromatin. Eur Phys J E Soft Matter 2006;19:241–9.

[9] Emanuel M, Radja NH, Henriksson A, Schiessel H. The physics behind the larger scale organization of DNA in eukaryotes. Phys Biol 2009;6:025008.

[10] Tark-Dame M, van Driel R, Heermann DW. Chromatin folding—from biology to polymer models and back. J Cell Sci 2011;124:839–45.

[11] Barbieri M, Scialdone A, Gamba A, Pombo A, Nicodemi M. Polymer physics, scaling and heterogeneity in the spatial organisation of chromosomes in the cell nucleus. Soft Matter 2013;9:8631–5.

[12] Nicodemi M, Pombo A. Models of chromosome structure. Curr Opin Cell Biol 2014;28:90–5.

[13] Rosa A, Everaers R. Structure and dynamics of interphase chromosomes. PLoS Comput Biol 2008;4:e1000153.

[14] Schram RD, Barkema GT, Schiessel H. On the stability of fractal globules. J Chem Phys 2013;138:224901.

[15] Nicodemi M, Prisco A. Thermodynamic pathways to genome spatial organization in the cell nucleus. Biophys J 2009;96:2168–77. http://dx.doi.org/10.1016/j.bpj.2008.12.3919.

[16] Barbieri M, Chotalia M, Fraser J, Lavitas L-M, Dostie J, Pombo A, et al. Complexity of chromatin folding is captured by the strings and binders switch model. Proc Natl Acad Sci USA 2012;109:16173–8. http://dx.doi.org/10.1073/pnas.1204799109.

[17] Wong H, et al. A predictive computational model of the dynamic 3D interphase yeast nucleus. Curr Biol 2012;22:1881–90.

[18] Grand RS, Pichugina T, Gehlen LR, Jones MB, Tsai P, Allison JR, et al. Chromosome conformation maps in fission yeast reveal cell cycle dependent sub nuclear structure. Nucleic Acids Res 2014;42:12585–12599.

[19] Scialdone A, Cataudella I, Barbieri M, Prisco A, Nicodemi M. Conformation regulation of the X chromosome inactivation center: a model. PLoS Comput Biol 2011;7:e1002229. PMID:22046112; http://dx.doi.org/10.1371/journal.pcbi.1002229.

[20] Chiariello AM, Annunziatella C, Bianco S, Esposito A, Nicodemi M. Polymer physics of chromosome large-scale 3D organisation. Sci Rep 2016;6:29775.

[21] Bohn M, Heermann DW. Diffusion-driven looping provides a consistent framework for chromatin organization. PLoS ONE 2010;5(8):e12218. http://dx.doi.org/10.1371/journal.pone.0012218.

[22] Tark-Dame M, Jerabek H, Manders EMM, van der Wateren IM, Heermann DW, van Driel R. Depletion of the chromatin looping proteins CTCF and cohesin causes chromatin compaction: insight into chromatin folding by polymer modelling. PLoS Comput Biol 2014;10(10):e1003877. http://dx.doi.org/10.1371/journal.pcbi.1003877.

[23] Brackley CA, Taylor S, Papantonis A, Cook PR, Marenduzzo D. Nonspecific bridging-induced attraction drives clustering of DNA-binding proteins and genome organization. PNAS 2013;110:E3605–E3611.

[24] Jost D, Carrivain P, Cavalli G, Vaillant C. Modeling epigenome folding: formation and dynamics of topologically associated chromatin domains. Nucleic Acids Res 2014;42:9553–61.

[25] Benedetti F, Dorier J, Burnier Y, Stasiak A. Models that include supercoiling of topological domains reproduce several known features of interphase chromosomes. Nucleic Acids Res 2014;42:2848–55.

[26] Le Tung BK, Imakaev Maxim V, Mirny Leonid A, Laub Michael T. High-resolution mapping of the spatial organization of a bacterial chromosome. Science 2013;342:731.

[27] Sanborn AL, Rao SS, Huang SC, Durand NC, Huntley MH, Jewett AI, et al. Chromatin extrusion explains key features of loop and domain formation in wild-type and engineered genomes. Proc Natl Acad Sci USA 2015;112:E6456–65.

[28] Naumova N, Imakaev M, Fudenberg G, Zhan Y, Lajoie BR, Mirny LA, et al. Organization of the mitotic chromosome. Science 2013;342:948–953.

[29] Cheng A, et al. A simple biophysical model emulates budding yeast chromosome condensation. eLife 2015;4:e05565.

[30] Baù D, et al. The three-dimensional folding of the α-globin gene domain reveals formation of chromatin globules. Nat Struct Mol Biol 2011;18:107–114.

[31] Duan Z, et al. A three-dimensional model of the yeast genome. Nature 2010;465:363–7.

[32] Serra F, Di Stefano M, Spill YG, Cuartero Y, Goodstadt M, Baù D, et al. Restraint-based three-dimensional modeling of genomes and genomic domains. FEBS Lett 2015;589:2987–95.

[33] Lesne A, Riposo J, Roger P, Cournac A, Mozziconacci J. 3D genome reconstruction from chromosomal contacts. Nat Methods 2014;11:1141–3.

[34] Hu M, Deng K, Qin Z, Dixon J, Selvaraj S, et al. Bayesian inference of spatial organizations of chromosomes. PLoS Comput Biol 2013;9(1):e1002893.

[35] Kalhor R, Tjong H, Jayathilaka N, Alber F, Chen L. Genome architectures revealed by tethered chromosome conformation capture and population-based modeling. Nat Biotechnol 2012;30:90–98.

[36] Amitai A, Holcman D. Diffusing polymers in confined microdomains and estimation of chromosomal territory sizes from chromosome capture data. PRL 2013;110:248105.

[37] Giorgetti L, et al. Predictive polymer modeling reveals coupled fluctuations in chromosome conformation and transcription. Cell 2014;157:950–63.

[38] Lupianez DG, et al. Disruptions of topological chromatin domains cause pathogenic rewiring of gene-enhancer interactions. Cell 2015;161(5):1012–1025.

[39] Hnisz D, Weintraub AS, Day DS, Valton AL, Bak RO, Li CH, et al. Activation of proto-oncogenes by disruption of chromosome neighborhoods. Science 2016;351:1454–8.

GLOSSARY

FISH (fluorescence in situ hybridization) a technique whereby fluorescent probes are bound to only specific genomic sequences on a chromosome having a high degree of complementarity. Combined with fluorescence microscopy it is employed to find the position of the probe.

Hi-C a method to map chromatin interactions genome-wide. Chromatin is crosslinked with formaldehyde, digested, and religated in a process whereby only DNA fragments that are bound together form ligation products. Sequencing will next provide information about which genomic sequences were physically in contact.

Monte Carlo methods a class of computational algorithms aiming to sample in an efficient way the large space of a set of variables of interest.

CHAPTER 10

From Chromosome Conformation Capture to Polymer Physics and Back: Investigating the Three-Dimensional Structure of Chromatin Within Topological Associating Domains

Guido Tiana[1], Edith Heard[2,3] and Luca Giorgetti[4]
[1]*University of Milano and Istituto Nazionale di Fisica Nucleare, Milano, Italy*
[2]*Institut Curie, PSL Research University, CNRS UMR3215, INSERM U934, Paris, France* [3]*Collège de France, Paris, France*
[4]*Friedrich Miescher Institute for Biomedical Research, Basel, Switzerland*

CONTENTS

INTRODUCTION

In mammals, the transcriptional control of many genes depends critically not only on promoter-proximal regulatory sequences, but also on long-range control elements such as enhancers. Such elements can be located up to megabases away from the gene that they control, and a large amount of evidence suggests that they regulate gene activity by looping out intervening DNA and associating in the three-dimensional nuclear space (reviewed for example in Ref. [1]). In some instances, enhancers act in a combinatorial fashion: the same enhancer can regulate several promoters at once, or at different times; also a given gene can be under the regulatory influence of several enhancers simultaneously [2]. Thus, understanding enhancer function requires a definition of the biochemical interactions that control enhancer–promoter specificity on the one hand; and on the other hand, the three-dimensional structure and the dynamics of chromosomes must be characterized, especially in the kilobase to megabase range which is where most promoter–enhancer interactions occur.

The rapid development of chromosome conformation capture (3C) and related techniques such as 4C, 5C, and Hi-C over the last two decades [3] has allowed a series of fundamental discoveries aimed at gaining such insights. These technologies have opened up unprecedented possibilities for the elucidation of how chromatin is organized in the cell nucleus. In 3C-based

Epigenetics and Systems Biology. DOI: http://dx.doi.org/10.1016/B978-0-12-803075-2.00010-6

methods, chromatin is crosslinked in cell nuclei, digested with a restriction enzyme and then religated in order to promote the formation of hybrid DNA molecules (Fig. 10.1A), which are then detected in most versions of the technique through high-throughput sequencing in order to determine which genomic sequences were found in three-dimensional proximity, and with what frequency. In the case of 5C (chromosome conformation capture carbon-copy [5]) and Hi-C [6], which are relevant to the topics discussed in this chapter, the interactions between many loci along a chromosome (5C) or between all genomic locations (Hi-C) can be detected (Fig. 10.1A). The results of these experiments can be plotted as a two-dimensional "interaction matrix," where the value of each entry (i,j) is proportional to the number of times a ligation product between sequences belonging to the genomic bins identified by the indexes i and j has been sequenced (Fig. 10.1B).

Together with other methods such as 4C (circularized 3C, where interactions of one locus vs. the entire genome can be detected) [7] and ChIA-PET (Chromatin Interaction Analysis by Paired-End Tag Sequencing, in which 3C is coupled to chromatin immunoprecipitation) [8], 5C and Hi-C experiments have been instrumental in revealing that the folding of mammalian chromosomes is highly hierarchical. Each chromosomal territory appears to be characterized by large-scale patterns of preferential associations into two so-called (A or B) compartments, genomic regions spanning several Mb of predominantly active or inactive chromatin, respectively, which tend to interact preferentially with other loci belonging to the same compartment [6]. At a smaller scale, sub-Mb contiguous blocks of preferential interactions exist, which are usually referred to as topological associating domains (TADs) [9,10]. TADs are often further characterized by the presence of sub-TAD structures that occasionally also define smaller domains dubbed "contact domains" [11]. All these folding layers result in a complex nested hierarchy of overlapping folding patterns that is readily visible in Hi-C maps (Fig. 10.1C).

Although all levels of chromosome folding may be relevant for transcriptional regulation, TADs are particularly interesting because their genomic size is comparable to the typical genomic distance between enhancers and promoters [11]. Indeed, genetic evidence has accumulated indicating that TADs constrain enhancer action on a limited set of promoters, possibly by increasing interactions between genomic elements within a domain and preventing high interactions across boundaries [12]. Moreover, structural variants that affect TAD boundaries have been shown to result in the creation of ectopic enhancer–promoter contacts that can cause oncogene activation [13], or ectopic gene activation leading to malformation syndromes [14]. It is therefore likely that three-dimensional structure of chromatin within TADs controls the spatial distance between enhancers and promoters, its fluctuations at the level single cells and its temporal dynamics, thus contributing to the control of gene expression [15].

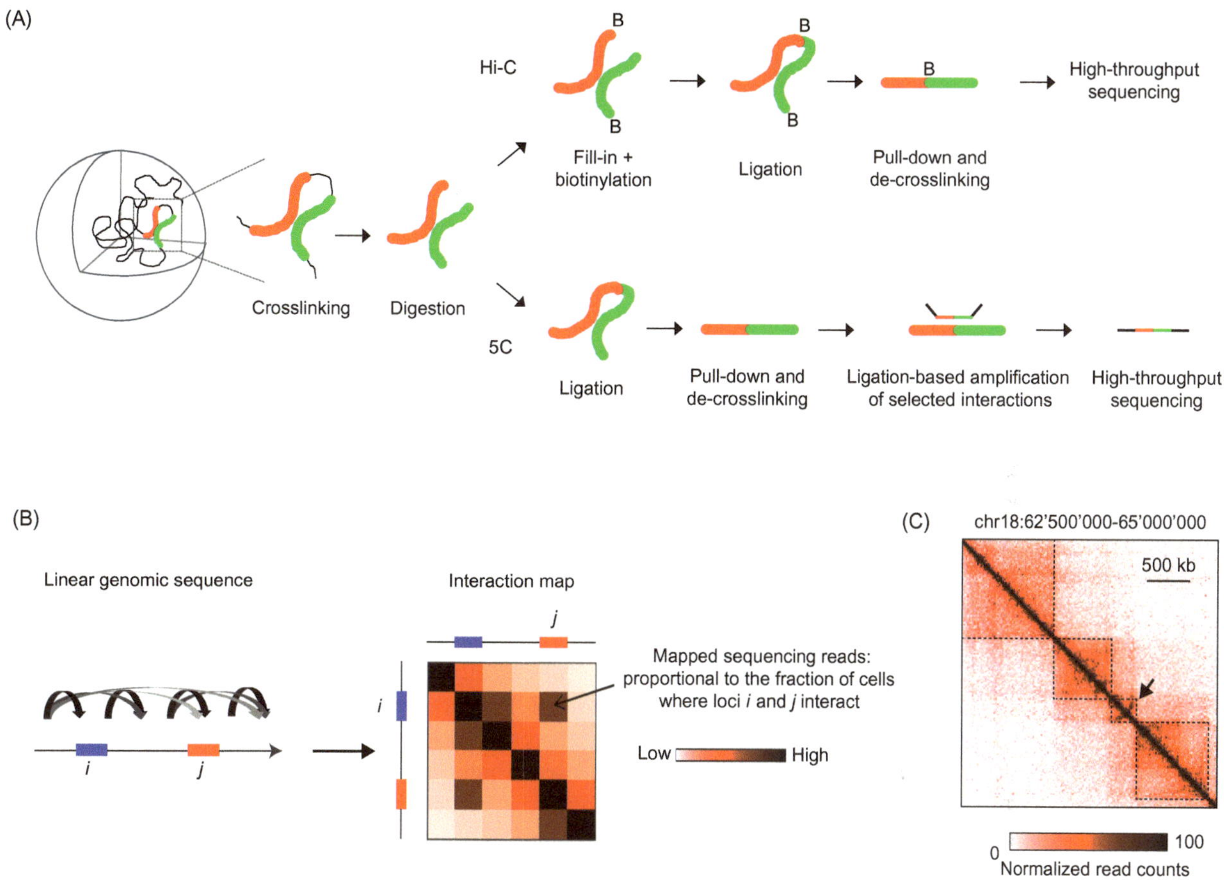

FIGURE 10.1 **(A) Schematics of 5C and Hi-C experiments.**
After crosslinking with formaldehyde, chromatin is digested in cell nuclei using a restriction enzyme. In Hi-C, restriction fragment ends are then repaired biotinylated and ligated, and following chromatin decrosslinking, pull-down is performed using streptavidin-coated beads that allows recovery of ligation products. Hybrid DNA molecules representing interactions between pairs of loci are then detected by high-throughput sequencing. In 5C, ligation is performed on nonrepaired restriction fragments immediately after digestion. after purification of ligation products and ligation-mediated amplification of sequence-specific oligonucleotides designed to target the subset of ligation events of relevance, PCR products representing a carbon copy of the original ligation events are then detected by high-throughput sequencing. (B) Both 5C and Hi-C convert the linear genomic sequence in a two-dimensional interaction matrix, which can be represented as a heat map where the color code is proportional to the number of read counts corresponding to each pair of genomic locations detected in the experiment. (C) Representative Hi-C results in a 3 Mb region of chromosome 18 in mouse embryonic stem cells, showing the complex overlap of nested folding patterns. Arrow indicates a prominent sub-TAD loop structure; dotted lines demarcate TAD boundaries. *Data from Giorgetti L, Lajoie B, Carter AC, Attia M, Zhan Y, Xu J, et al. Structural organization of the inactive X chromosome in the mouse. Nature 2016 in press, binned at 20 kb.*

5C and Hi-C datasets can be used quantitatively to investigate chromosome structure, given that each element in an interaction maps can be interpreted as being proportional to the contact probabilities of the associated chromosomal loci [16]. A natural consequence of this interpretation is that the complicated nested hierarchies of interaction patterns observed in contact maps, including TADs and their internal substructures, arise from a large number of alternative, and possibly highly dynamic, conformations of the chromatin fiber [17]. However, the population-based nature of 3C-based experiments does not grant immediate access to either the single-cell conformations of the chromatin fiber, or their dynamics. Alternative approaches are therefore needed to study chromosome conformation at the level of single cells. On the one hand, single-cell techniques such as DNA fluorescence in situ hybridization (DNA FISH) or live-cell imaging of DNA loci can be used to investigate both the cell-to-cell and temporal variability of chromosome conformation in single cells [18,19]. However, these techniques can only inform on the positions and/or dynamics of a limited number of chromosomal loci simultaneously. On the other hand, Hi-C can be performed in single cells [20], but its limited resolution has not so far enabled determination of single conformations inside single TADs.

How can then one derive realistic single-cell trajectories of the chromatin fiber starting from 5C (or Hi-C) interaction maps? Reconstructing three-dimensional molecular structures from two-dimensional interaction datasets is a classical inverse problem that is recurrently encountered in different fields of biophysics. One example is protein structure determination based on two-dimensional nuclear magnetic resonance spectroscopy techniques, such as 2D-NOESY [21] where spatial proximity between atoms in the polypeptide chain is detected through nuclear spin cross-relaxation. When analyzing 2D-NOESY data, proximity maps are converted in three-dimensional protein conformations using physical models of the peptide chain, and searching specifically for those conformations that reproduce the experimental maps. Several studies have recently shown that similar physical methods can be used to reconstruct the three-dimensional conformations of the chromatin fiber starting from 5C or Hi-C datasets [22–28]. Unlike the field of protein structure determination however, the quantitative laws that link signal intensity in 5C or Hi-C matrices to spatial proximity between genomic locations are unknown; different assumptions and approaches have therefore been used to reconstruct chromosomal structures (reviewed in Ref. [29]). Here we will briefly describe the general principle underlying some of these "data-driven" models and then focus on the approach we developed [30,31] to reconstruct the single-cell conformations of the chromatin fiber inside single TADs, which was inspired by methods that were created to derive the highly variable structures of intrinsically disordered proteins [32].

FROM 5C/HI-C MAPS TO THREE-DIMENSIONAL STRUCTURES

The major problem that one faces when trying to deconvolve 5C/Hi-C maps into three-dimensional conformations of the chromatin fiber with the help of physical models, is that the experimental data is averaged over the ensemble of conformations that the fiber can assume in the nuclei of the millions of cells that are used in the experiment. To understand this point, consider an idealized case where the structure of the fiber is identical in all cells. In this case the 5C/Hi-C map would be very simple (Fig. 10.2A) and contain only two possible values: either zero, or a positive number (proportional to the number of sequencing read counts). Inverting the problem, and converting the 5C/Hi-C map into the single structure that originated it, would also be relatively simple as it would reduce to the construction of a polymer model whose pairs of loci corresponding to the nonzero elements of the contact map are in contact. On the other hand, if the conformations of the fiber differ from cell to cell, the 5C/Hi-C map represents an average of the many possible signals produced by the various conformations, as sketched in Fig. 10.2B. In this case, the inverse problem (deducing the set of conformations from the population-averaged map) is far from simple, unless it is treated using simplified "mean-field" approaches (see Section "Description of the Data-driven Polymer Model" below). However, this is a far more likely scenario than the previous, as a large number of single-cell experiments performed, e.g. by DNA FISH have revealed that chromosome conformation is highly variable at the level of single cells, at all scales of chromosomal organization [9,33,34].

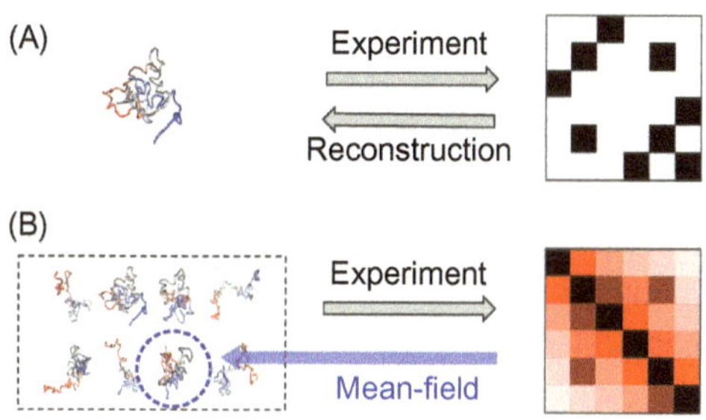

FIGURE 10.2 A sketch of the problems associated with structure reconstruction based on 5C/Hi-C data.

(A) If the experimental signal came from a single conformation, the 5C/Hi-C map would be a binary matrix and the structural reconstruction would be straightforward. On the other hand, if the signal comes from an ensemble of conformations as in (B), structural reconstruction is a complicated problem. In a mean-field approximation, one assumes that there is a single conformation that is responsible for most of the experimental signal. The structural reconstruction problem then reduces to finding this this conformation, neglecting all the others.

The first step that is taken when using physical models to deconvolve 3C-based experiments into three-dimensional conformations is to assume that the behavior of the chromatin fiber can be described in terms of a polymer model, represented as a sequence of beads arranged along a string (Fig. 10.3). This is a standard representation in polymer physics and provides a good approximation provided that the internal structure to each bead does not change significantly over time. In order to simulate specific attractions or repulsions between the various parts of the chromatin fiber, beads are allowed to attract or repel each other through attractive or repulsive interaction potentials (such as the one sketched in Fig. 10.3). Given that the beads-on-a-string model is a coarse-grained representation of the chromatin fiber where each bead typically represents thousands of base pairs, interaction potentials must be considered as "effective" energies that summarize the interactions between all subparts of one bead (e.g., mediated by all DNA-binding proteins that are bound to the sequences inside a bead), as well as the steric repulsion between different regions of the fiber.

A fundamental assumption of data-driven models is that the polymer chain is at thermodynamic equilibrium. This hypothesis can certainly be questioned when studying large-scale properties of chromosomes, which in metazoans are so long that the associated equilibration times are expected to be much longer than the typical duration of a cell cycle [35]. However, at the scale of single TADs the conformational dynamics might be fast enough that the chromatin fiber is able to sample many conformations during one cell cycle. Although so far there are no available measurements of chromatin conformational dynamics of chromatin within TADs, which would require that motion of at least a pair of chromosomal loci be measured in the same TAD in living cells, this seems to be supported by existing measurements of the mobility of single chromosomal loci [36,37]. In the equilibrium approximation, the millions of cells on which the 5C/Hi-C experiment is performed are considered to represent the thermodynamic ensemble in the polymer model; in particular, experimental averages over the cell population can be compared to averages over the thermodynamic ensemble. This is a key concept, since it allows comparison of the 5C/Hi-C datasets to quantities extracted from polymer simulations.

As the chromatin fiber is (at least locally) supposed to be at equilibrium, the probability distribution of the different conformations that it can assume are governed by the Boltzmann distribution: The most likely conformations are those with the lowest interaction energies. The number of conformations displaying minimal energy depends on the particular interaction potentials between pairs of beads. If a single, low-energy conformation exists, the single polymer conformations in the equilibrium ensemble will often be similar to it. On the other hand, if the arrangement of interaction potentials along the chromatin fiber is such that many conformations of the polymer with comparably low energies exist, the conformations of the simulated chromatin fiber ensemble will fluctuate among them.

In order to deconvolve the three-dimensional conformations of the chromatin fiber from 5C or Hi-C data, the conformations in the Boltzmann ensemble must then be compared to the experimental data. This is the step where different methods diverge. In the model, we recently implemented [30,38] inspired by structure reconstruction of disordered proteins [32], as well as in more recent studies [39], each single conformation in the Boltzmann ensemble is converted in a "single-cell" contact map. Individual maps are then averaged, and compared to experimental 5C or Hi-C map (see the detailed description of the model in Section "Description of the Data-driven Polymer Model" below). This allows information on the single conformations that gave rise to one particular population-averaged map to be maintained.

Another class of models [22,26–28,40,41] assumes instead the existence of an *average* conformation that is responsible for most of the experimental signal. In physical terms, this is called a "mean-field approximation," and allows an average three-dimensional structure that is essentially an approximate view of the overall structure of the fiber in the region of interest to be obtained (Fig. 10.2B). In these approaches, 3C-based counts are transformed into restraints on the population-averaged distance between pairs of beads each representing large amounts of genomic sequence, and a computational approach is used to find the structure that satisfies such restraints as much as possible. The main limitation of such a procedure is that despite the diversity of the ensemble of conformations which produced the 5C/Hi-C map, the procedure results in a single, average structure. Structural fluctuations around the mean structure in these studies do not represent the thermodynamic fluctuations of the polymer but are rather a measure of the accuracy with which the mean-field restraints can be satisfied.

However, DNA FISH experiments have shown that the cell-to-cell differences in the 3D distances between pairs of chromosomal loci can be extensive, comparable with their mean distances [9,30,33,34]. This suggests that to have a more realistic idea of the conformational properties of chromosomes, and in particular at the level of TADs, it is necessary to go beyond the mean-field approximation. In the next sections, we describe in detail the method we developed [30,38] that, starting from 5C data, was originally aimed at reconstructing the conformation of the chromatin fiber within two neighboring TADs containing the promoters of *Tsix* and *Xist*, two key regulators of X-chromosome inactivation in female mouse embryonic stem cells (ESCs) [9].

DESCRIPTION OF THE DATA-DRIVEN POLYMER MODEL

In order to build a coarse-grained model of the chromatin fiber spanning the *Xist* and *Tsix* TADs in the X-inactivation center (Xic) region, we mapped the genomic sequence of the locus on a series of beads, each representing an amount of sequence equal to the resolution of experimental 5C maps (3 kb). In our model, each pair of beads i and j can attract or repel each other through an interaction potential u_{ij} that depends on their mutual distance d_{ij} and is shaped like a step function (Fig. 10.3):

$$u_{ij} = \begin{cases} \infty & \text{if } d_{ij} < R_{\mathrm{HC}} \\ B_{ij} & \text{if } R_{\mathrm{HC}} < d_{ij} < R \\ 0 & \text{if } d_{ij} > R \end{cases} \qquad (10.1)$$

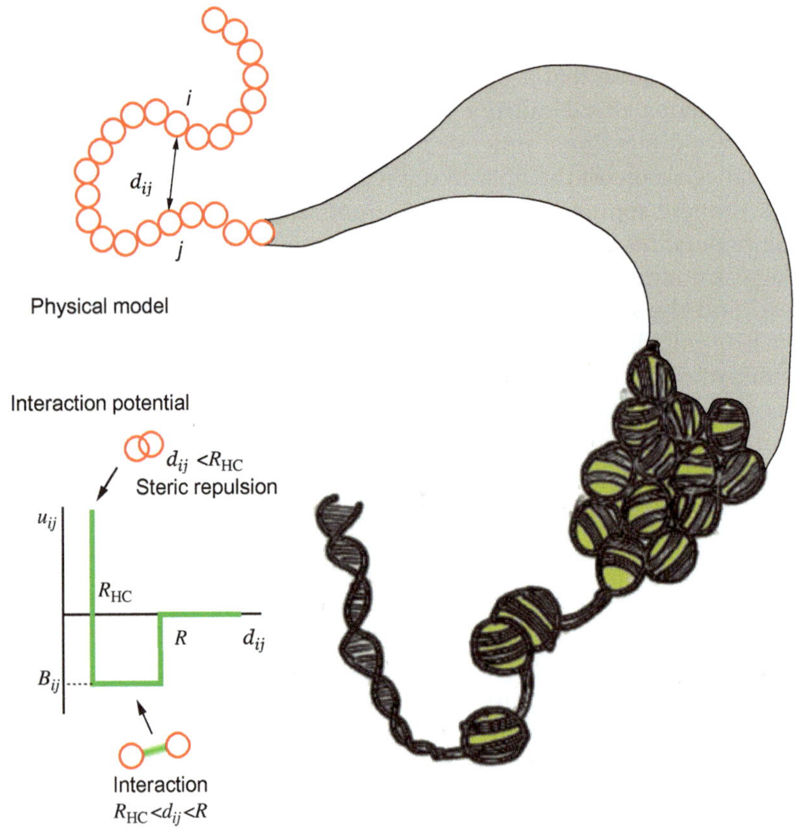

FIGURE 10.3 The chromatin fiber is modeled as a chain of beads on an inextensible string (red beads). Each pair of beads i and j interact through an interaction potential that depends on their mutual three-dimensional distance d_{ij}, the interaction energy B_{ij} depends on the specific pair of beads. Here u_{ij} is the interaction energy between tw beads, R is the interaction range, R_{HC} is the hard-core repulsion radius.

where R_{HC} is a "hardcore" radius preventing beads from crossing each other, and R the interaction distance between two beads. Thus, the total energy U of the chain in any given configuration is given by the sum of all possible pair interactions u_{ij}

$$U = \sum_{i<j} u_{ij} \quad (10.2)$$

The total energy defines the probabilities for the various configurations to occur in the equilibrium ensemble as described above.

Note that in this model, if two beads are closer than the hard-core radius R_{HC} their interaction energy is infinite. This leads to such a large increase in energy that any associated conformation is effectively absent from the Boltzmann ensemble of the model polymer (i.e., is prohibited). This implements the basic fact that the chromatin fiber cannot cross itself. If the two beads are located farther than R_{HC} but closer than the interaction distance R, they interact with an energy B_{ij} which depends on the specific pair i and j, whereas if they are further away they do not interact. The distance R can be thought of as the range of contact between the DNA-bound proteins that could mediate specific interactions between DNA loci and also defines the range at which two beads can be "crosslinked" together and thus define a contact in the model conformations. The physics of this polymer model, and in particular the various conformations belonging in its Boltzmann ensemble and their relative probability of occurrence, are entirely specified by the energy matrix B_{ij} that quantifies to which extent beads i and j attract or repel each other. Operatively, given a certain energy matrix B_{ij}, the Boltzmann ensemble of conformations can be generated using a computational scheme called Monte Carlo algorithm [42], which allows simulation of the correct thermodynamic fluctuations of the chain among all possible conformations. Different energy matrices B_{ij} result in ensemble of conformations that favor the probability of some conformations rather than others.

In order to deconvolve the ensemble of structures from a 5C map, our method tries to find the "optimal" energies B_{ij} with which the model produces a population-averaged contact map that is as similar as possible to the experimental 5C map (Fig. 10.4). For this, we developed an algorithm, which we dubbed MonteGrappa (described in detail in Ref. [38]). On the one hand, MonteGrappa allows the efficient sampling of the conformational space in the Monte Carlo part of the algorithm, in order to produce the highest possible number of different conformations of the chromatin fiber given a matrix B_{ij}. On the other hand, it enables the efficient exploration of the space of interaction energies B_{ij} for rapid convergence toward an optimal choice that maximizes the agreement with the 5C data. At the end of the procedure, one obtains a specific energy matrix B_{ij} that gives rise to conformations, the contacts of which, when averaged over the entire Boltzmann ensemble, are very

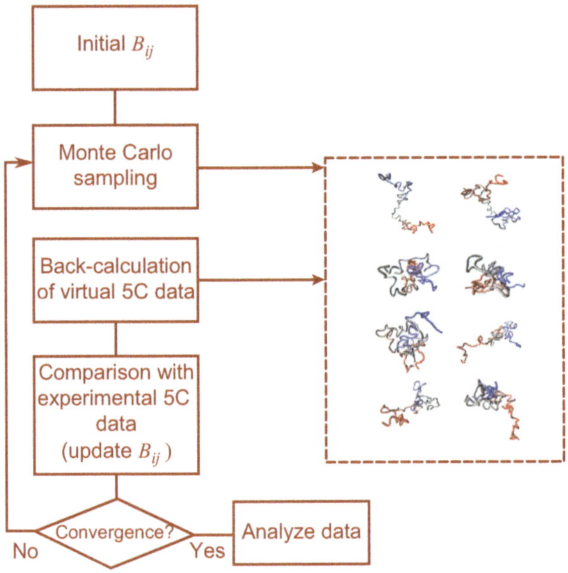

FIGURE 10.4 The workflow implemented in the MonteGrappa algorithm [38] to obtain the optimal interaction matrix *Bij* which determines the physics of the polymer model. Starting from a random choice of the initial set of energies B_{ij}, a thermodynamic sampling of the conformations of the polymer is performed with a Monte Carlo algorithm. From the ensemble of sampled conformations, a virtual 5C map is back-calculated and compared with the actual 5C data. The procedure is iterated until optimal agreement is reached. The associated ensemble of conformations, which is compatible with the experimental data, is then analyzed.

similar to the experimental 5C map. The ensemble of sampled conformations can then be analyzed to obtain properties that are not accessible directly from the 5C/Hi-C map, such as conformational fluctuations [30] or correlations among different types of contacts [36].

In the case of the *Tsix* and *Xist* TADs in mouse ESCs, we obtained structural models that are in agreement not only with the 5C map that was used to derive them, but most importantly, that agreed with results that obtained using high-resolution 3D DNA FISH, representing single-cell information using an alternative technique [19]. In these experiments, we accurately measured the three-dimensional distances between several pairs of loci within the *Tsix* TAD and compared the average distances and their cell-to-cell variability with those predicted by the computational model. We found that the combination of potential energies B_{ij} that produced the best agreement with the 5C contact matrix also predicted the full FISH results with striking accuracy [19].

It should be noted that the polymer model is dimensionless, in the sense that it does require an explicit assumption of how many nanometers each

3 kb bead occupies in physical space. All distances in the model are in fact expressed in units of the bead size, which can assume any numerical value. However, the latter will determine the effective spatial dimensions of the polymer conformations and is therefore key for comparing the simulations with the experiments. In Ref. [30], we decided to extract this value a posteriori as the scale factor that is able to convert the three-dimensional distances predicted by the model with the high-resolution DNA FISH experiments mentioned above. Interestingly, this lead to the estimation that a bead of 3 kb of chromatin at the Xic has a diameter of approximately 30 nm, corresponding to an effective contact and cross-linking range (parameter *R*) of approximately 80 nm, compatible with the idea that protein complexes mediate the interactions between the various parts of the chromatin fiber.

These results show that using a thermodynamic model of the chromatin fiber, which is based on the assumption that the fiber is locally at equilibrium, allows the reconstruction of a realistic model for chromosome conformation inside TADs, including its cell-to-cell variability.

WHAT CAN BE LEARNED FROM DATA-DRIVEN, POLYMER-BASED RECONSTRUCTIONS OF CHROMOSOME STRUCTURE?

Realistic polymer models can be used to investigate conformational properties that cannot be easily accessed experimentally. For example, the model reconstruction of the *Tsix* and *Xist* TADs showed that the observed 5C contacts arise from a wide variety of different underlying structures despite the presence of well-demarcated boundaries in the 5C map (Fig. 10.5A and B). Boundaries between adjacent TADs do not arise because the two contiguous regions of the chromatin fiber form two stable "globules" but rather emerge as statistical patterns in a strongly fluctuating chromosomal neighborhood (Fig. 10.5B). Similarly, even the strongest sub-TAD contacts that arise in the *Tsix* TAD (Fig. 10.5C) do not represent stable loops but rather stochastic events that occur more frequently than others within the structural ensemble (and thus the cell population). This has important implications, as it suggests that even in regions of relatively highly organized chromosome structure such as TADs that contain multiple sub-TAD loops [43], loop structures might occur only in a relatively small fraction of cells at the same time.

In the case of the *Tsix* TAD, only slightly more than half of the conformations that the model predicted are compactly folded and rich in contacts between even the most interacting loci, such as the *Xite* locus, the *Chic1* gene and the *Linx* noncoding RNA (Fig. 10.4D). These three loci tend to colocalize together in the more compact TAD conformations, whereas they are more

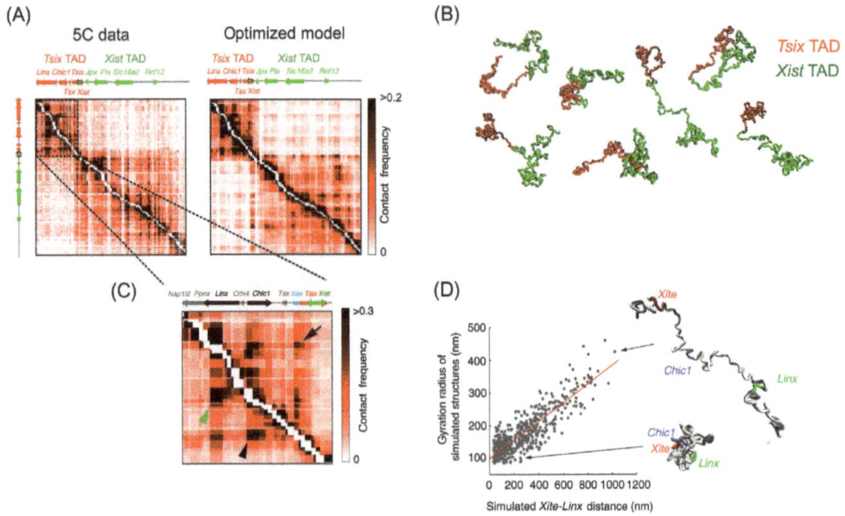

FIGURE 10.5 (A) The 5C interaction map in the region of the TADs of Tsix and Xist in mouse ES cells (left), compared with those back-calculated from the model (right).
(B) A zoom-in of the Tsix TAD with the long-range interactions between *Xite*, *Chic1*, and *Linx* indicated by arrows. (C) A set of representative conformations of the polymer model, showing that the *Tsix* and *Xist* TADs arise from highly variable and partially intermingling conformations. (D) A scatter plot of the radius of gyration of the polymer model of the *Tsix* TAD versus the distance between the *Xite* and *Linx* loci. Each dot corresponds to one conformation in the structural ensemble. The distance between the long-range interacting elements *Xite* and *Linx* tends to be small when the conformation of the entire TAD is compact, and vice versa. *Adapted from Giorgetti L, Galupa R, Nora EP, Piolot T, Lam F, Dekker J, et al. Predictive polymer modeling reveals coupled fluctuations in chromosome conformation and transcription. Cell 2014;157(4)950–63.*

separate from each other when the TAD is in an overall more elongated state. This finding, which was validated by high-resolution DNA FISH, might have important implications for the way in which the expression level of *Tsix* itself is regulated at the level of single cells, as its TAD contains several potential enhancer regions, whose reciprocal distance as well as their distance from promoters within the TAD are modulated by conformational fluctuations. In fact, by performing sequential RNA-DNA FISH experiments, we detected a correlation between the overall compactness of the TAD and the transcription level of *Tsix* at the level of single cells [30].

Despite substantial cell-to-cell differences, chromosome structure inside single TADs is not random. Careful analysis of the structural ensembles that were derived for the *Tsix* and *Xist* TADs [36] showed that the folding of the chromatin fiber is much less structured than a protein for example, which is characterized by a unique, well-defined conformation. However,

TAD conformations are more structured than a random coil where no specific interactions are imposed on a freely fluctuating polymer. In particular, frequently interacting sites such as *Xite*, *Chic1*, and *Xite* in the *Tsix* TAD are found to correspond to mutually attractive interactions B_{ij} [36]. These interactions favor the stochastic appearance of compact conformations involving the entire TAD, and together with weaker interactions between other loci along the fiber in the TAD, determine the overall fluctuations of the chromatin fiber in the region.

It is however noteworthy that the interaction energies between frequently interacting loci are not strong enough to stabilize compact conformations. Thus, compact and elongated conformation of the entire TAD are not separated by important energy barriers and the transition from one state to the other could, at least in principle, occur frequently. Predicting the time scale of structural fluctuations based on a model derived from population-averaged, static datasets such as 5C is however not possible unless some dynamical properties of the system are known. To be able to study the kinetic properties of the fiber within the *Tsix* TAD, we set out to measure the mobility of a genomic locus that lies within this TAD in living cells. To do this, we exploited a mouse ESC line where an array of Tet operator (TetO) DNA-binding sites has been inserted into the 3′ untranslated region of the *Chic1* gene [44]. These have also been engineered to express the Tet repressor (TetR), a bacterial DNA-binding protein that binds TetO sites with high specificity, fused to Green fluorescent protein (GFP). The array of TetO sites can then be visualized under a microscope as a diffraction-limited spot in the nucleus upon binding of TetR-GFP [36] (Fig. 10.6A). Although the dynamics of this locus had already been measured over longtime scales (comparable to one cell cycle) in Ref. [44], we measured its mobility at much shorter time scales and extracted the instantaneous diffusion coefficient of a locus within the *Tsix* TAD which we determined to be approximately $4000\,nm^2\,s^{-1}$ [36]. By feeding our polymer model with this experimentally determined value, we were able to predict that the time needed for the entire *Tsix* TAD to fluctuate among the many available conformations is likely to be in the order of a few hundred minutes (Fig. 10.6B). In this case, sub-TAD chromosomal contacts and possible enhancer–promoter interactions in particular, are likely to be dynamic events that can occur and disassemble several times during a single-cell cycle. If proven, this implies that enhancer–promoter contacts can assemble and disassemble within TADs several times during a single-cell cycle.

Another important implication of our model derived from the *Tsix/Xist* locus in ESCs is that interactions between sites that are located *internally* to each single TAD seem to be sufficient to simultaneously (1) create boundaries between adjacent TADs and (2) drive the internal structures that determine sub-TAD contacts between regulatory regions. This was inferred by comparing

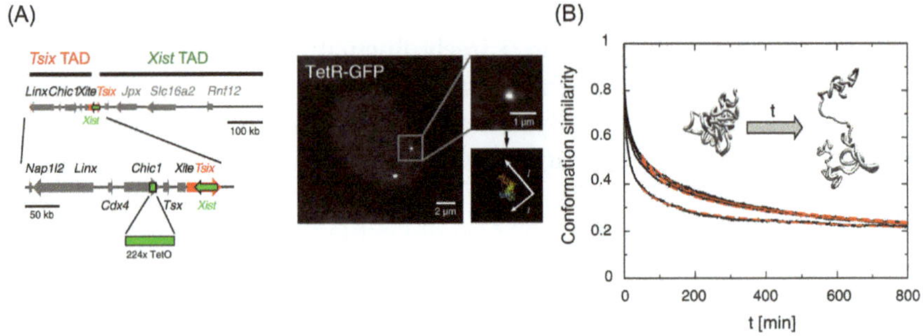

FIGURE 10.6 (A) Left: Scheme showing the genomic position of the Tet operator (TetO) array, which was inserted in the 3′UTR region of the *Chic1* gene.
Right: Upon binding of TetR-GFP, the TetO array can be visualized in live cells and single-particle tracking was performed to characterize its motion. (B) The time it takes for the chromatin fiber within the *Tsix* TAD to fold from an initial configuration into an unrelated one is of the order of few hundred minutes based on dynamical simulations performed with the diffusion coefficient measured at the *Chic1* locus. Four independent simulations performed starting from different initial conformations are shown (black line), together with their fit with an exponential function (red line). *Adapted from Tiana G, Amitai A, Pollex T, Piolot T, Holcman D, Heard E, et al. Structural fluctuations of the chromatin fiber within topologically associating domains. Biophys J. 2016;110(6):1234–45.*

the predictions of the model to experiments performed in ESCs carrying deletions of either the *Tsix/Xist* TAD boundary, or a group of DNA-binding sites for the CCCTC-binding factor (CTCF) internal to the *Tsix* TAD [30]. The fact that the model correctly predicts the experimental outcome of these deletions implies that sub-TAD attractive interactions between strongly attracting loci contribute to folding each TAD independently, which in turn establishes physical separation across the boundary between adjacent TADs. Interestingly, several lines of experimental evidence have recently accumulated showing that these attractive interactions might be due to mechanisms that largely depend on CTCF binding to DNA. CTCF is the only known insulator protein characterized so far in vertebrates and has long been associated to the formation of chromosomal loops together with its binding partner cohesin [45]. Indeed, sub-TAD "loops" and structures seem to arise as a consequence of effective attractive interactions between two loci where CTCF binds to DNA motifs that are arranged in a "head-to-head" orientation [43,46,47] (i.e., in a convergent orientation of the asymmetric binding motif). Although the exact molecular mechanisms by which these interactions arise is unclear [45], our analysis of model predictions and experiments suggests that TADs might well be established by overlapping patterns of nested CTCF-associated "loop" structures [43]. Importantly, deletion, inversion, and inactivation of CTCF motifs associated to sub-TAD loop structures has been shown to result not only in altered patterns of interactions within the corresponding loop domain, but also in structural perturbations that extend across the boundaries with adjacent domains [46,48]. It will be interesting to see if (and to what extent) the appearance of strong boundaries between adjacent TADs can be explained in terms of a peculiar density and arrangement of CTCF binding sites.

WHAT ARE THE MOLECULAR MECHANISMS BEHIND INTERACTION ENERGIES?

An important aspect of "data-driven" physical models of the chromatin fiber such as the one we developed is that by definition, the model remains agnostic concerning the molecular origin of the forces or interaction potentials that are "learned" from the 5C/Hi-C dataset. In particular in our model, we make no assumptions to link the mechanisms that lead to the formation of attractions or repulsions on the real chromatin fiber, with the elements of the energy matrix B_{ij}. In the case of the two TADs that we simulated, the specific attractive mutual interactions that emerged often corresponded to clustered CTCF sites such as those located within the *Xite*, *Chic1*, and *Linx* loci [30,36]. However, these "spontaneously" emerge during the unbiased optimization of energies B_{ij}, as they contribute to producing an optimal agreement with the 5C map. It is also interesting to notice that by definition, interactions in our model are described through the contact potentials u_{ij} described above and sketched in Fig. 10.3, where interactions between distal beads occur by direct interaction through classical "looping" (i.e., by physical association in the three-dimensional space).

A relevant question, however, is what could be the molecular determinants of the interactions between chromosomal loci. To investigate whether and how specific molecular mechanisms can give rise to the structural features that can be observed in 5C/Hi-C experiments, and notably TADs and sub-TAD interactions, it is possible to use another class of physical models, which can be referred to as "de novo" or "mechanism-driven" models (reviewed in Ref. [29]). These models start from a hypothetical physical or biological principle and test whether this principle is able to reproduce the features observed in experiments. Thus, contrary to data-driven approaches, they *do not* take Hi-C or 5C maps as an input, but rather try to predict it from first principles.

One possible mechanism creating specific chromosomal structure might be that interactions between proteins or protein complexes bound to specific DNA sequence locations mediate looping interactions. This is the implicit assumption of the model we implemented (see Section "Description of the Data-driven Polymer Model" above). However, a literal interpretation of this model at the molecular level would require a very large numbers of different DNA-binding factors, which are able to interact with each other in a combinatorial way to give rise to the various B_{ij} interaction energies. So far, only a few proteins have been suggested to be able to mediate interactions between distal chromosome loci. The best known are CTCF and cohesin [45], the depletion of which has been shown to result in reduction of TAD and sub-TAD structure [49]. Although it is currently unclear how CTCF and cohesin participate in chromosomal structure, CTCF has been reported to form dimers or multimers, which could drive looping interactions between distal sites [50]. In addition, certain subunits of cohesin have been shown to interact in vitro with CTCF [51]. Other mechanisms are likely to contribute to mediating long-range three-dimensional looping interactions, such as Mediator complex, that could be implicated in stabilizing interactions between enhancers and promoters [52]. However, with such a small number of candidate DNA-binding factors able to mediate interactions, it is difficult to explain how well-defined and complex structures such as TADs and their internal interactions can be created by direct looping. In fact, polymer models mimicking the presence of one [53], two [54,55], or a few [56] types of direct interactions between specific loci can produce TAD-like domains. However, these domains do not appear in the equilibrium ensemble of the polymer models but occur as a transient long-lived phase, which eventually leaves place to a less structured globule with no domain structure. Thus, in models that suppose that a few types of direct interactions between distal sites exist, the presence of domains reminiscent of TADs is observed as a transient out-of-equilibrium effect, and suggests that TADs can be observed because the associated structural states exist at least for the time of a cell cycle.

The idea that an entire chromosome (which is typically tens of Mb long in mammals) does not have enough time to equilibrate during a cell cycle in ESC (which lasts for about 16 hours) is based on simple physical arguments [35]. However, at the genomic length scale of TADs (few hundred kb) such an argument might not hold true given that the mobility of single chromosomal loci can be relatively fast, with microscopic diffusion coefficients in the order of thousands of $nm^2 s^{-1}$ [36], which could be too fast to guarantee that out-of-equilibrium phenomena are relevant at this scale.

Another mechanism has been proposed for the formation of TADs in the presence of a few types of specific interactions, without resorting to out-of-equilibrium effects. The "string and binders" model assumes that different loci can interact by simultaneously binding one or few diffusing proteins, which have a mild affinity for the fiber and are present in finite amount in the nucleus [39,57,58]. Despite the apparent similarity, there is one fundamental difference with the models described above. In the "direct interaction" models, the DNA-binding sites for factors that mediate looping interactions are assumed to always be bound (i.e., the nuclear concentration of the factor is sufficiently high to ensure approximately 100% occupancy of specific binding sites). Thus, one can regard the chromatin locus itself as an effective interaction unit: If two loci are close in space, they will always attract each other. In the "string and binders" model, if the nuclear concentration of binder proteins is very large, the result is identical to that of direct interaction between loci. If it is very low, the chromatin fiber is essentially a disordered coil because of the absence of attracting interactions. But an intermediate concentration regime exists, in which the fiber displays TAD-like structures [57].

Another mechanism proposed to account for TAD formation is the "loop extrusion" model [59–61] where effective attractive interactions between CTCF-bound chromosome loci occur through an active, energy-consuming mechanism by which an unknown factor binds randomly to chromatin and actively extrudes loops until the process is stopped by a CTCF molecule bound in the correct orientation [59,61]. Model calculations showed that this mechanism can give rise to TADs and sub-TAD loop signals between convergent CTCF sites in virtually generated Hi-C maps. Moreover, this model predicts important outcomes observed in Hi-C experiments, such as why two loop domains merge when a CTCF site is lost, and where a new boundary will be created [48]. Cohesin was proposed as a possible loop-extrusion factor, because it contains an ATPase domain and it topologically embraces the chromatin fiber. However, it is unclear how cohesin could actively extrude loops, and what molecular mechanism would allow CTCF, and not other transcription factors or the polymerase itself to stop the motion of cohesin [45].

A final model proposed to explain the emergence of TADs and sub-TAD structures, which is somewhat distinct from all the previous models, is based on the fact that in the presence of transcription, DNA can become supercoiled as a consequence of the torsional stress imposed on the chromatin fiber by active polymerases [62]. A simple model implementing supercoiling to the stretch of chromatin located between two "boundary elements" was shown to be capable of producing structures reminiscent of TADs [63]. This is a general model based on polymer physics, which does not provide predictions on the detailed structure of the chromatin fiber of well-defined cell types, but can indeed coexist with all the other models, fostering and modulating the formation of TADs.

Irrespective of the specific molecular mechanisms that might mediate interactions between chromosomal loci, it is interesting to notice that all the "de novo" physical models described above result in structures that are only marginally stable, and that undergo large fluctuations among alternative conformations. Thus, whatever the molecular mechanism, the effective interaction energies between chromosomal loci in these models are expected to be in the range of few k_BT, where k_B is the Boltzmann constant and T is the experimental temperature (usually 37°C). This is in the range of the numerical values we obtained for the potentials B_{ij} that we optimized in order to match the 5C maps in the *Tsix/Xist* region.

CONCLUSIONS AND PERSPECTIVES

Although the molecular mechanisms that give rise to chromosomal structures are still incompletely characterized, we are convinced that the use of data-driven physical methods can be highly informative to derive realistic and predictive coarse-grained structural models of the chromatin fiber at the scale of TADs. As illustrated by the case of the *Tsix/Xist* locus to which the method we developed was initially applied, such models are able to accurately describe the structure of the chromatin fiber, including its cell-to-cell differences and to predict potential outcomes of experiments performed in mutant cells carrying deletions that impact chromosomal structure. Although in the case presented here the structural reconstruction of TADs was based on 5C datasets, the recent development of even more high-resolution Hi-C [43,64] has meant that such datasets could be similarly exploited in the future. For example in Capture-C [65] and Capture Hi-C [66], oligonucleotide-based capture can be used to enrich for a specific subset of the interactions present in 3C or Hi-C libraries, which can then be sequenced and mapped with high genomic coverage. It will thus be of great interest to combine these new high-resolution datasets with various types of polymer-driven modeling as described above. This should enable an understanding of how the complex hierarchical organization of chromosomes into nested loops, domains, TADs and groups of TADs arises and what these structures corresponds to in single cells.

References

[1] Spitz F. Gene regulation at a distance: from remote enhancers to 3D regulatory ensembles. Semin Cell Dev Biol 2016;75:57–67.

[2] Lagha M, Bothma JP, Levine M. Mechanisms of transcriptional precision in animal development. Trends Genet 2012;28(8):409–16.

[3] Denker A, de Laat W. The second decade of 3C technologies: detailed insights into nuclear organization. Genes Dev 2016;30(12)):1357–82.

[4] Giorgetti L, Lajoie B, Carter AC, Attia M, Zhan Y, Xu J, et al. Structural organization of the inactive X chromosome in the mouse. Nature 2016;535:575–579.

[5] Dostie J, Richmond TA, Arnaout RA, Selzer RR, Lee WL, Honan TA, et al. Chromosome conformation capture carbon copy (5C): a massively parallel solution for mapping interactions between genomic elements. Genome Res 2006;16(10):1299–309.

[6] Lieberman-Aiden E, Berkum NL, van Williams L, Imakaev M, Ragoczy T, Telling A, et al. Comprehensive mapping of long-range interactions reveals folding principles of the human genome. Science 2009;326(5950):289–93.

[7] Simonis M, Klous P, Splinter E, Moshkin Y, Willemsen R, de Wit E, et al. Nuclear organization of active and inactive chromatin domains uncovered by chromosome conformation capture-on-chip (4C). Nat Genet 2006;38(11):1348–54.

[8] Fullwood MJ, Liu MH, Pan YF, Liu J, Xu H, Mohamed YB, et al. An oestrogen-receptor-α-bound human chromatin interactome. Nature 2009;462(7269):58–64.

[9] Nora EP, Lajoie BR, Schulz EG, Giorgetti L, Okamoto I, Servant N, et al. Spatial partitioning of the regulatory landscape of the X-inactivation centre. Nature 2012;485(7398):381–5.

[10] Dixon JR, Selvaraj S, Yue F, Kim A, Li Y, Shen Y, et al. Topological domains in mammalian genomes identified by analysis of chromatin interactions. Nature 2012;485(7398):376–80.

[11] Smallwood A, Ren B. Genome organization and long-range regulation of gene expression by enhancers. Curr Opin Cell Biol 2013;25(3):387–94.

[12] Symmons O, Uslu VV, Tsujimura T, Ruf S, Nassari S, Schwarzer W, et al. Functional and topological characteristics of mammalian regulatory domains. Genome Res 2014;24(3):390–400.

[13] Hnisz D, Weintraub AS, Day DS, Valton A-L, Bak RO, Li CH, et al. Activation of proto-oncogenes by disruption of chromosome neighborhoods. Science 2016;351:1454–8.

[14] Lupiáñez DG, Kraft K, Heinrich V, Krawitz P, Brancati F, Klopocki E, et al. Disruptions of topological chromatin domains cause pathogenic rewiring of gene-enhancer interactions. Cell 2015;161(5):1012–25.

[15] van Arensbergen J, van Steensel B, Bussemaker HJ. In search of the determinants of enhancer–promoter interaction specificity. Trends Cell Biol 2014;24(11):695–702.

[16] Fudenberg G, Mirny LA. Higher-order chromatin structure: bridging physics and biology. Curr Opin Genet Dev 2012;22(2):115–24.

[17] Junier I, Spill YG, Marti-Renom MA, Beato M, le Dily F. On the demultiplexing of chromosome capture conformation data. FEBS Lett 2015;589(20, Part A):3005–13.

[18] Chen B, Gilbert LA, Cimini BA, Schnitzbauer J, Zhang W, Li G-W, et al. Dynamic imaging of genomic loci in living human cells by an optimized CRISPR/Cas system. Cell 2013;155(7):1479–91.

[19] Giorgetti L, Piolot T, Heard E. High-resolution 3D DNA FISH using plasmid probes and computational correction of optical aberrations to study chromatin structure at the sub-megabase scale. In: Nakagawa S, Hirose T, editors. Nuclear bodies and noncoding RNAs: Methods and Protocols. New York: Springer; 2015. p. 37–53. Available from: http://dx.doi.org/10.1007/978-1-4939-2253-6_3. [accessed 12.04.16]

[20] Nagano T, Lubling Y, Stevens TJ, Schoenfelder S, Yaffe E, Dean W, et al. Single-cell Hi-C reveals cell-to-cell variability in chromosome structure. Nature 2013;502(7469):59–64.

[21] Wüthrich K. The way to NMR structures of proteins. Nat Struct Mol Biol 2001;8(11):923–5.

[22] Baù D, Sanyal A, Lajoie BR, Capriotti E, Byron M, Lawrence JB, et al. The three-dimensional folding of the α-globin gene domain reveals formation of chromatin globules. Nat Struct Mol Biol 2011;18(1):107–14.

[23] Serra F, Di Stefano M, Spill YG, Cuartero Y, Goodstadt M, Baù D, et al. Restraint-based three-dimensional modeling of genomes and genomic domains. FEBS Lett 2015;589(20, Part A):2987–95.

[24] Umbarger MA, Toro E, Wright MA, Porreca GJ, Baù D, Hong S-H, et al. The Three-dimensional architecture of a bacterial genome and its alteration by genetic perturbation. Mol Cell 2011;44(2):252–64.

[25] Tanizawa H, Iwasaki O, Tanaka A, Capizzi JR, Wickramasinghe P, Lee M, et al. Mapping of long-range associations throughout the fission yeast genome reveals global genome organization linked to transcriptional regulation. Nucleic Acids Res 2010;38(22):8164–77.

[26] Marbouty M, Le Gall A, Cattoni DI, Cournac A, Koh A, Fiche J-B, et al. Condensin- and replication-mediated bacterial chromosome folding and origin condensation revealed by Hi-C and super-resolution imaging. Mol Cell 2015;59(4):588–602.

[27] Lesne A, Riposo J, Roger P, Cournac A, Mozziconacci J. 3D genome reconstruction from chromosomal contacts. Nat Methods 2014;11(11):1141–3.

[28] Varoquaux N, Ay F, Noble WS, Vert J-P. A statistical approach for inferring the 3D structure of the genome. Bioinformatics 2014;30(12):i26–33.

[29] Imakaev MV, Fudenberg G, Mirny LA. Modeling chromosomes: beyond pretty pictures. FEBS Lett 2015;589(20PartA):3031–6.

[30] Giorgetti L, Galupa R, Nora EP, Piolot T, Lam F, Dekker J, et al. Predictive polymer modeling reveals coupled fluctuations in chromosome conformation and transcription. Cell 2014;157(4):950–63.

[31] Tiana G, Sutto L, Broglia RA. Use of the Metropolis algorithm to simulate the dynamics of protein chains. Phys Stat Mech Its Appl 2007;380:241–9.

[32] Norgaard AB, Ferkinghoff-Borg J, Lindorff-Larsen K. Experimental parameterization of an energy function for the simulation of unfolded proteins. Biophys J 2008;94(1):182–92.

[33] Hakim O, Sung M-H, Voss TC, Splinter E, John S, Sabo PJ, et al. Diverse gene reprogramming events occur in the same spatial clusters of distal regulatory elements. Genome Res 2011;21(5):697–706.

[34] Amano T, Sagai T, Tanabe H, Mizushina Y, Nakazawa H, Shiroishi T. Chromosomal dynamics at the Shh locus: limb bud-specific differential regulation of competence and active transcription. Dev Cell 2009;16(1):47–57.

[35] Rosa A, Everaers R. Structure and dynamics of interphase chromosomes. PLoS Comput Biol 2008;4(8):e1000153.

[36] Tiana G, Amitai A, Pollex T, Piolot T, Holcman D, Heard E, et al. Structural fluctuations of the chromatin fiber within topologically associating domains. Biophys J 2016;110(6):1234–45.

[37] Lucas JS, Zhang Y, Dudko OK, Murre C. 3D trajectories adopted by coding and regulatory DNA elements: first-passage times for genomic interactions. Cell 2014;158(2):339–52.

[38] Tiana G, Villa F, Zhan Y, Capelli R, Paissoni C, Sormanni P, et al. MonteGrappa: an iterative Monte Carlo program to optimize biomolecular potentials in simplified models. Comput Phys Commun 2015;186:93–104.

[39] Brackley CA, Brown JM, Waithe D, Babbs C, Davies J, Hughes JR, et al. Predicting the three-dimensional folding of cis-regulatory regions in mammalian genomes using bioinformatic data and polymer models. Genome Biol 2016;17:59.

[40] Kalhor R, Tjong H, Jayathilaka N, Alber F, Chen L. Genome architectures revealed by tethered chromosome conformation capture and population-based modeling. Nat Biotechnol 2012;30(1):90–8.

[41] Duan Z, Andronescu M, Schutz K, McIlwain S, Kim YJ, Lee C, et al. A three-dimensional model of the yeast genome. Nature 2010;465(7296):363–7.

[42] Metropolis N, Rosenbluth AW, Rosenbluth MN, Teller AH, Teller E. Equation of state calculations by fast computing machines. J Chem Phys 1953;21(6):1087–92.

[43] Rao SSP, Huntley MH, Durand NC, Stamenova EK, Bochkov ID, Robinson JT, et al. A 3D map of the human genome at kilobase resolution reveals principles of chromatin looping. Cell 2014;159(7):1665–80.

[44] Masui O, Bonnet I, Le Baccon P, Brito I, Pollex T, Murphy N, et al. Live-cell chromosome dynamics and outcome of X chromosome pairing events during ES cell differentiation. Cell 2011;145(3):447–58.

[45] Merkenschlager M, Nora EP. CTCF and cohesin in genome folding and transcriptional gene regulation. Annu Rev Genomics Hum Genet 2016;17:17–43.

[46] de Wit E, Vos ESM, Holwerda SJB, Valdes-Quezada C, Verstegen MJAM, Teunissen H, et al. CTCF binding polarity determines chromatin looping. Mol Cell 2015;60(4):676–84.

[47] Vietri Rudan M, Barrington C, Henderson S, Ernst C, Odom DT, Tanay A, et al. Comparative Hi-C reveals that CTCF underlies evolution of chromosomal domain architecture. Cell Rep 2015;10(8):1297–309.

[48] Sanborn AL, Rao SSP, Huang S-C, Durand NC, Huntley MH, Jewett AI, et al. Chromatin extrusion explains key features of loop and domain formation in wild-type and engineered genomes. Proc Natl Acad Sci 2015;112(47):E6456–65.

[49] Zuin J, Dixon JR, van der Reijden MIJA, Ye Z, Kolovos P, Brouwer RWW, et al. Cohesin and CTCF differentially affect chromatin architecture and gene expression in human cells. Proc Natl Acad Sci 2014;111(3):996–1001.

[50] Pant V, Kurukuti S, Pugacheva E, Shamsuddin S, Mariano P, Renkawitz R, et al. Mutation of a single CTCF target site within the H19 imprinting control region leads to loss of Igf2 imprinting and complex patterns of de novo methylation upon maternal inheritance. Mol Cell Biol 2004;24(8):3497–504.

[51] Xiao T, Wallace J, Felsenfeld G. Specific sites in the C terminus of CTCF interact with the SA2 subunit of the cohesin complex and are required for cohesin-dependent insulation activity. Mol Cell Biol 2011;31(11):2174–83.

[52] Kagey MH, Newman JJ, Bilodeau S, Zhan Y, Orlando DA, van Berkum NL, et al. Mediator and cohesin connect gene expression and chromatin architecture. Nature 2010;467(7314):430–5.

[53] Mirny LA. The fractal globule as a model of chromatin architecture in the cell. Chromosome Res 2011;19(1):37–51.

[54] Scolari VF, Lagomarsino MC. Combined collapse by bridging and self-adhesion in a prototypical polymer model inspired by the bacterial nucleoid. Soft Matter 2015;11(9):1677–87.

[55] Nazarov LI, Tamm MV, Avetisov VA, Nechaev SK. A statistical model of intra-chromosome contact maps. Soft Matter 2015;11(5):1019–25.

[56] Jost D, Carrivain P, Cavalli G, Vaillant C. Modeling epigenome folding: formation and dynamics of topologically associated chromatin domains. Nucleic Acids Res 2014;42(15):9553–61.

[57] Barbieri M, Chotalia M, Fraser J, Lavitas L-M, Dostie J, Pombo A, et al. Complexity of chromatin folding is captured by the strings and binders switch model. Proc Natl Acad Sci 2012;109(40):16173–8.

[58] Brackley CA, Taylor S, Papantonis A, Cook PR, Marenduzzo D. Nonspecific bridging-induced attraction drives clustering of DNA-binding proteins and genome organization. Proc Natl Acad Sci 2013;110(38):E3605–11.

[59] Fudenberg G, Imakaev M, Lu C, Goloborodko A, Abdennur N, Mirny LA. Formation of chromosomal domains by loop extrusion. Cell Rep 2016;15(9):2038–49.

[60] Nasmyth K. Disseminating the genome: joining, resolving, and separating sister chromatids during mitosis and meiosis. Annu Rev Genet 2001;35(1):673–745.

[61] Alipour E, Marko JF. Self-organization of domain structures by DNA-loop-extruding enzymes. Nucleic Acids Res 2012;40(22):11202–12.

[62] Naughton C, Avlonitis N, Corless S, Prendergast JG, Mati IK, Eijk PP, et al. Transcription forms and remodels supercoiling domains unfolding large-scale chromatin structures. Nat Struct Mol Biol 2013;20(3):387–95.

[63] Benedetti F, Dorier J, Burnier Y, Stasiak A. Models that include supercoiling of topological domains reproduce several known features of interphase chromosomes. Nucleic Acids Res 2014;42(5):2848–55.

[64] Franke M, Ibrahim DM, Andrey G, Schwarzer W, Heinrich V, Schöpflin R, et al. Formation of new chromatin domains determines pathogenicity of genomic duplications. Nature

[Internet] 2016 Oct 5 [cited 2016 Oct 6];advance online publication. Available from: http://www.nature.com/nature/journal/vaop/ncurrent/full/nature19800.html.

[65] Davies JOJ, Telenius JM, McGowan SJ, Roberts NA, Taylor S, Higgs DR, et al. Multiplexed analysis of chromosome conformation at vastly improved sensitivity. Nat Methods 2016;13(1):74–80.

[66] Martin P, McGovern A, Orozco G, Duffus K, Yarwood A, Schoenfelder S, et al. Capture Hi-C reveals novel candidate genes and complex long-range interactions with related autoimmune risk loci. Nat Commun 2015;6:10069.

GLOSSARY

3C (Chromosome conformation capture) a class of molecular biology techniques allowing one to detect physical interactions between genomic loci. 3C-based methods consists in cross-linking cells or tissues with formaldehyde, digesting chromatin in isolated nuclei with a restriction enzyme, and subsequently religating the resulting crosslinked restriction fragments. Decrosslinking and DNA purification allows one to detect ligation products corresponding to loci that were found to occur in spatial proximity by PCR (3C), ligation-assisted amplification followed by high-throughput sequencing (5C), or direct next generation sequencing (Hi-C).

Boltzmann ensemble according to thermodynamics, a system of interacting elements (such as a polymer) at fixed temperature fluctuates among all its possible conformations, spending in each conformation of this ensemble a fraction of time proportional to the exponential of minus its energy (this fraction is called Boltzmann probability).

Diffusion coefficient for a system which moves regularly, for example along a straight trajectory, velocity is a measure of how fast it moves. For molecules in a solvent, which move irregularly due to the random scattering with the solvent, the diffusion coefficient measures something analogous to velocity, by determining how quickly a particle spreads around through the square of the walked distance over time.

DNA FISH (fluorescence in situ hybridization) cytogenetic technique in which fluorescent DNA probes are hybridized to chromosomes in fixed and permeabilized cells following slight denaturation of the chromatin fiber, allowing to determine the three-dimensional positions of target loci using fluorescence microscopy.

Mean-field approximation it consists in approximating the average of a complex function of the conformations of a system (e.g., their contact probability) with the value of the function evaluated in a single, representative conformation.

TAD (Topologically associating domain) submegabase region of a chromosome showing preferential physical interactions with respect to neighboring regions. Boundaries of TADs have been shown to be largely conserved during evolution and differentiation. TADs are thought to constrain enhancer action by increasing interaction frequencies with promoters within one genomic domain, and preventing deleterious interaction with neighboring regions.

LIST OF ACRONYMS AND ABBREVIATIONS

3C chromosome conformation capture
5C chromosome conformation capture carbon-copy
CTCF CCCTC-binding factor
TAD topologically associating domain

CHAPTER 11

A Combination Approach Based on Live-Cell Imaging and Computational Modeling to Further Our Understanding of Chromatin, Epigenetics, and the Genome

Kayo Hibino[1,2], Kazunari Kaizu[3], Koichi Takahashi[3] and Kazuhiro Maeshima[1,2,*]

[1]*National Institute of Genetics, Mishima, Shizuoka, Japan*

[2]*Sokendai (Graduate University for Advanced Studies), Mishima, Shizuoka, Japan*

[3]*RIKEN Quantitative Biology Center, Suita, Osaka, Japan*

CONTENTS

INTRODUCTION: GENOMIC DNA, NUCLEOSOMES, AND CHROMATIN

How is 2-m genomic DNA organized three-dimensionally within a cell, and how does it behave within living cells? Inside cells, DNA is wrapped around a core histone octamer that consists of the histone proteins H2A, H2B, H3, and H4 and forms a nucleosome [1,2]. Each nucleosome particle is held together by linker DNA (~20–80 bp; ~6.6–27 nm), which makes up repetitive motifs of ~200 bp and is also called a "10-nm fiber" [1–3] (Fig. 11.1). The 10-nm fiber associated with the various nonhistone proteins is traditionally described as "chromatin."

The 10-nm fiber has long been assumed to be folded into regular 30-nm chromatin fibers [4–8] and further into hierarchical helical structures [9,10] (Fig. 11.1). However, recent evidence from electron microscopy, X-ray scattering, chromosome conformation capture, and super-resolution imaging suggests that such regular fibers do not generally exist in cells [11–21], and that 10-nm fibers are instead folded irregularly [4] (Fig. 11.1). This irregularly folded structure implies a physically dynamic state of chromatin. This state may play a role in the regulation of RNA transcription and DNA replication, which is also affected by epigenetic modifications on chromatin.

To investigate local chromatin structure and dynamics, our group has developed a novel system for single nucleosome imaging in a whole-cell nucleus using

*Corresponding address: kmaeshim@nig.ac.jp or kayo_hibino@nig.ac.jp

Epigenetics and Systems Biology. DOI: http://dx.doi.org/10.1016/B978-0-12-803075-2.00011-8

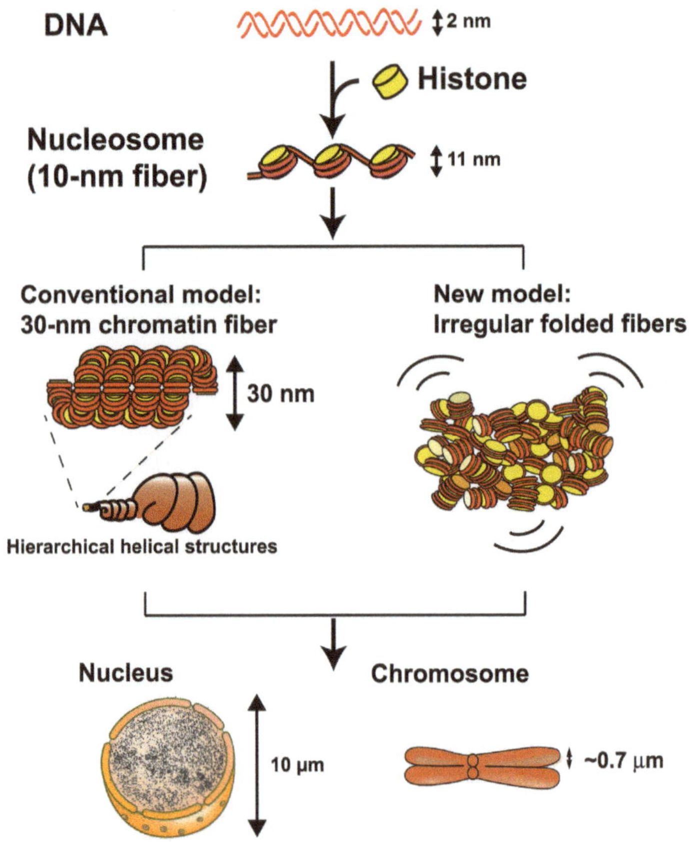

FIGURE 11.1 **Genomic DNA, histones, the nucleosome, and chromatin.**
A long DNA (deoxyribonucleic acid) molecule (red) is wrapped around basic core histone octamers (yellow) to form a nucleosome fiber or 10-nm fiber and further organized in a nucleus or chromosome. The 10-nm fiber has long been assumed to be folded into regular 30-nm chromatin fibers and further into hierarchical helical structures (conventional model). However, recent evidence suggests that such regular fibers do not generally exist in cells, and that the 10-nm fibers are folded irregularly (new model). *The illustrations were modified from Maeshima K, Ide S, Hibino K, Sasai M. Liquid-like behavior of chromatin. Curr Opin Genet Dev 2016;37:36–45.*

super-resolution microscopy (photoactivated localization microscopy) [22]. Combining this approach with computational modeling (e.g., the reconstruction of the chromatin environment *in silico*) makes it even more powerful, as modeling allows us to predict the behavior of the proteins under conditions that are either difficult to observe directly or difficult to generate experimentally. In turn, the imaging method provides the information needed to reconstruct the chromatin environment *in silico*. Thus, imaging and computational modeling are complementary techniques for investigating local chromatin structure and dynamics.

EXPERIMENTAL TECHNIQUES: SINGLE-NUCLEOSOME IMAGING IN LIVING CELLS USING SUPER-RESOLUTION MICROSCOPY

To perform single-nucleosome imaging in living mammalian cells, we needed to fluorescently label a small number of nucleosomes among the ~3×10^7 nucleosomes in a single nucleus [23]. For this purpose, photoactivatable-Green fluorescent protein (PA-GFP) [24,25] fused with histone H4 was expressed in HeLa cells at a very low level (2–8 × 10^5 molecules in a single nucleus) [23,26]. Usually PA-GFP has almost no fluorescence when excited by 488-nm light, and fluorescence increases 100-fold upon ~400-nm laser activation. Unexpectedly, we found that a small amount of PA-GFP-H4 was activated spontaneously without the laser activation (Fig. 11.2A) [23,26]. For nucleosome imaging, we used oblique illumination (or HILO) microscopy [26,27]. This system can illuminate a limited thin layer inside a nucleus, allowing single-nucleosome images to be obtained with low background (Fig. 11.2B). Using the oblique illumination microscopy, spontaneously activated PA-GFP-H4 nucleosomes in the stable cell lines were observed as bright dots (Fig. 11.2C). The number of nucleosome signals in the each image frame was ~100 so that the individual signals did not typically overlap. During ~1000 frames acquisition, ~100,000 of nucleosome signals were detected. This observed number was sufficient to investigate trends of local chromatin structure and dynamics.

We recorded the nucleosome signals in living cells as movies at video rate (30 frames^{-1}). The precise center of individual nucleosome signals in an image frame was determined by fitting them to an assumed Gaussian point spread function, with typically less than 10-nm optical precision [28]. After determining the precise position of each nucleosome in every frame, we tracked the positions of identical molecules frame by frame (Fig. 11.2D). Surprisingly, we found a large degree of movement or fluctuation among the nucleosomes: the average displacement (movement) during a period of 30 milliseconds (ms) was about 60 nm, which is approximately sixfold larger than the diameter of the nucleosome (Fig. 11.2E). Because the displacements of purified enhanced GFP (EGFP) molecules immobilized on the glass surface or the cross-linked nucleosomes in the glutaraldehyde-fixed cells were much smaller than those observed in living cells, the results indicated that the majority of the displacement resulted from the movement of nucleosomes in living cells and not from instability of the microscope system, which could induce apparent movements of the samples fixed on the glass surface [23,26].

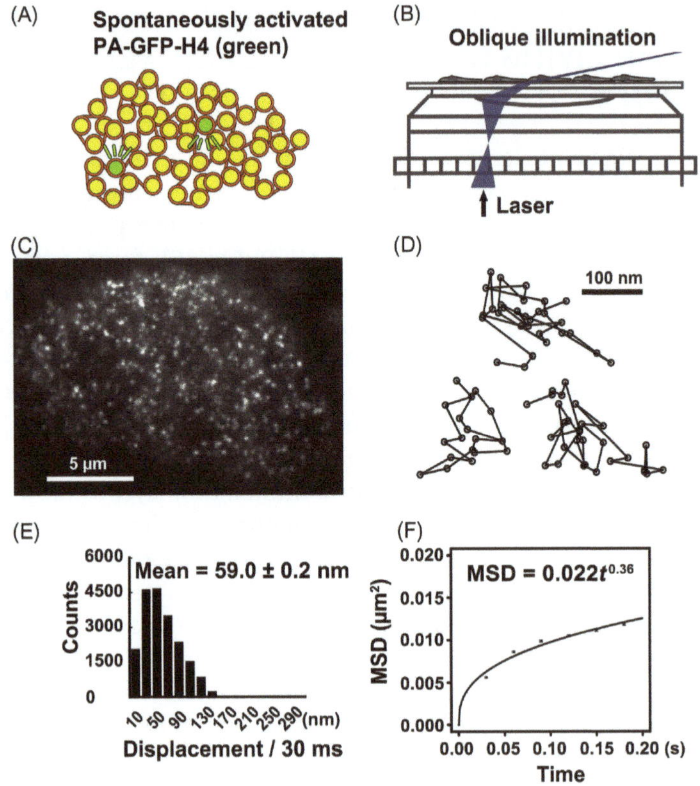

FIGURE 11.2 Single nucleosome imaging.

(A) To perform single nucleosome imaging in living cells, a small number of histone H4 proteins were fluorescently labeled using photoactivatable GFP (PA-GFP-H4). (B) Oblique illumination microscopy (HILO system [27]) was carried out using a Nikon laser TIRF microscope Ti with a Sapphire 488-nm laser [26]. (C) Single molecules of PA-GFP-H4 were observed as bright dots using oblique illumination microscopy. A free software, u-track, was used for single nucleosome tracking. Dots were fitted to an assumed Gaussian point spread function to determine the precise centers of signals at a high resolution [28]. (D) Representative trajectories of the tracked single nucleosomes. (E) Displacement (movement) histograms of single nucleosomes during a period of 30 ms ($n = 35{,}000$ in seven cells). (F) Plots of the mean-square displacements (MSDs) of single nucleosomes from 0 to 0.18 s in interphase chromatin. The plots were fitted as an anomalous diffusion curve. *The data in panels (C–F) were reproduced from Nozaki T, Kaizu K, Pack CG, Tamura S, Tani T, Hihara S, et al. Flexible and dynamic nucleosome fiber in living mammalian cells. Nucleus 2013;4(5):349–56.*

To analyze local nucleosome movement in more detail, we plotted the mean square displacement (MSD) values of the nucleosomes in interphase chromatin (Fig. 11.2F). The MSD plot fitted well with the anomalous diffusion model: $\text{MSD} = 0.022t^{0.36}$. This result supported the restricted nucleosome movement model in which nucleosome movement is restricted to within a certain range, in a manner analogous to a "dog on a leash."

MODELING: MONTE CARLO SIMULATION OF CHROMATIN DYNAMICS

What is the biological relevance of such local nucleosome movement in living cells? How can the movement contribute to the regulation of RNA transcription or DNA replication? To address these questions, it is essential to compare the chromatin environment with and without nucleosome fluctuation. It is, however, difficult to modulate the degree of nucleosome fluctuation experimentally. In addition, the number of nucleosomes in a single nucleus is too great ($\sim 10^7$ nucleosomes/pL_{-1}) to simultaneously observe individual nucleosomes along with the behavior of the proteins of interest. Computational modeling can predict the behavior of proteins under conditions that are either difficult to observe directly by imaging or difficult to generate experimentally. Here, we used the Metropolis Monte Carlo method [29,30], which can reproduce movements of objects such as nucleosomes by making a random movement. In particular, we carried out behavioral simulations of individual diffusing proteins under various nucleosome concentrations with and without their fluctuation (Fig. 11.3). We also discuss the biological relevance of nucleosome fluctuation in the "scanning and targeting genomic DNA" process.

For the *in silico* reconstruction of the chromatin environment, we used the physical parameters obtained from single-nucleosome imaging analyses and other available data [23,26]. All molecules, including nucleosomes, were represented as hard spherical bodies. Their diffusive motions were calculated without long-range potential interactions. Spherical particles of 10-nm diameter, which mimic nucleosomes, were placed in the simulation at concentrations from 0.1 to 1 mM (red spheres in Fig. 11.3A and B). A concentration higher than 0.5 mM likely corresponds to dense regions or heterochromatin in the cell. The spheres of 13-nm hydrodynamic diameter (green spheres) represented a model for diffusing protein in the environment and moved according to a random walk with a predefined diffusion coefficient (Fig. 11.3B) [23]. In the environment with the immobilized nucleosomes (red spheres) at a concentration of 0.1 mM corresponding to less dense chromatin regions, the diffusing proteins (green spheres) moved around freely (Fig. 11.4A, left). However, with the fixed 10-nm nucleosomes at a concentration of 0.5 mM, which corresponds to dense or heterochromatin environments, the diffusing proteins were confined to a small area around their starting positions (Fig. 11.4A, center).

We next constructed a model with fluctuating nucleosomes. In this model, nucleosomes fluctuate, but their movements are restricted to a certain range. In this dynamic environment, apparently free diffusion of the model proteins (green spheres) was observed with 0.5-mM nucleosomes (red spheres)

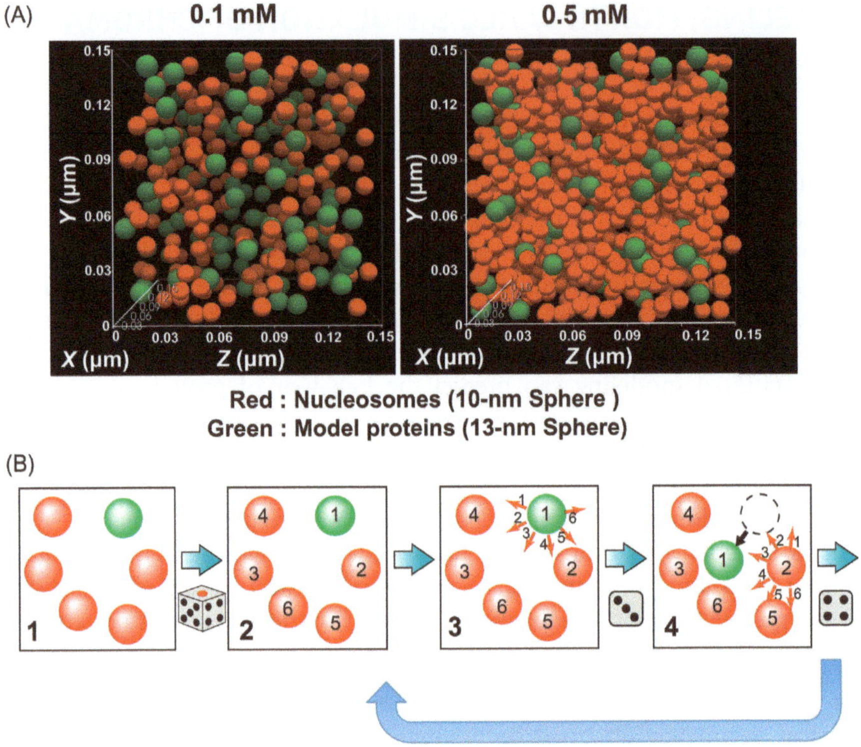

FIGURE 11.3 **Reconstruction of the chromatin environment *in silico* [23,26].**
(A) The nucleosome is represented as a 10-nm red sphere and fixed randomly in a restricted space, but in a manner that avoids any overlap, at a concentration of 0.1 mM (left) or 0.5 mM (right, corresponding to condensed chromatin domain). A model protein is represented as a 13-nm sphere (green). (B) A simple scheme of the simulation procedure: We first decided the order of the spheres to be moved at random (Steps 1 and 2). The first sphere was then moved randomly (Step 3). Its displacement followed the three-dimensional normal distribution with the standard deviation ($SD^2 = 6Ddt$). We turned to the next sphere (Step 4) and repeated this process a number of times. Finally, we traced the movement of the model protein. *Illustrations reproduced from Hihara S, Pack CG, Kaizu K, Tani T, Hanafusa T, Nozaki T, et al. Local nucleosome dynamics facilitate chromatin accessibility in living mammalian cells. Cell Rep 2012;2(6):1645–56 and Nozaki T, Kaizu K, Pack CG, Tamura S, Tani T, Hihara S, et al. Flexible and dynamic nucleosome fiber in living mammalian cells. Nucleus 2013;4(5):349–56.*

(Fig. 11.4A, right). This result suggests that in a dynamic chromatin environment proteins can move freely, even with 0.5-mM nucleosomes. In addition to the 13-nm diameter model proteins (green spheres), we also showed that a model protein of 21-nm diameter, which is twice as large as the nucleosome, can move freely in 0.5-mM fluctuating nucleosomes [23]. Strikingly, a maximum displacement (or fluctuation) of 10–20 nm for the nucleosomes was sufficient for the model proteins to diffuse freely in the crowded environment (Fig. 11.4B). Furthermore, we also found that the local movement was

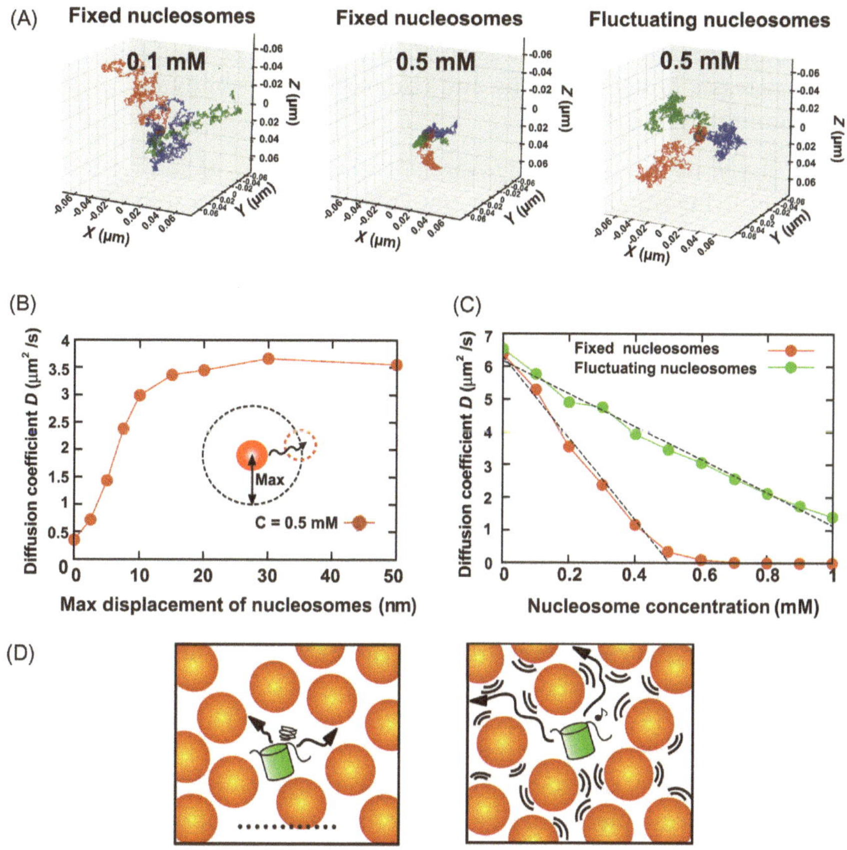

FIGURE 11.4 Monte Carlo computer simulation results.

(A) Tracing patterns of three model proteins (13-nm spheres) under various conditions. At 0.1-mM fixed nucleosomes (10-nm spheres), the model proteins (13-nm spheres) move around freely (left). However, at 0.5-mM fixed nucleosomes (10-nm spheres), the model proteins (13-nm spheres) are unable to move far from their starting points (center). In the environment with fluctuation of 0.5-mM nucleosomes, model proteins can move around freely (right). Each nucleosome behaves like "a dog on a leash." The leash length is 20 nm. The three different temporal trajectories of the model proteins for 0.2 ms are indicated in blue, green, and red. (B) Terminal diffusion coefficients of the model proteins with 0.5-mM nucleosomes and various "dog leash" lengths (maximum displacement of nucleosomes). Note that a 10–20-nm displacement of the nucleosomes allows the model proteins to diffuse quite freely. (C) Terminal diffusion coefficients of the model proteins at various nucleosome concentrations, which are fixed (red) or fluctuating (green). Note that, even at the relatively low nucleosome concentration range of 0.1–0.4 mM, local fluctuation facilitated movement of the model proteins. The "dog leash" (maximum 10-nm sphere displacement) length is 20 nm. (D) Cartoons show that a protein (green) is stacked in a confined space of fixed nucleosomes (left), and the protein is able to move freely with nucleosome fluctuation (right).
Data and illustrations reproduced from Hihara S, Pack CG, Kaizu K, Tani T, Hanafusa T, Nozaki T, et al. Local nucleosome dynamics facilitate chromatin accessibility in living mammalian cells. Cell Rep 2012;2(6):1645–56 and Nozaki T, Kaizu K, Pack CG, Tamura S, Tani T, Hihara S, et al. Flexible and dynamic nucleosome fiber in living mammalian cells. Nucleus 2013;4(5):349–56.

effective even at nucleosome concentrations lower than 0.5 mM (Fig. 11.4C). These findings suggest that local fluctuation (10–20 nm) of nucleosomes drastically increases protein mobility and chromatin accessibility in the chromatin environment (Fig. 11.4D).

In this study, we uncovered the novel, local dynamics of individual nucleosomes in living human cells. Our results show that nucleosome fluctuation facilitates chromatin accessibility, which is advantageous for "scanning and targeting genomic DNA" reactions in various processes, including RNA transcription and DNA replication [3,23,26].

ANOTHER RESULT FROM THE MODEL: PHYSICAL SIZE OF TRANSCRIPTION COMPLEXES

A similar modeling approach provided a clue to another interesting question regarding transcriptional regulation: why are transcription complexes so large? For instance, gene-specific transcription factors that scan the genome and bind regulatory elements are relatively small (~50 kDa; ~5 nm diameter). Mediators, RNA polymerases, nucleosome remodelers, and histone modifiers are large protein complexes 1–3 MDa in size (~15–25 nm diameter) (Fig. 11.5A) [31]. As recent findings suggest that chromatin forms numerous condensed domains (e.g., topologically associating domains) (Fig. 11.6A) [32–34], and the size exclusion effect of such chromatin domains may limit the accessibility of DNA sequences to transcription complexes, the physical size of transcription complexes may be significant for transcriptional regulation in chromatin domains [31].

First, we evaluated the physical size limit of transcription factors that can enter condensed chromatin domain using the Monte Carlo method. In the simulation, transcription factors and nucleosomes were represented as diffusing spherical particles with diameters of 5–30 nm and mobile spherical particles of 10-nm hydrodynamic diameter. The nucleosomes were placed in the right half of the simulation space ("chromatin domain") at a concentration of 0.5 mM (Fig. 11.5B). In this model, nucleosomes are mobile, but their movements are restricted.

Fig. 11.5C shows the fraction of each sphere type localized in the chromatin domain as a function of time. After ~1 ms (arrow in Fig. 11.5C), ~40% of the 5-nm spheres moved around within the chromatin domain. The 10-nm spheres had a similar tendency with much lower (~15%) efficiency. However, we rarely found spheres >15 nm in size (15–30-nm spheres) in the chromatin domain. Rather, these larger spheres were confined outside the chromatin domains. These results indicate that chromatin domains have a potential barrier for spheres with diameters >15 nm (Fig. 11.5B and C). We also found that

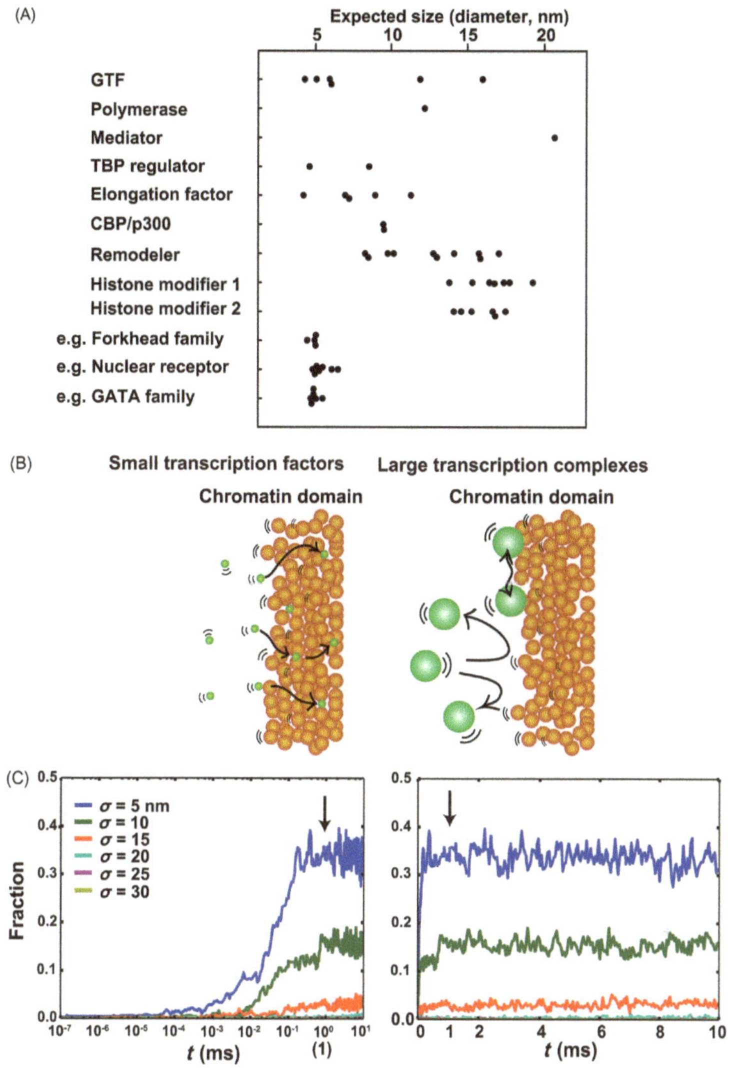

FIGURE 11.5 Expected size of transcription factors and complexes and simulation scheme.
(A) Expected diameters of the transcription factors and complexes were calculated on the basis of available sequence data and formula and shown as a scatter plot. Note that, while gene-specific transcription factors such as the Forkhead family, nuclear receptors, and GATA family are relatively small (~50 kDa; ~5 nm diameter), Mediators, RNA polymerases, nucleosome remodelers, and histone modifiers are large (~15–25 nm diameter). Histone modifier 1 and 2 are histone methylation and acetylation enzymes, respectively. (B) Simulation scheme to study the behavior of transcription factors in the chromatin domain. Simulation volume is ~215 × ~215 × ~215 nm. The transcription factors and nucleosomes are represented as diffusing spherical particles with diameters of 5 nm (green spheres in left) or 30 nm (green spheres in right) and spherical particles of 10-nm diameter (yellow spheres), respectively. The mobile nucleosomes were placed in the right half of the simulation space ("chromatin domain") at a concentration of 0.5 mM. Spheres of various sizes were placed and moved around freely in the left half of the space, which is free from nucleosomes. Note that the small green spheres can move in the chromatin domain, but the large green spheres cannot. (C) Fractions of spheres of various sizes localized in the mobile chromatin domain. Time series (log-scale at left; linear-scale at right) of the fraction of each sphere type within the chromatin domain (nucleosomes at 0.5 mM; fluctuation size is 20 nm). For each sphere type (5, 10, 15, 20, 25, and 30-nm diameters), 50 spheres were initially placed in the nonchromatin domain, avoiding overlap [left half of the space in (B)]. The initial fraction was 0. Each plot was averaged over 10 independent trials. The plots show that equilibrium is reached after 1 ms (arrow). Because of the exclusion volume of nucleosomes, maximum fraction values varied depending on sphere size. The maximum fraction value is ~0.4 for the 5-nm spheres. *Data and illustrations reproduced from Maeshima K, Kaizu K, Tamura S, Nozaki T, Kokubo T, Takahashi K. The physical size of transcription factors is key to transcriptional regulation in the chromatin domains. J Phys: Condens Matters 2015;27:064116. (10 pp) with modifications.*

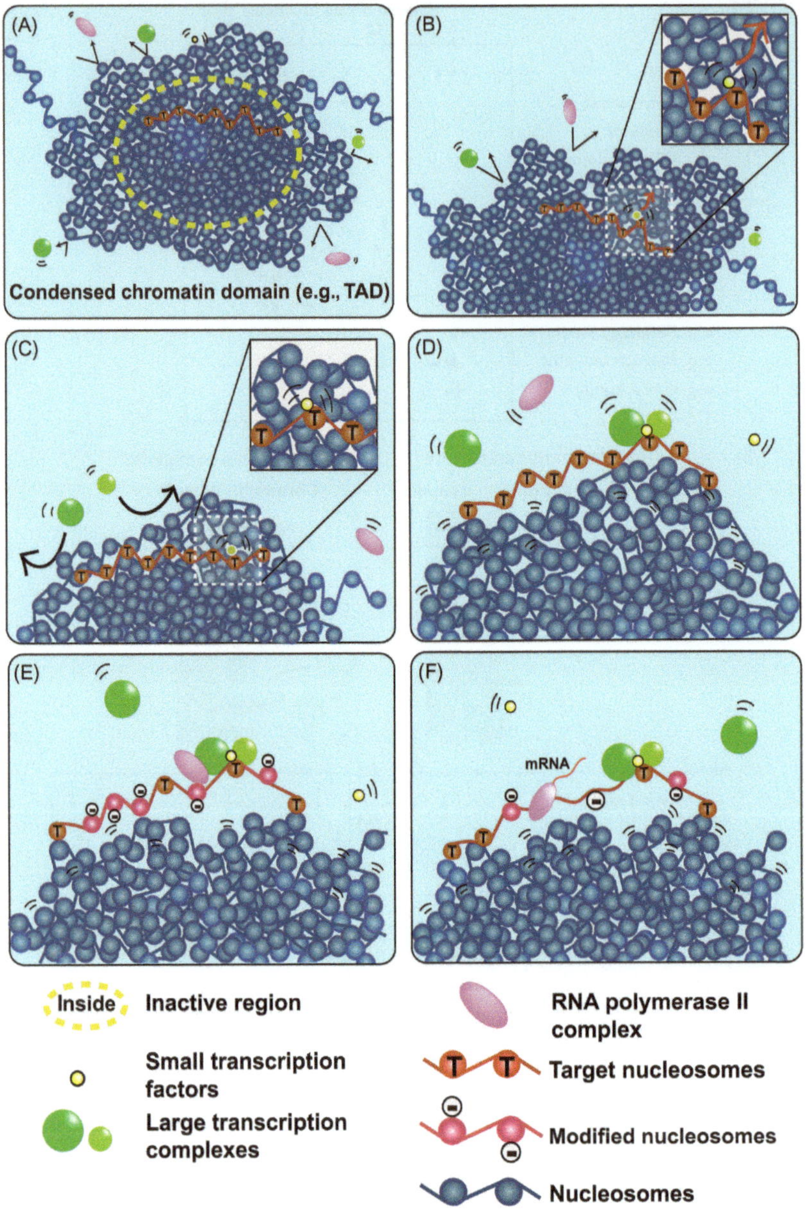

FIGURE 11.6 The "buoy" model for transcriptional regulation [31].
(A) Condensed chromatin domains (e.g., TADs) are shown as blue nucleosomes. Transcription is allowed only on the domain surfaces and is prevented inside the chromatin domains (inside the region marked with a yellow line). (B) and (C) Small transcription factors (yellow) can move in the condensed chromatin domain, but the large transcription complexes (green) cannot. The regions marked with white squares are enlarged in the insets. The small transcription factor bound to the target region (red nucleosomes) can recruit large transcription complexes when it is relocated to the domain surface. (D) Binding of large transcription complexes (green) keeps the transcriptional region (red nucleosomes) on the domain surface like a "buoy." In the absence of binding of large complexes and lacking the buoy function, the regions will go back into the domains, owing to their highly dynamic chromatin nature, and become transcriptionally inactive. (E) and (F) Histone acetylation by histone modifiers (modified nucleosomes) or nucleosome eviction by nucleosome remodelers increases the negative charge of the regions (marked with "–"), and regions are then isolated from the domain surface by repulsion force. The repulsion force further stabilizes active regions. *Illustrations reproduced from Maeshima K, Kaizu K, Tamura S, Nozaki T, Kokubo T, Takahashi K. The physical size of transcription factors is key to transcriptional regulation in the chromatin domains. J Phys: Condens Matters 2015;27:064116. (10 pp) with modifications.*

this potential barrier was effective not only at a highly crowded 0.5-mM concentration of nucleosomes, but also under the 0.3-mM concentration, which is a relatively open chromatin state [31]. Note that, as mentioned above (Fig. 11.4), once proteins enter the condensed domains, they can move around with the help of nucleosome fluctuation. However, the proteins outside the domains tend to remain there because they have more space. We assume that this tendency generates a barrier to the chromatin domain (Fig. 11.5B).

Based on these results, as well as available size information and knowledge on the transcription factors, we presented a "buoy" model for transcriptional regulation in the chromatin domains (Fig. 11.6). Once a large transcription complex binds its target site on the condensed chromatin domain, it can hold the target region on the domain surface like a buoy and keep the region transcriptionally active. This situation seems like the "Brazil nut effect," where the largest particles rise to the surface of a granular material containing a mixture of particles that vary in size. Further details of this model follow (see also Ref. [31]).

Small gene-specific transcription factors such as Forkhead family factors, nuclear receptors, and GATA family factors (Fig. 11.5A), can dive into chromatin domains and scan the genome for their target sequences (Fig. 11.6B). When the target gene has a chance of being exposed on the domain surface (Fig. 11.6C), these factors can then recruit large protein complexes such as Mediators, nucleosome remodelers, histone modifiers, and large RNA polymerases. This step depends largely on local nucleosome movement (fluctuation) within the domains, presumably driven by Brownian motion. Owing to their large size, such complexes prevent the site from moving back to the inner region of chromatin domains; they act as a buoy to keep the bound regions at the domain surface (Fig. 11.6D–F). Indeed, transcription seems to occur outside the chromatin domains [35–37]. We propose that buoyancy is provided by factors (or complexes) that mediate the transcription process itself and that this buoyancy mechanism may maintain transcriptional competency and facilitate transcription (Fig. 11.6D–F) [31]. This size-dependent specialization of target-scanning and surface-tethering functions could provide novel insight into the mechanisms of other DNA processes, such as replication and repair/recombination.

PERSPECTIVE

We developed a single-nucleosome imaging system combined with computational modeling. This system can directly reveal chromatin dynamics at the single nucleosome level, as well as shed light upon their biological relevance. Recently we showed that such dynamics can provide a structural information on how the chromatin is organized in living cells [38]. One intriguing avenue for further research is the spatial and temporal regulation of chromatin dynamics: what is the difference between euchromatin and heterochromatin in terms of dynamics? What kinds of protein factors are involved in the regulatory process? How can various epigenetic modifications, such as histone acetylation/methylation and DNA methylation, spatially and temporally control chromatin dynamics in living cells? How can such modifications give rise to functional outputs of genome information? The novel system described here provides the means for investigating these important questions.

Our system also allows cellular structures to be observed at a sufficiently high resolution to visualize chromatin domains in living cells. Super-resolution live-cell imaging with dual-color labeling of chromatin domains and specific transcription factors/complexes is possible using this system. This would permit us to verify where the domains are located within nuclei, how transcription factors behave inside and around the domains, and the validity of our proposed buoy model.

References

[1] Luger K, Mader AW, Richmond RK, Sargent DF, Richmond TJ. Crystal structure of the nucleosome core particle at 2.8 Å resolution. Nature 1997;389(6648):251–60.

[2] Olins DE, Olins AL. Chromatin history: our view from the bridge. Nat Rev Mol Cell Biol 2003;4(10):809–14.

[3] Maeshima K, Imai R, Tamura S, Nozaki T. Chromatin as dynamic 10-nm fibers. Chromosoma 2014;123:225–37.

[4] Maeshima K, Ide S, Hibino K, Sasai M. Liquid-like behavior of chromatin. Curr Opin Genet Dev 2016;37:36–45.

[5] Finch JT, Klug A. Solenoidal model for superstructure in chromatin. Proc Natl Acad Sci U S A 1976;73(6):1897–901.

[6] Woodcock CL, Frado LL, Rattner JB. The higher-order structure of chromatin: evidence for a helical ribbon arrangement. J Cell Biol 1984;99(1 Pt 1):42–52.

[7] Schalch T, Duda S, Sargent DF, Richmond TJ. X-ray structure of a tetranucleosome and its implications for the chromatin fibre. Nature 2005;436(7047):138–41.

[8] Robinson PJ, Fairall L, Huynh VA, Rhodes D. EM measurements define the dimensions of the "30-nm" chromatin fiber: evidence for a compact, interdigitated structure. Proc Natl Acad Sci U S A 2006;103(17):6506–11.

[9] Sedat J, Manuelidis L. A direct approach to the structure of eukaryotic chromosomes. Cold Spring Harb Symp Quant Biol 1978;42(Pt 1):331–50.

[10] Belmont AS, Sedat JW, Agard DA. Three-dimensional approach to mitotic chromosome structure: evidence for a complex hierarchical organization. J Cell Biol 1987;105(1):77–92.

[11] McDowall AW, Smith JM, Dubochet J. Cryo-electron microscopy of vitrified chromosomes in situ. Embo J 1986;5(6):1395–402.

[12] Eltsov M, Maclellan KM, Maeshima K, Frangakis AS, Dubochet J. Analysis of cryo-electron microscopy images does not support the existence of 30-nm chromatin fibers in mitotic chromosomes in situ. Proc Natl Acad Sci U S A 2008;105(50):19732–7.

[13] Nishino Y, Eltsov M, Joti Y, Ito K, Takata H, Takahashi Y, et al. Human mitotic chromosomes consist predominantly of irregularly folded nucleosome fibres without a 30-nm chromatin structure. EMBO J 2012;31(7):1644–53.

[14] Joti Y, Hikima T, Nishino Y, Kamada F, Hihara S, Takata H, et al. Chromosomes without a 30-nm chromatin fiber. Nucleus 2012;3:404–10.

[15] Fussner E, Strauss M, Djuric U, Li R, Ahmed K, Hart M, et al. Open and closed domains in the mouse genome are configured as 10-nm chromatin fibres. EMBO Rep 2012;13(11):992–6.

[16] Chen C, Hwa Lim H, Shi J, Tamura S, Maeshima K, Surana U, et al. Budding yeast chromatin is dispersed in a crowded nucleoplasm. Mol Biol Cell 2016;27(21):3357–68.

[17] Eltsov M, Sosnovski S, Olins AL, Olins DE. ELCS in ice: cryo-electron microscopy of nuclear envelope-limited chromatin sheets. Chromosoma 2014;123(3):303–12.

[18] Hsieh TH, Weiner A, Lajoie B, Dekker J, Friedman N, Rando OJ. Mapping nucleosome resolution chromosome folding in yeast by micro-C. Cell 2015;162(1):108–19.

[19] Ricci MA, Manzo C, Garcia-Parajo MF, Lakadamyali M, Cosma MP. Chromatin fibers are formed by heterogeneous groups of nucleosomes in vivo. Cell 2015;160(6):1145–58.

[20] Sanborn AL, Rao SS, Huang SC, Durand NC, Huntley MH, Jewett AI, et al. Chromatin extrusion explains key features of loop and domain formation in wild-type and engineered genomes. Proc Natl Acad Sci U S A 2015;112(47):E6456–65.

[21] Maeshima K, Rogge R, Tamura S, Joti Y, Hikima T, Szerlong H, et al. Nucleosomal arrays self-assemble into supramolecular globular structures lacking 30-nm fibers. EMBO J 2016;35:1115–32.

[22] Betzig E, Patterson GH, Sougrat R, Lindwasser OW, Olenych S, Bonifacino JS, et al. Imaging intracellular fluorescent proteins at nanometer resolution. Science (New York, NY) 2006;313(5793):1642–5.

[23] Hihara S, Pack CG, Kaizu K, Tani T, Hanafusa T, Nozaki T, et al. Local nucleosome dynamics facilitate chromatin accessibility in living mammalian cells. Cell Rep 2012;2(6):1645–56.

[24] Patterson GH, Lippincott-Schwartz J. A photoactivatable GFP for selective photolabeling of proteins and cells. Science (New York, NY) 2002;297(5588):1873–7.

[25] Wiesmeijer K, Krouwels IM, Tanke HJ, Dirks RW. Chromatin movement visualized with photoactivable GFP-labeled histone H4. Differentiation 2008;76(1):83–90.

[26] Nozaki T, Kaizu K, Pack CG, Tamura S, Tani T, Hihara S, et al. Flexible and dynamic nucleosome fiber in living mammalian cells. Nucleus 2013;4(5):349–56.

[27] Tokunaga M, Imamoto N, Sakata-Sogawa K. Highly inclined thin illumination enables clear single-molecule imaging in cells. Nat Methods 2008;5(2):159–61.

[28] Lippincott-Schwartz J, Patterson GH. Photoactivatable fluorescent proteins for diffraction-limited and super-resolution imaging. Trends Cell Biol 2009;19(11):555–65.

[29] Metropolis N, Rosenbluth AW, Rosenbluth MN, Teller AH, Teller E. Equation of state calculations by fast computing machines. J Chem Phys 1953;21(6):1087. (6 p).

[30] Morelli MJ, ten Wolde PR. Reaction Brownian dynamics and the effect of spatial fluctuations on the gain of a push-pull network. J Chem Phys 2008;129(5):054112.

[31] Maeshima K, Kaizu K, Tamura S, Nozaki T, Kokubo T, Takahashi K. The physical size of transcription factors is key to transcriptional regulation in the chromatin domains. J Phys: Condens Matters 2015;27:064116. (10 pp).

[32] Dekker J, Marti-Renom MA, Mirny LA. Exploring the three-dimensional organization of genomes: interpreting chromatin interaction data. Nat Rev Genet 2013;14(6):390–403.

[33] Smallwood A, Ren B. Genome organization and long-range regulation of gene expression by enhancers. Curr Opin Cell Biol 2013;25(3):387–94.

[34] Dekker J, Heard E. Structural and functional diversity of Topologically Associating Domains. FEBS Lett 2015;589(20 Pt A):2877–84.

[35] Fakan S, van Driel R. The perichromatin region: a functional compartment in the nucleus that determines large-scale chromatin folding. Semin Cell Dev Biol 2007;18(5):676–81.

[36] Markaki Y, Gunkel M, Schermelleh L, Beichmanis S, Neumann J, Heidemann M, et al. Functional nuclear organization of transcription and DNA replication: a topographical marriage between chromatin domains and the interchromatin compartment. Cold Spring Harb Symp Quant Biol 2010;75:475–92.

[37] Niedojadlo J, Perret-Vivancos C, Kalland KH, Cmarko D, Cremer T, van Driel R, et al. Transcribed DNA is preferentially located in the perichromatin region of mammalian cell nuclei. Exp Cell Res 2011;317(4):433–44.

[38] Shinkai S, Nozaki T, Maeshima K, Togashi Y. Dynamic nucleosome movement tells Structural information of topological chromatin domains in human cells. PLOS Comput Biol 2016;12:e1005136.

GLOSSARY

Diffusion coefficient proportionality constant to the squared velocity of the diffusing particles or polymer, which depends on the temperature, viscosity of the fluid, and their size according to the Stokes–Einstein relation.

DNA transactions genomic DNA-related cellular functions, such as RNA transcription, DNA replication, and DNA repair/recombination.

Heterochromatin chromatin regions which are generally more tightly packed and have less active genes. By contrast, euchromatin is more open and transcriptionally active regions.

Hydrodynamic diameter the diameter calculated from the diffusional properties of the particle or polymer based on the Stokes–Einstein relation.

Mean square displacement (MSD) a measure of the deviation over time between the position of a particle and some reference position, and a common measure of the spatial extent of random motion.

Metropolis Monte Carlo method computational algorithm that performs a numerical integration by making a random movement and evaluating whether the movement is acceptable based on the change in potential energy.

Super-resolution microscopy a type of light microscopy which can collect images with a resolution higher than the diffraction limit (~200 nm).

Transcription factors proteins that bind to specific DNA sequences and can control the on-off status of certain genes; that is, the condition in which genetic information is transcribed from DNA to RNA.

LIST OF ACRONYMS AND ABBREVIATIONS

DNA Deoxyribonucleic acid

PA-GFP Photoactivatable-Green fluorescent protein

EGFP Enhanced GFP

MSD Mean-square displacement

CHAPTER 12

Capturing Chromosome Structural Properties From Their Spatial and Temporal Fluctuations

Aurélien Bancaud, Olivier Gadal and Kerstin Bystricky
LAAS, CNRS, University of Toulouse, Toulouse, France

CONTENTS

INTRODUCTION

Twenty years ago, the first whole genome, the DNA sequence of the baker's yeast *Saccharomyces cerevisiae*, was sequenced [1]. Many others followed, from bacteria to vertebrates and plants, whose genomes comprise between one million to several dozens of billion nucleotides. These genomes were unveiled close to completeness opening the door to pan-genomic comparisons and, although debated, personalized medical diagnostics [2–6]. This one-dimensional (1D) information is precious, but understanding the intricacies of genome function requires exploring additional levels. One of them, the Epigenome, a refined "dress-code" of the DNA fiber in eucaryotes, brought us one step closer to comprehending how nuclear functions are regulated. How the epigenome itself is organized and operational requires of us to place it into a 3D context. Spatio-temporal dynamics of the genome, the *4D Nucleome*, have been shown to respond to endogenous and exogenous cues: physiological (a striking example is that in photoreceptor rod cells heterochromatin zones are inverted to accommodate nocturnal vision [7]), hormonal (hormone stimulation alters general chromatin compaction in human cancer cells [8]) or metabolic (yeast nuclear size and gene expression reorganizes when changing the carbon source [9]) to cite just a few examples. 3D position or chromatin fiber conformation further influences cellular functions, notably gene expression, recombination and repair are less efficient near the nuclear periphery where compaction levels are presumably greater than elsewhere in the nucleus [10–13]. Transcriptional silencing is thought to largely depend on formation of denser structures as seen for entire genomes (spermatozoa or erythrocytes [14,15]), whole chromosomes (the Barr body representing X-inactivation in mammals [16]), chromosome domains (chromocenters in mice and plants [17,18]), or specific loci (mating type loci in yeast [19,20] or imprinted loci in mammals [21,22]). Numerous examples

Epigenetics and Systems Biology. DOI: http://dx.doi.org/10.1016/B978-0-12-803075-2.00012-X

exist where changes in 4D nucleome features have been correlated with dedifferentiation, genome instability, and disease [23–26].

Therefore, elucidation of physical mechanisms governing chromosome folding and their interplay with genome functions are the subject of intense research [27]. High resolution chromosome conformation capture (3C)-based methods (see Ref. [28] for a recent review), which become more and more popular for exploring 3D organization of genomes, have been instrumental in determining the folding of chromosomes at length scales spanning 10 kb up to Gb. They provide contact frequency maps that are indicative of the occurrence at which sequences are ligated together after crosslinking [29]. High frequencies correlate strongly with spatial proximity but are prone to fixation artifacts which can be normalized to some extent [30–32], and others that call for additional information usually obtained with direct imaging observation [33–35]. These imaging studies have been essentially based on physical distance measurements between chromosome loci separated by given genomic distances, and they provided some of the first insights on the folding principles of chromosomes at length scales from 100 nm to a few μm ([36,37], see Ref. [38] for a review).

Chromatin loci also exhibit fast spatio-temporal dynamics, which have been extensively documented since the development of labeling methods based on bacterial operator systems by the Murray lab in 1997 [39]. Chromosomes undergo random fluctuations but also infrequent directional movements [40] associated with transcription activation [41]. Regulation of chromosome dynamics is influenced by the spatial organization of genes during transcription [42–45] and DNA repair [46–49]. At larger scales, chromosome reorganization events detected during development [50] could be correlated to transcription activation [51]. More recently, approaches, in particular displacement correlation spectroscopy based on time-resolved image correlation analysis [52] to map chromatin dynamics simultaneously across the whole nucleus have been developed to infer chromosome large-scale displacements. The modern view is that chromosomes are increasingly dynamic at every length scale from the molecular level—where the rapid turnover of chromatin associated proteins has been observed long ago by photo-bleaching techniques [53]—to the level of chromosome territories. However, the physical mechanisms that govern these fluctuations remain elusive at any length scale and raise questions about the exact contribution of active forces [54], which may bring chromatin to a far-from-equilibrium state.

In this review, we focus on technologies for chromosome loci analysis and discuss the insights on chromosome properties in relation to their biological functions that can be gained applying them. We will first overview methods to label single or multiple chromosome loci in living cells, then discuss the techniques to perform particle tracking and high-throughput localization imaging, systems biology approaches for automated detection and statistical analysis, and the models to analyze the data. We finally discuss whether or not chromatin motion and quantitative positional analysis are relevant to investigate chromosome properties and propose some experiments to gain confidence and/or refine the structural insights gained with models.

CHROMOSOME LABELING TECHNOLOGIES

Chromatin dynamics can be analyzed at multiple scales in space and time using different fluorescence-based imaging methods. At great length scales, the movements of genome domains were followed by confocal microscopy using fluorescently tagged histones [55,56], generally performed after partial photo-bleaching or photo-activation. The minimal area that can be studied with this approach is defined by the size of the laser spot used to photo-bleach or photo-convert fluorescent histones [57], which encompasses several Mb to Gb of chromatin. Live cell tracking of telomeres over several minutes corroborated the notion of self-contained chromosome domains in yeast [58] and metazoan cells [59]. 3C-based technologies revealed the existence of topologically associated domains (TADs) of smaller dimensions in the range 0.1 to 10 Mb [60], within which DNA–DNA interactions are more frequent as compared to the bulk of the genome. Elucidating their origin and structural bases requires new microscopy techniques, including collections of specific DNA probes to label a target region [61], which are so far limited to analysis in fixed cells. At smaller length scales, visualization of chromosome loci can be performed without sequence specificity by incorporating injected or transfected fluorescent nucleotides during replication [62], or indirectly, by expressing photo-convertible fluorescent histones [63]. The former technique achieves excellent signal-to-noise ratio (SNR, see more below) when labeling "large" loci of ~50 kb [64], whereas the latter permits analysis of trajectories even of single nucleosomes. Visualization of DNA at specific genomic loci requires labeling techniques that create a fluorescent focus detectable above background levels. Several systems useful for live cell imaging have been developed (Table 12.1), for a review see Ref. [83]. FROS (fluorescent repressor operator system, [75]) is based on insertion of numerous repeats of bacterial operator sequences (150–250 bp per unit) to which fluorescent repressor fusion proteins bind. The FROS system includes TetR/tetO [75], LacI/lacO [84], and λcl/λOp [85]. Although widely used and very powerful, genome biology may be affected by FROS insertion, and biological consequences should be tested before and after insertion [86]. FROS insertion can promote heterochromatin formation at a nearby silenced locus [87] and block DNA or RNA polymerase progression. Operator repeats should therefore be inserted

Table 12.1 Methods for Tracking Chromatin Motion in Living Cells

Method	Origin (*n* = Number of Systems in use)	Sequences Labeled	Reference
Fluorescent dNTPs	dNTP (*n* = 4), BrdU	Bulk labeling; whole genome single cell injections or "rubbing"; transient labeling	[65]
Histones	H2B-GFP, etc. (*n* > 4)	Bulk labeling; whole genome photoactivatable fusion allows activating individual chromosomes, chromosome territories	[52,63,66,67]
Chromatin associated proteins	Rap1-GFP (yeast), TRF1-GFP (mammalian), Rad52-GFP, mch-53BP1 others	Visualization of repetitive sequences (telomeres, centromeres) or protein accumulation at double strand breaks	[59,68–70]
MS2/MCP	mRNA *n* = 2 (MS2, PP7)	Visualization of specific nascent mRNA transcripts bearing repetitive stem loop sequences (genome editing); indirect, close to gene	[71–73]
Fluorescent antibody fragments (Fab)	Antibody (unlimited but synthesis laborious)	Labeling of histones or proteins associated with repetitive gene arrays; positional information	[74]
FROS (fluorescent repressor operator system)	*n* = 3; LacO/lacl (*E. coli* chromosome); TetO/TetR (Tn10 transposon tetracycline gene); LO/LCi (phage lambda); potentially other bacteria operons (i.e., lexA; CuO/CymR; ycnKJI)	Visualization of specific DNA sites by inserting arrays of operator sites (genome editing)	[39,75,76]
TALE (transcription activator like effector proteins)	*n* = 2 but potentially ∞ Xanthomonas oryzae transcription activator	Specific natural repetitive sequences (i.e., telomeres, centromeres)	[77,78]
CRISPR/dCas9	*n* = 3; Streptococcus pyrogenes (Sp), Neisseria meningitides (Nm), and Streptococcus thermophilus (St1)	Repetitive sequences or numerous sites recognized by small guide RNAs (*n* > ~70); transient labeling	[79,80]
SunTag	Antibody to a multimerized protein scaffold	Visualization of specific DNA sites by inserting a protein binding site (genome editing)	[81]
ANCHOR (ParB/INT)	*n* = 3 (potentially more) bacterial chromosome partition systems (Burkholderiaecae)	Visualization of specific DNA sites by inserting a short nonrepetitive sequence (genome editing)	[82], Germier et al. in preparation

as far as possible from regulatory elements, but still close enough to be considered part of the tagged and tracked locus to reduce possible side effects. This can be 100–800 bp in 3′ and 5′ of transcribed genes or several kb when studying telomeres.

Taking advantage of protein oligomerization rather than DNA sequence repeats, Suntag [81], based on antibody recognition of amplified protein segments, and ANCHOR (ParB/INT), based on bacterial partitioning complexes ([82]; Germier et al. submitted, Mariame et al. submitted) are less invasive, yet stable new labeling techniques available for use in yeast and most metazoan cells. Unfortunately, quantitative comparison of the different labeling technologies in terms of SNR and motion is largely missing. Different sizes (number of repeats) of lacO arrays have been used to distinguish two DNA loci in yeast (e.g., Refs. [88,89]), but only qualitatively. Several groups have implemented reduced size operator arrays and assessed their suitability for detecting position or even live cell tracking, yet most of these analyses based on SNR are not published. A very careful analysis of the impact of photo-bleaching on yeast cell viability with respect to fluorescent intensity, and imaging quality was performed by the Sedat lab using the OMX microscope [90]. They demonstrated that 1% of usual excitation intensity can suffice to track loci using the OMX and applying denoising algorithms. The ANCHOR1 and ANCHOR2 (ParB1—ParB2/INT) systems have a lower SNR than the lac and tet FROS, on average, in yeast, but spots can be tracked as efficiently using a spinning-disk (Nipkow-disk) confocal system. Improved resolution and SNR for point-like signals is achieved because only a small fraction of the sample is visible at any given instant through a series of pinholes on a rapidly rotating disk. The ANCHOR3 (Germier et al. submitted, Audibert et al. unpublished) system offers greatly improved SNR in both yeast and mammalian cells.

The use of CRISPR (Clustered Regularly Interspaced Short Palindromic Repeats) and inactivated or dead CRISPR-associated (dCas9) protein fused to fluorescent proteins is very promising to label multiple sites transiently. Sites are specifically targeted by short guide RNAs complementary to genomic CRISPR sequences [79,80] or TALE [77,78]. Here also, little quantitative data are available. The first study [79] showed that dCas9-GFP labeled, naturally occurring repetitive regions colocalize with corresponding FISH probes in fixed cells. They quantified movement of telomeres visualized by dCas9-GFP targeted to telomere specific guide RNAs and compared it to the one recorded using the telomere protein TRF1 fused to GFP. The detected motion was heterogeneous in both cases. Fluorescence intensity relative to the determined diffusion coefficient was plotted to exclude any microscopy artifacts. Combining dCas9 labeling with the MS2 RNA-tagging system improves the capabilities for multicolor labeling, but the SNR is still limited [91]. Recently, the Pederson group demonstrated that several dCas9 from different bacteria allowed visualization of multiple spots that could also be tracked, but quantitation here also needs optimization: intensity or efficiency of the orthologous systems with respect to other labeling techniques have yet to be estimated [80].

SPATIAL DYNAMICS: REAL-TIME IMAGING AND TRACKING OF SINGLE CHROMOSOME LOCI

Imaging Experiments

Chromosome loci tracking is carried out with different fluorescence microscopes. Citing only the main setups including (multiphoton) confocal, spinning-disk, wide-field, or total-internal reflection microscopes (see Refs. [92–94] for review). Depending on the application, the microscopy technique is chosen to track objects in 2D or 3D with optimal temporal resolutions; the choice of the light source as well as the photo-detector (cameras or diodes) is also guided by the requirement of the experiment. The exposure time of the camera is usually defined by tuning the excitation power to maximize the brightness of the fluorescent object by maintaining an acceptable level of photo-bleaching throughout the experiment. Whenever single (or a limited number of) loci are tracked, it is desirable to reduce the size of the frame of the camera (or region of interest, ROI), particularly for confocal laser scanning microscope confocal and sCMOS cameras (frame reduction is more complex for EM-CCD cameras, only achieved with an optical mask placed ahead of the sensor) to reduce the transfer time in between two consecutive images. Hence, for a given exposure time, the interframe interval can be optimized to reduce the consequences of photo-bleaching. The key parameter defining the quality of a recording is SNR of the target fluorescent region, as described in the following section.

A Few Facts on the Signal-to-Noise Ratio in Single Particle Experiments

The brightness of a fluorescent locus is estimated according to its SNR (see examples in the upper panel of Fig. 12.1A). This quantity is defined by the intensity of the locus corrected for the background signal ($A = 4$ and 13 in the two upper graphs of Fig. 12.1A) divided by the noise level in the background denoted by Δ in Fig. 12.1A. This level is inferred from the standard deviation of the background signal. High SNR allows one to track an object with minimal localization error with a low false-detection rate, and thus to retrieve the maximal load of information out of a given experiment. In order to illustrate this statement, let us focus on the analysis of typical trajectories recorded for molecular transactions occurring in the nucleoplasm. We use the temporal evolution of the mean square displacement (MSD), which measures

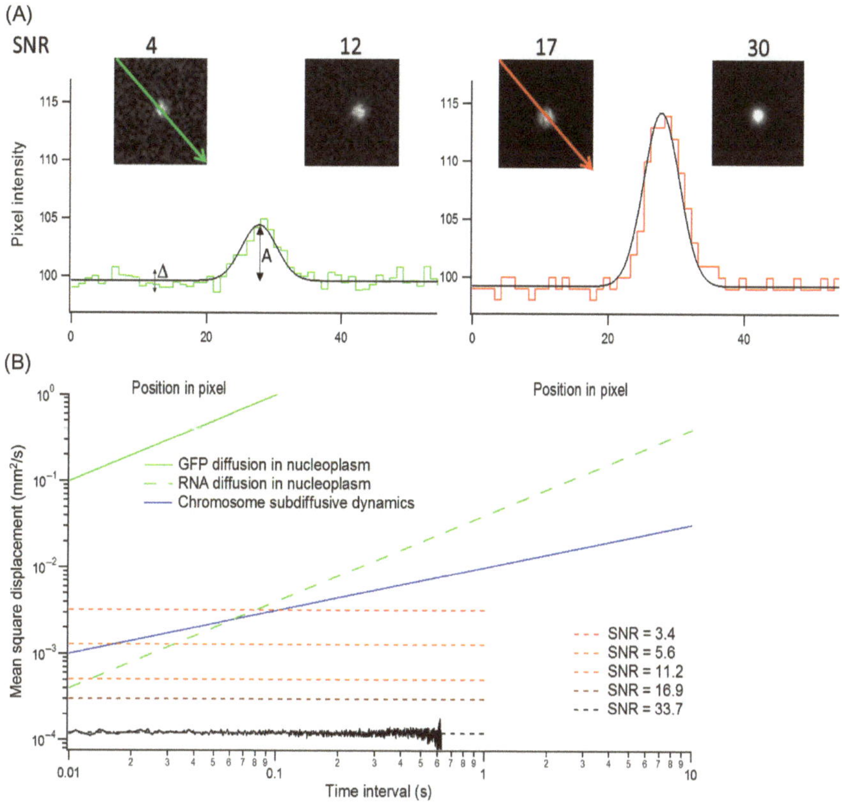

FIGURE 12.1 Particle tracking and positioning error.
(A) The upper panel shows four fluorescence micrographs of 0.4 μm nanoparticles imaged with a wide-field microscope equipped with a 100× objective (NA = 1.4). The different SNR are obtained by exposing the nanoparticles to variable doses of illumination. The two associated graphs correspond to the intensity vs pixel position along the respective green and red arrows drawn on the two micrographs. Note that the color of the datasets is associated to the arrows and the black lines are Gaussian fits of width 3.7 pixels. (B) The graph represents the temporal variation of the MSD for three typical situations in the nucleoplasm. In green, we consider free diffusion of GFP [95] or RNA [96], and in blue the subdiffusive response detected for chromosome loci [97]. The five horizontal lines correspond to the lower limits of detection for variable values of the SNR, as indicated in inset. Note that these horizontal lines are deduced from simulations of fixed particles, which produce noisy black traces, as represented for SNR = 33.7.

the average distance traveled by a target after a given time lag, as readout of these experiments (see Section 2.3 for more information). In Fig. 12.1B, we report diffusive MSD responses for GFP or RNA molecules in the nucleoplasm (green lines), as well as the typical subdiffusive behavior detected for chromatin loci in yeast (blue line). Depending on the SNR of the object, tracking accuracy is limited spatially to a lower limit, which is presented by the five horizontal lines of the plot. It is not possible to detect spatial fluctuations smaller than these limits, due to positioning incertitude. For a locus characterized by an SNR of 3.4, chromosome spatial fluctuations or RNA diffusion cannot be sampled for time scales below 0.1 s, but GFP diffusion is perfectly well characterized for this signal quality. In order to track chromatin down to 10 ms, it is necessary to increase the SNR to ~10, by e.g., increasing the power of the light source, augmenting the gain in photodetectors, and adjusting the size of the frames.

Tracking and Trajectory Reconstruction

DNA locus motion is analyzed using single particle tracking (SPT) techniques. SPT is based on the ability to accurately identify the center of a fluorescent spot at each time point of time-lapse imaging (2D or 3D) and connect those positions to reconstruct trajectories. Spatial resolution of SPT is not limited by the diffraction limit of ~250 nm for GFP visualization, but by the SNR that depends on the number of photons detected and the noise generated by microscopic setup and detector. Automated detection based on Gaussian fitting is typically appropriate for high SNR, which have been rigorously compared in Ref. [98], but other detection methods [85,99] might be more robust and should be explored when SNR cannot be increased.

The standard approach to assess the level of localization precision consists in tracking simulated spots at a fixed position with a registered level of SNR. Simulated spots are generated using Matlab algorithms (or equivalent), in which we assign a Gaussian profile for the particle characteristics and add a Poisson noise to every pixel for background noise approximation. The size of the particle's Gaussian profile is dependent on the microscopy setup as well as the camera's pixel size, especially for objects smaller than the pixel size. These parameters should be chosen to match experimental conditions. Note that the localization precision can also be measured experimentally by tracking fluorescent beads chemically bound to coverslips or with chemically fixed FROS tagged cells, but artifacts associated to mechanical drift of the microscope during the acquisition should not be overlooked in this case. Simulated images are thus more relevant to assess the absolute positioning ("ground-truth" error), although subtle sources of errors in localization due to motion blur during camera acquisition (called dynamic errors; see Ref. [100] for a review) are often neglected.

Regarding observation of chromosomal loci in living yeast, genes can be tracked in 2D with a resolution below 10 nm, i.e., with an SNR larger than ~30, for time intervals 200 ms using FROS labeling (data not shown). It is often noted that chromosome loci move in the nuclear volume in 3D, and several techniques have been set up to track their motion along the three spatial directions, including, to cite but a few, micro-mirrors [101], double-helix point spread function microscope [102], astigmatic microscope [66], or confocal microscopy [93]. So far, however, the advantages of 3D over 2D tracking remain unclear, because the acquisition of images in 3D reduces the temporal resolution and/or decreases the tracking accuracy, hence may limit the quality of the data retrieved from tracking experiments. Furthermore, the organization of chromosomes with respect to the focal plane of the objective is expected to be random (effects of gravity are a priori negligible), so the movement of chromosomes in the focal plane of the objective is similar to that in the third dimension. Therefore, the information gathered from 3D tracking should essentially be identical to that collected in 2D, and the complications associated to 3D microscopy should not be rewarded with more accurate physical insights.

Data Analysis

The diffusion coefficient D and the velocity v of the measured trajectories can be retrieved from the analysis of the MSD:

$$MSD_{1D}(\tau) = \langle (x(t+\tau) - x(t))^2 \rangle_t = 2D\tau + v^2\tau^2 \quad (12.1)$$

with τ the time interval, x the position of the tracer, and t the time variable for moving average.

The MSD corresponds to the average distance traveled after a given time lag τ. Note that other statistical functions may be relevant for the analysis of random processes, including the velocity correlation function [102,103] or the angle distribution between consecutive steps [104]. The motion of an isolated object diffusing in a viscous fluid is known to be associated to a linear temporal variation of the MSD with time, as reported in Eq. (12.1). Yet this model obviously constitutes an oversimplification, which can hardly be applied to probes moving in the nucleoplasm. In fact, the presence of obstacles with a structured or random spatial distribution [105] leads to deviation from the standard model with the MSD scaling as a power of time. This anomalous diffusive behavior, which has recently gained in popularity for chromosome fluctuation analysis in living cells (see Table 12.2 and Ref. [111]), is characterized with the following equation:

$$MSD_{1D}(\tau) = 2\Gamma\tau^{\alpha} \quad (12.2)$$

with α the anomaly exponent and Γ the amplitude of the MSD (units of $\mu m^2\ t^{-\alpha}$). Notably pioneering studies of chromosome dynamics have been initially analyzed with confined random walk models, i.e., normal diffusion in the small time limit and a plateau in the long time limit (see Ref. [112] for

Table 12.2 Overview on Recent Studies on Chromosome Dynamics

	Study	Anomaly Exponent	Time Scale (s)	Interpretation
Bacteria	[106]	0.4	1–200	Rouse in viscoelastic media
	[103]	0.4	1–200	"Superthermal" response
	[107]	0.4	0.1–10	Fractional Langevin motion
Yeast	[39]	1 and plateau	12–500	Confined random motion Reptation for diffusion coefficient
	[44]	0.4–0.5	4–100	Subdiffusion
	[101]	0.4	0.2–4	Subdiffusion
	[108]	0.5 ± 0.07	0.1–200	Rouse model
	[97]	0.52 ± 0.07	0.02–500	Rouse model
	[102]	0.7	0.1–10	Fractional Brownian motion
	[109]	0.7 and 0.5	0.2–100	Rouse model
Metazoan	[40]	ND	0.01–15	Two time regimes for "diffusive" responses
	[68]	0.3, 0.5, 1.2	0.01–1000	Reptation
	[110]	0.4–0.7	0.5–100	Subdiffusion
	[65]	0.5	0.5–20	Subdiffusion

a review). It is conceptually difficult to discriminate between confined random walk and anomalous diffusion based on fitting the MSD curve, because both models rely on two-fitting parameters. However, it should be noted that models of polymer motion are nearly systematically associated to anomalous diffusion [113].

Importantly, many mathematical and/or physical approaches have been developed to delineate between the most popular models of anomalous diffusion. Although normal diffusion originates from random collision with the solvent whatever the time scale, anomalous diffusion is associated to some degree of memory over time [112]. For instance, the combination of random fluctuations with attachment of a polymer segment to its neighboring sites is associated to history effects in the trajectory of monomers because elastic forces hold back the monomer toward its previous position. More generally, models with short memory include continuous-time random walk and obstructed diffusion; on the other hand, fractional Brownian motion and fractional Levy stable motion correspond to a process with a long memory. Various tests have been established to differentiate between these classes of models, see e.g.,

Ref. [114]. Interestingly, these tests have been applied to study anomalous diffusion of telomeres [115] or yeast chromosomes [102], and they showed that chromosomes loci dynamics were consistent with fractional Brownian dynamics, which are very common for polymers. This result strongly supports the fact the analysis of the statistical properties of chromosome loci trajectories can be used to probe structural information of chromosomes.

The anomaly exponents describing the trajectories chromosome loci in yeast, bacteria, or mammalian cells (Table 12.2) have been consistently measured close to 0.5. This behavior is very common for polymers for it corresponds to the Rouse regime. The Rouse model is expected to be relevant if the dynamics of a polymer is driven by elastic interactions along the chain with negligible long-range interactions mediated by solvent-flux [113]. This approximation is consistent with the molecular environment of the nucleoplasm as the level of crowding [116,117] should screen out long-range hydrodynamic flows. Think about spinning a fork in a spaghetti bowl: The fluid flow is dissipated at the vicinity of the cutlery. The Rouse model describes the chain as a series of beads connected by elastic springs, and it relies on two molecular parameters of the polymer, namely the monomer friction coefficient ζ, and the length each segment which is defined by the Kuhn length b. Note that each Kuhn segment are associated with ball-joint connections, so that the orientation between consecutive monomers is random. Therefore, the longer the Kuhn length, the stiffer the polymer. The expression for the MSD in 1D reads

$$MSD_{1D}(\tau) = \left(\frac{4b^2k_BT}{3\pi\zeta}\right)^{0.5} \times \tau^{0.5} \tag{12.3}$$

with k_B and T the Boltzmann constant and the temperature, respectively. Note that the friction coefficient is related to the Kuhn length according to $\zeta = 3\pi\eta b$ with η the viscosity for translational diffusion of monomers. The fact that the dynamics of monomers is determined by the polymer elastic properties implies that motion analysis offers a unique window to characterizing chromosome structural properties in vivo. The amplitude of the MSD has been estimated to $\Gamma \sim 0.01\,\mu m^2 s^{-0.5}$ for yeast chromosomes [97]. Given that nucleoplasmic viscosity spans 4 to 7 mPa s [95,118,119], we deduce that the Kuhn length is ~1 nm. This value of the Kuhn length expected to be on the order of 30 to 200 nm according to models and experiments [120] is much smaller than expected for a 10-nm fiber. This surprisingly high flexibility raises concerns about the consistency of this model and suggesting that additional physical effects, e.g., active forces out of equilibrium [103], may be considered. This step forward in the modeling remains a key issue for the structural description of chromosomes and for the understanding of how chromosome spatio-temporal dynamics are connected to their biological function in vivo.

MAPPING CHROMOSOME CONFORMATIONAL DYNAMICS BY SINGLE/MULTI LOCI POSITIONING

Folding and physical data of chromatin conformation can be inferred from position of one or several specific DNA loci with respect to nuclear landmarks or relative to each other. Bulk labeling methods using fluorescent histones such as H2B-GFP or fluorescent dNTPs cannot inform on the path the underlying chromosome fiber follows. Increased labeling density and new, bright fluorescent dyes may ultimately, using super-resolution imaging techniques, yield a picture of this path [121] and systems biology approaches can inform on the probabilistic distribution of highly mobile DNA loci in vivo.

In yeast, two high-throughput approaches based on automated, high precision tracking of single or multiple labeled sites have proven powerful in providing detailed information on the folding principles of chromosomes in the nucleus at steady state ([76,108,122], Figs. 12.2 and 12.3). They are described in the following paragraphs.

Single Locus Localization in the Nucleus in a Large Population of Cells

Investigations of gene localization are intrinsically limited by the diffraction blur of optical microscopy. Diffraction limits the spatial resolution to 200–500 nm at best, a major handicap considering the size of yeast nuclei of 2 μm in diameter. For multiple-point localization, this limitation can only be overcome with super-resolution detection, which so far remains a semi-automated and time-consuming technique, allowing the sampling of ~100 cells. A method coined "gene map" analysis [122] combines fully automated cell nuclei detection and single gene locus super-resolution 3D tracking (Fig. 12.2A). 3D coordinates of single gene loci tagged using tetO FROS with respect to two nuclear landmarks, the nucleolus and the nuclear center, are automatically determined with a dedicated software. The positions can then be assembled in a single probabilistic representation of the gene's localization in nuclear space (Fig. 12.2B). The resolution of this gene localization is significantly better than the diffraction limit, and only limited by the SNR and the number of cells analyzed. Most genomic loci are confined to "gene territories" which can be remodeled during transcriptional activation [122,123]. Overlay of different gene maps can distinguish preferential areas in which the respective DNA sites roam (Fig. 12.2C). These areas can be compared between different conditions or used to trace the path of a chromosome. For instance, by systematically inserting labels along chromosome XII, with 10 outside the rDNA and 3 in the rDNA region, the organization of chromosome XII, the largest yeast chromosome with ~2 Mb has been determined [108]. The territory occupied by each tagged locus is relatively large

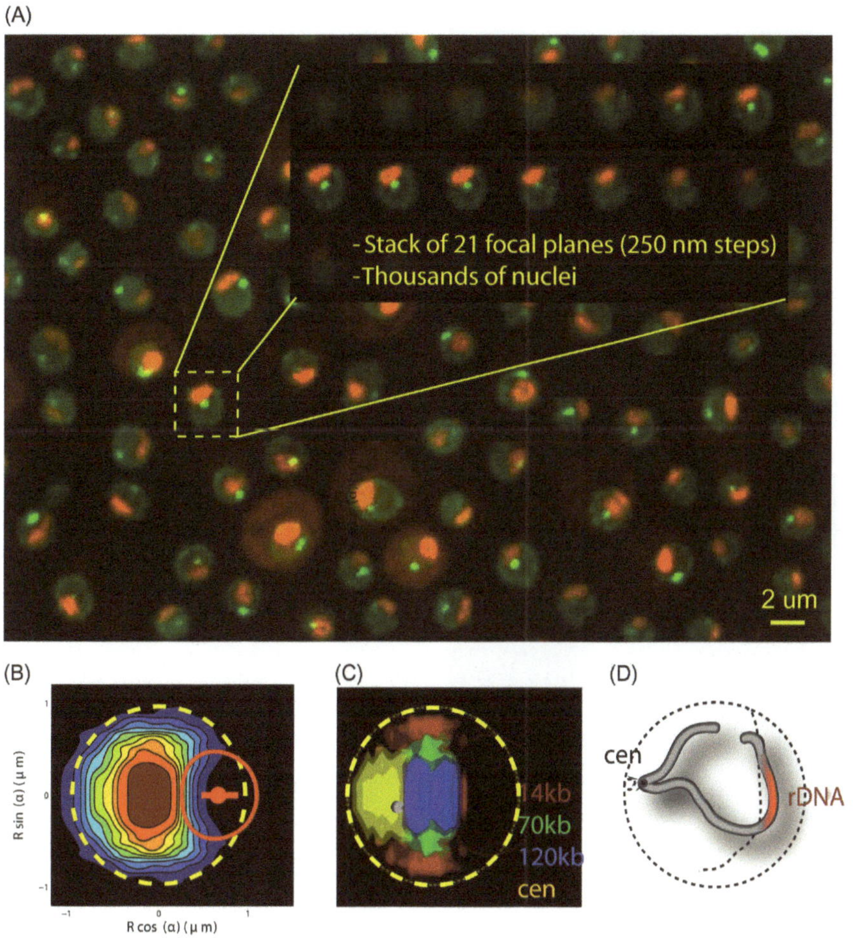

FIGURE 12.2 A systems biology approach to study gene position.

(A) Yeast nuclei were imaged in 3D acquiring 21 *z* planes at 250 nm intervals using a spinning-disk microscope. The nuclear envelope (green) and the nucleolus (red) were labeled by expressing endogenously tagged fusion proteins GFP-Nup49 (nucleoporin; green) and Nop1-mcherry, respectively. The DNA locus of interest is labeled by inserting an array of 256 TetO and expressing TetR-GFP. Excitation time was 200 ms. An example of a field of ~100 nuclei is shown. (B) An example of a gene map showing the probabilistic distribution of the imaged DNA locus from thousands of nuclei. Only round shaped nuclei are analyzed. (C) Gene maps obtained from different yeast strains in which distinct DNA loci were labeled [here loci along yeast chromosome XII at positions 14, 70, and 120 kb from the left telomere and the centromere (cen)] can be overlaid. (D) The relative position of the gene maps allows reconstructing the path of chromosomes with respect to nuclear landmarks.

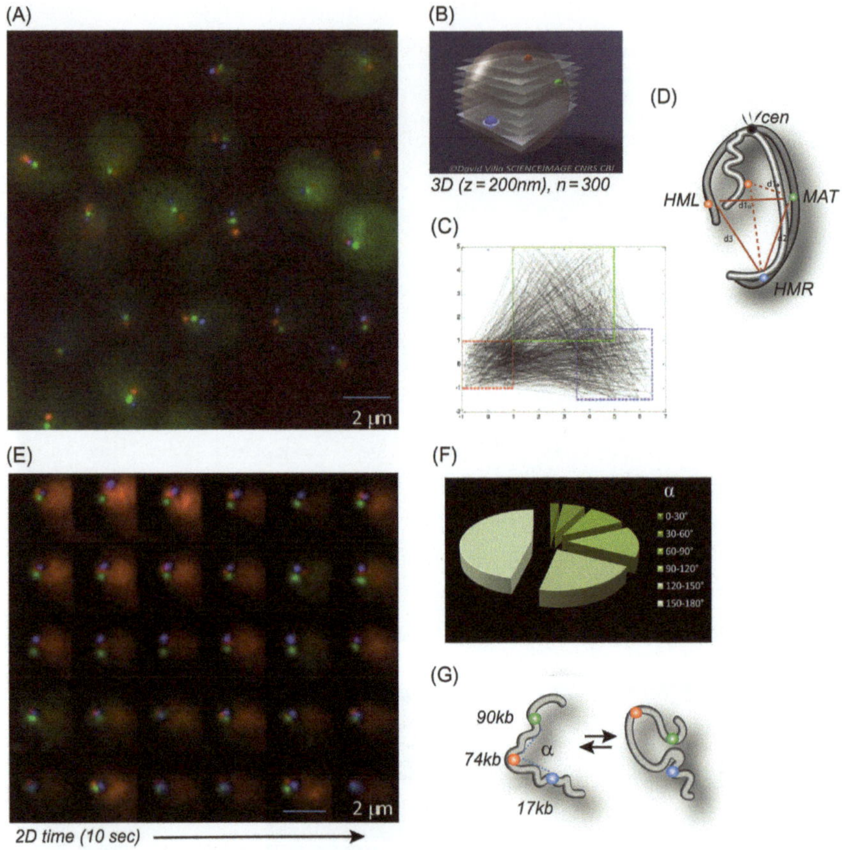

FIGURE 12.3 Multiloci labeling to determine chromosome conformation and folding.
(A) Live yeasts were imaged in 3D acquiring 21 z stacks separated by 250 nm intervals using a wide-field microscope. Three distinct loci were labeled simultaneously by inserting arrays of LacO, TetO, and λO (next to the three mating type loci, *HML*, *MAT*, and *HMR*, on chromosome III), and visualized by coexpressing CFP-LacI, TetR-mcherry, and λI-YFP. Excitation times were 300–400 ms. (B) The center of gravity of each spot and the distance between the three spots was determined automatically from hundreds of nuclei. (C) An inverse mathematical modeling approach was used to determine the areas in which each DNA locus can roam when linked to the 2 others (survival zones in red, blue, and green contours). (D) Relative positions of the survival zones can be compared between two cell types. (E) A triple labeled single yeast cell was imaged in 2D in real-time, 1 image/10 seconds over 5 minutes. Here, FROS arrays were inserted along the same chromosome arm 17, 74, and 90 kb from the left telomere of chromosome III. (F) Distributions of angles formed at the central red locus formed between the blue and green one is shown from $n = 350$ cells. All angles, as represented in panel (G) and (H), are possible, suggesting that the red locus can move with respect to the two others without constraint.

compared to the overall volume of the nucleus and maps of two genes distant by 100 kb in genomic distance along the same chromosome arm readily overlap. Snapshots of a tagged locus in many different cells can inform on its 3D area of constraint, which also correlates with the average region explored by a locus through random fluctuations [108]. The resulting data can be quantified with physics models, in which chromosomes are treated as polymer chains with volume exclusion and nuclear constraints defined by the nuclear envelope and the nucleolus [124,125]. This approach convincingly confirmed the Rabl-like organization of the yeast nucleus, an organization based on regrouping centromeres near the spindle pole body (a centrosome like structure embedded in the nuclear envelope) and clustering subsets of telomeres on the opposite end of the nucleus. Yet it did not allow determining chromatin structural parameters (see Ref. [38] for a review). The gene map method is readily applicable to fixed cells and DNA loci identified by FISH. Future improvements may permit analysis of relative positions and interdependence of several sites simultaneously, but it is so far difficult owing to the lack of high intensity fluorophores required for labeling multiple loci simultaneously.

Multilocus Localization for Chromosome Conformation Analysis

In both yeast and mammalian cells, the regulation of position and associated dynamic behavior has been shown to be influenced by replication [92], transcription [44] as well as DNA repair [46]. Two genomic sites can be labeled simultaneously to assess their relative position, in order to retrieve the propensity of two telomeres to coalesce [126], or to measure the relationship of their physical distance with respect to their separation on the linear chromosome, a measure useful to model the chromatin fiber (for a review see Ref. [127]). All these studies have so far been based on FROS labeling which allows one to distinguish two sites by either different fluorescent intensities achieved using one short and one long array (128 vs 256 repeats [88]), or two orthologous systems (lacO and tetO [128]). Note that a third site can be tagged using the bacterial lambda binding sequence λ FROS ([76,129]; Fig. 12.3A). Relative positions and angles can then be calculated by determining the position of three sites within the same nucleus. This approach has been instrumental in determining folding of chromosome III (320 kb) in yeast. Chromosome III folding appeared to be cell type specific, although preferential perinuclear localization was indistinguishable between the two cell types when analyzing individually tagged loci near the telomeres (*HML* and *HMR*) [47]. Analysis of chromosome structure based on 3-point localization relies on dedicated software that automatically computes the 3D coordinates of each of the three loci from hundreds of nuclei (Fig. 12.3B). Distances and angles from the triangle can then be assembled in probabilistic representation of their relative position. Generated heat-maps inform on constraint between two loci relative to a third one [76]. An inverse mathematical modeling approach has also been developed to generate survival zones for the three loci by fitting the experimental data iteratively. Survival zones define the area within which a locus can be positioned when physically bound to two other on the same chromosome fiber. In the case of chromosome III, it was possible to show that loci on the left arm of the chromosome had less freedom of position relative to the right arm of the chromosome in the alpha cell type than in the **a** cell type (Fig. 12.3C and D). Given the inability of the left arm of chromosome III to efficiently contact loci on the right one, chromosome conformation appears to relate to the mechanisms of homology search, the limiting step in homologous recombination. The geometrical "3 loci" algorithm is readily applicable to analyze position of three distinct sites at any distance in fixed and living cells.

Real-Time Tracking of Multiple Loci

Fitting the mean distance between two chromosomal sites with a polymer model based on a worm-like chain provides physical parameters of the chromatin fiber, one of which is the persistence length, or the Kuhn length (see above) divided by 2, a measure of polymer flexibility [128]. This value suggested that, in situ, the fiber adopted a rather stiff conformation compatible with a ~30 nm fiber. As the mere existence of such a fiber is more and more debated [97,130], models have to be refined and additional parameters have to be computed. Chromatin is highly mobile and most loci explore a rather large subnuclear space. Hence, by using the mean distance motion and variability of measured distances are not taken into account. It is therefore highly desirable to combine multi-locus tracking to refine the analysis of distance averages. For instance, the distance between two telomeres of the same chromosome shows a general mean around 1 μm [126], but telomeres of several short chromosomes move in a coordinated fashion ("dancing in tune") over a 5 min interval, and telomeres of distinct chromosomes do not [58]. Correlated movement can be observed in 1- s intervals in 2D, but only in 3- min intervals in 3D. The duration of contact between two distinct loci can also be determined: for example the heterochromatic *HML* and *HMR* loci stay together on average 10 s [131]. Simultaneous tracking of three or more discrete loci in close vicinity (<100 kb or less) has the greatest potential to describe fiber folding. An example in 2D is shown in Fig. 12.3E–G, where the red labeled central locus, 20 and 50 kb separation to the green and blue one, respectively, on the same chromosome arm, moves with unconstrained amplitude, the three spots readily forming angles between 0 and 180 degree. Hence, the chromatin fiber in a yeast interphase cell can bend around itself over distances of <20 kb within less than 5 min. Although such a study seems to be well-suited to derive insights on the flexibility of the fiber, it remains to be shown that static data (gene map or multi-locus data) can be bridged with fluctuation measurements. These questions clearly define an area of future research of genome biophysics which is essential to further elucidate the relationship between chromosome structural properties and their biological functions, as described in the following paragraph.

OUTLOOK—FUTURE DEVELOPMENTS

The goal is now to elucidate how chromosome structural properties enable local or global reorganization of the nuclear architecture with respect to nuclear and cellular processes. This task amounts to define questions of systems biology where technology and physics can bring novel insights into nuclear organization. Its accomplishment will rely on technological developments in order to thoroughly sample the parameter "space" of molecular reactions involving chromosomes—as well as in physics—to build consistent and predictive chromosome models.

Technological Pipeline

Although the throughput of SPT has significantly improved over the past 10 years, much effort remains to be carried out for screening multiple conditions and multiple colors. Let us first focus on multicolor fluorescence microscopy. Tracking two (or more) loci simultaneously in real-time requires compromises between frequency, resolution, and duration [83]: in 3D, the number of focal planes reduces the number of time frames; in 2D, one spot may move out of focus during the acquisition. The choice between 2D and 3D acquisition is hence subtle and the microscopy setup to be employed is not straightforward. For instance, while confocal microscopes allow simultaneous use of several laser sources, spinning-disk or wide-field microscope impose the choice between alternating wavelengths during stack acquisition or acquiring a stack for each wavelength. The first solution slows frequency depending on the speed change in light excitation and filter (wheel filters take ~30 ms to switch), and the second reduces positional precision in particular for fast moving spots (if one stack takes 1 s to complete, the second stack will be acquired after 1 s). Nevertheless confocal microscopy is not a fast acquisition technique, and scanning photo-multiplicators require reduction of the frame size, i.e., the number of pixels scanned (ROI) to gain enough speed for live cell tracking.

Automated quantitative high-throughput microscopy (HTM) techniques are rapidly progressing. Confocal fluorescence micrographs of multiple fields can be captured from each microplate well of 96 or 384-well plates [121,132], or from custom made microfluidic systems [133] generating massive amounts of data which will require custom Matlab algorithms and Image J plugins for automated, multiparametric analyses. Yet, while statistical analysis of single locus trajectories is now well-established with, e.g., the MSD, the mathematical tools to use for multiple loci need to be developed. This task involves challenging computational and statistical approaches able to treat very noisy data from imaging very small nuclear segments.

Chromosome Structure and Dynamics and Their Link to Genome Transactions

In textbook models of DNA processing, chromosomes have been assumed to be immobile and proteins to scan the DNA backbone infinitely much faster [134–136]. According to this target search mechanism, the folding of chromosomes plays an essential role to guide multitarget binding [137]. The analysis of 3C data with polymer looping models seems to provide an excellent solution to better understand the orchestration of the expression of multiple genes simultaneously. Interestingly, the "immobility" hypothesis does not hold for chromatin repair pathways where distant chromosome sites have to interact. In this context, the physics of chromatin motion and its interplay with chromatin structure remain to be clarified. Moreover it is widely believed that genes "move" upon activating transcription, but whether the change in localization is a cause or a consequence of transcription and concomitant to mRNA production is not known. So far a few real-time examples have shown a change in position as mRNA is produced [41,42], but other studies of transcription activation have rather pointed to a reduced mobility [138]. Also, it is well known that RNA polymerase forms clusters at active genes which are rather immobile [71], but quantitative studies of DNA motion as transcription proceeds need to be developed. Improving labeling and imaging techniques with shorter inserts, better SNR, greater resolution and speed will eventually allow tracking short-lived events, such as enhancer–promoter contacts. Together with consistent molecular modeling of chromatin, this convergence of physics and biology will clarify many aspects of the multiple regulation mechanisms of the genome.

References

[1] Mewes HW, et al. Overview of the yeast genome. Nature 1997;387:7–65.

[2] Richards S. It's more than stamp collecting: how genome sequencing can unify biological research. Trends Genet 2015;31:411–21.

[3] Vernikos G, Medini D, Riley DR, Tettelin H. Ten years of pan-genome analyses. Curr Opin Microbiol 2015;23:148–54.

[4] Paterson AH, Freeling Michael, Tang H, Wang X. Insights from the comparison of plant genome sequences. Annu Rev Plant Biol 2010;61:349–72.

[5] Baltimore D. Our genome unveiled. Nature 2001;409:814–16.

[6] Venter JC, Adams MD, Myers EW, et al. The sequence of the human genome. Science 2001;291:1304–51.

[7] Solovei I, et al. Spatial preservation of nuclear chromatin architecture during three-dimensional fluorescence in situ hybridization (3D-FISH). Exp Cell Res 2002;276:10–23.

[8] Dily FL, et al. Distinct structural transitions of chromatin topological domains correlate with coordinated hormone-induced gene regulation. Genes Dev 2014;28:2151–62.

[9] Taddei A, Gasser SM. Structure and function in the budding yeast nucleus. Genetics 2012;192:107–29.

[10] Fraser P, Bickmore W. Nuclear organization of the genome and the potential for gene regulation. Nature 2007;447:413–17.

[11] Therizols P, et al. Chromatin decondensation is sufficient to alter nuclear organization in embryonic stem cells. Science 2014;346:1238–42.

[12] Lemaître C, et al. Nuclear position dictates DNA repair pathway choice. Genes Dev 2014;28:2450–63.

[13] Meister P, Taddei A. Building silent compartments at the nuclear periphery: a recurrent theme. Curr Opin Genet Dev 2013;23:96–103.

[14] Gilbert N, et al. Chromatin architecture of the human genome: gene-rich domains are enriched in open chromatin fibers. Cell 2004;118:555–66.

[15] Fussner E, Ahmed K, Dehghani H, Strauss M, Bazett-Jones DP. Cold Spring Harb Symp Quant. Biol 2010;75:245–249.

[16] Shamsuddin AKM, Tang C-K. Barr bodies in testis with Klinefelter syndrome. Urology 1980;15:74–6.

[17] Schöfer C, Weipoltshammer K. Gene dynamics and nuclear architecture during differentiation. Differentiation 2008;76:41–56.

[18] Fransz P, de Jong JH, Lysak M, Castiglione MR, Schubert I. Interphase chromosomes in Arabidopsis are organized as well defined chromocenters from which euchromatin loops emanate. PNAS 2002;99:14584–9.

[19] Nasmyth KA. The regulation of yeast mating-type chromatin structure by SIR: an action at a distance affecting both transcription and transposition. Cell 1982;30:567–78.

[20] Weiss K, Simpson RT. High-resolution structural analysis of chromatin at specific loci: *Saccharomyces cerevisiae* silent mating type locus HMLα. Mol Cell Biol 1998;18:5392–403.

[21] Pombo A, Dillon N. Three-dimensional genome architecture: players and mechanisms. Nat Rev Mol Cell Biol 2015;16:245–57.

[22] Lahbib-Mansais Y, et al. Expressed alleles of imprinted IGF2, DLK1 and MEG3 colocalize in 3D-preserved nuclei of porcine fetal cells. BMC Cell Biol 2016;17:35.

[23] Chen H, et al. Functional organization of the human 4D Nucleome. PNAS 2015;112:8002–7.

[24] Cremer T, et al. The 4D nucleome: evidence for a dynamic nuclear landscape based on co-aligned active and inactive nuclear compartments. FEBS Lett 2015;589:2931–43.

[25] Sexton T, Cavalli G. The role of chromosome domains in shaping the functional genome. Cell 2015;160:1049–59.

[26] Tashiro S, Lanctôt C. The international nucleome consortium. Nucleus 2015;6:89–92.

[27] Gasser SM. Nuclear architecture: past and future tense. Trends Cell Biol 2016;26:473–5.

[28] Denker A, de Laat W. A long-distance chromatin affair. Cell 2015;162:942–3.

[29] Ay F, Noble WS. Analysis methods for studying the 3D architecture of the genome. Genome Biol 2015;16:183.

[30] Belmont AS. Large-scale chromatin organization: the good, the surprising, and the still perplexing. Curr Opin Cell Biol 2014;26:69–78.

[31] Cournac A, Marie-Nelly H, Marbouty M, Koszul R, Mozziconacci J. Normalization of a chromosomal contact map. BMC Genomics 2012;13:436.

[32] Yaffe E, Tanay A. Probabilistic modeling of Hi-C contact maps eliminates systematic biases to characterize global chromosomal architecture. Nat Genet 2011;43:1059–65.

[33] Shachar S, Voss TC, Pegoraro G, Sciascia N, Misteli T. Identification of gene positioning factors using high-throughput imaging mapping. Cell 2015;162:911–23.

[34] Williamson I, et al. Spatial genome organization: contrasting views from chromosome conformation capture and fluorescence in situ hybridization. Genes Dev 2014;28:2778–91.

[35] Giorgetti L, Heard E. Closing the loop: 3C versus DNA FISH. Genome Biol 2016;17:215.

[36] Mateos-Langerak J, et al. Spatially confined folding of chromatin in the interphase nucleus. Proc Natl Acad Sci U S A 2009;106:3812–17.

[37] Yokota H, van den Engh G, Hearst JE, Sachs RK, Trask BJ. Evidence for the organization of chromatin in megabase pair-sized loops arranged along a random walk path in the human G0/G1 interphase nucleus. J Cell Biol 1995;130:1239–49.

[38] Huet S, et al. Relevance and limitations of crowding, fractal, and polymer models to describe nuclear architecture. Int Rev Cell Mol Biol 2014;307:443–79.

[39] Marshall W, et al. Interphase chromosomes undergo constrained diffusional motion in living cells. Curr Biol 1997;7:930–9.

[40] Levi V, Ruan Q, Plutz M, Belmont AS, Gratton E. Chromatin dynamics in interphase cells revealed by tracking in a two-photon excitation microscope. Biophys J 2005;89:4275–85.

[41] Chuang C-H, et al. Long-range directional movement of an interphase chromosome site. Curr Biol 2006;16:825–31.

[42] Tumbar T, Belmont AS. Interphase movements of a DNA chromosome region modulated by VP16 transcriptional activator. Nat Cell Biol 2001;3:134–9.

[43] Taddei A, et al. Nuclear pore association confers optimal expression levels for an inducible yeast gene. Nature 2006;441:774–8.

[44] Cabal GG, et al. SAGA interacting factors confine sub-diffusion of transcribed genes to the nuclear envelope. Nature 2006;441:770–3.

[45] Chubb JR, Boyle S, Perry P, Bickmore WA. Chromatin motion is constrained by association with nuclear compartments in human cells. Curr Biol 2002;12:439–45.

[46] Lisby M, Mortensen UH, Rothstein R. Colocalization of multiple DNA double-strand breaks at a single Rad52 repair center. Nat Cell Biol 2003;5:572–7.

[47] Bystricky K, et al. Regulation of nuclear positioning and dynamics of the silent mating type loci by the yeast Ku70/Ku80 complex. Mol Cell Biol 2009;29:835.

[48] Nagai S, et al. Functional targeting of DNA damage to a nuclear pore-associated SUMO-dependent ubiquitin ligase. Science 2008;322:597–602.

[49] Lemaître C, Soutoglou E. Double strand break (DSB) repair in heterochromatin and heterochromatin proteins in DSB repair. DNA Repair (Amst) 2014;19:163–8.

[50] Avivi Y, et al. Reorganization of specific chromosomal domains and activation of silent genes in plant cells acquiring pluripotentiality. Dev Dynam 2004;230:12–22.

[51] Chambeyron S, Bickmore WA. Chromatin decondensation and nuclear reorganization of the HoxB locus upon induction of transcription. Genes Dev 2004;18:1119–30.

[52] Zidovska A, Weitz DA, Mitchison TJ. Micron-scale coherence in interphase chromatin dynamics. PNAS 2013;110:15555–60.

[53] Misteli T. Protein dynamics: implications for nuclear architecture and gene expression. Science 2001;291:843–7.

[54] Bruinsma R, Grosberg AY, Rabin Y, Zidovska A. Chromatin hydrodynamics. Biophys J 2014;106:1871–81.

[55] Gerlich D, Ellenberg J. 4D imaging to assay complex dynamics in live specimens. Nat Cell Biol 2003(Suppl):S14–19.

[56] Walter J, Schermelleh L, Cremer M, Tashiro S, Cremer T. Chromosome order in HeLa cells changes during mitosis and early G1, but is stably maintained during subsequent interphase stages. J Cell Biol 2003;160:685–97.

[57] Liu J, Vidi P-A, Lelièvre SA, Irudayaraj JMK. Nanoscale histone localization in live cells reveals reduced chromatin mobility in response to DNA damage. J Cell Sci 2015;128:599–604.

[58] Bystricky K, Laroche T, Van Houwe G, Blaszczyk M, Gasser SM. Chromosome looping in yeast. JCB 2005;168:375.

[59] Kepten E, Weron A, Bronstein I, Burnecki K, Garini Y. Uniform contraction-expansion description of relative centromere and telomere motion. Biophys J 2015;109:1454–62.

[60] Dixon JR, et al. Topological domains in mammalian genomes identified by analysis of chromatin interactions. Nature 2012;485:376–80.

[61] Wang S, et al. Spatial organization of chromatin domains and compartments in single chromosomes. Science 2016:aaf8084.

[62] Zink D, Sadoni N, Stelzer E. Visualizing chromatin and chromosomes in living cells. Methods 2003;29:42–50.

[63] Hihara S, et al. Local Nucleosome Dynamics Facilitate Chromatin Accessibility in Living Mammalian Cells. Cell Rep 2012;2:1645–56.

[64] Lacroix J, et al. Analysis of DNA Replication by Optical Mapping in Nanochannels Small 2016;12:5963–5970.

[65] Lebeaupin T, Sellou H, Timinszky G, Huet S. Chromatin dynamics at DNA breaks: what, how and why? AIMS Biophys 2015;2:458–75.

[66] Récamier V, et al. Single cell correlation fractal dimension of chromatin. Nucleus 2014;5:75–84.

[67] Gerlich D, et al. Global chromosome positions are transmitted through mitosis in mammalian cells. Cell 2004;112:751–64.

[68] Bronstein I, et al. Transient anomalous diffusion of telomeres in the nucleus of mammalian cells. Phys Rev Lett 2009;103:18102.

[69] Dimitrova N, Chen Y-CM, Spector DL, de Lange T. 53BP1 promotes non-homologous end joining of telomeres by increasing chromatin mobility. Nature 2008;456:524–8.

[70] Wanat JJ, et al. Csm4, in collaboration with Ndj1, mediates telomere-led chromosome dynamics and recombination during yeast meiosis. PLoS Genet 2008;4:e1000188.

[71] Darzacq X, et al. Imaging transcription in living cells. Annu Rev Biophys 2009;38: 173–96.

[72] Bertrand E, Chartrand PS, Shenoy SM, Singer RH, Long RM. Localization of ASH1 mRNA particles in living yeast. Mol Cell 1998;2:437–45.

[73] Hocine S, Raymond P, Zenklusen D, Chao JA, Singer RH. Single-molecule analysis of gene expression using two-color RNA labeling in live yeast. Nat Meth 2013;10:119–21.

[74] Stasevich TJ, et al. Regulation of RNA polymerase II activation by histone acetylation in single living cells. Nature 2014;516:272–5.

[75] Straight AF, Belmont AS, Robinett CC, Murray AW. GFP tagging of budding yeast chromosomes reveals that protein–protein interactions can mediate sister chromatid cohesion. Curr Biol 1996;6:1599–608.

[76] Lassadi I, Kamgoué A, Goiffon I, Tanguy-le-Gac N, Bystricky K. Differential chromosome conformations as hallmarks of cellular identity revealed by mathematical polymer modeling. PLoS Comput Biol 2015;11:e1004306. http://dx.doi.org/10.1371/journal.pcbi.1004306.

[77] Ma H, Reyes-Gutierrez P, Pederson T. Visualization of repetitive DNA sequences in human chromosomes with transcription activator-like effectors. PNAS 2013;110:21048–53.

[78] Miyanari Y, Ziegler-Birling C, Torres-Padilla M-E. Live visualization of chromatin dynamics with fluorescent TALEs. Nat Struct Mol Biol 2013;20:1321–4.

[79] Chen B, et al. Dynamic imaging of genomic loci in living human cells by an optimized CRISPR/Cas system. Cell 2013;155:1479–91.

[80] Ma H, et al. Multicolor CRISPR labeling of chromosomal loci in human cells. Proc Natl Acad Sci 2015;112:3002–3007.

[81] Tanenbaum ME, Gilbert LA, Qi LS, Weissman JS, Vale RD. A protein-tagging system for signal amplification in gene expression and fluorescence imaging. Cell 2014;159:635–46.

[82] Saad H, et al. DNA dynamics during early double-strand break processing revealed by non-intrusive imaging of living cells. PLoS Genet 2014;10:e1004187.

[83] Bystricky K. Chromosome dynamics and folding in eukaryotes: insights from live cell microscopy. FEBS Lett 2015;589:3014–22.

[84] Robinett CC, et al. In vivo localization of DNA sequences and visualization of large-scale chromatin organization using lac operator/repressor recognition. J Cell Biol 1996;135:1685.

[85] Lassadi I, Bystricky K. Tracking of Single and Multiple Genomic Loci in Living Yeast Cells. Methods Mol Biol 2011;745:499–522. Available from: http://dx.doi.org/10.1007/978-1-61779-129-1_29.

[86] Loïodice I, Dubarry M, Taddei A. Current protocols in cell biology. John Wiley & Sons, Inc.; 2014. ⟨http://onlinelibrary.wiley.com/doi/10.1002/0471143030.cb2217s62/abstract⟩.

[87] Dubarry M, Loïodice I, Chen CL, Thermes C, Taddei A. Tight protein–DNA interactions favor gene silencing. Genes Dev 2011;25:1365–70.

[88] Simon P, Houston P, Broach J. Directional bias during mating type switching in Saccharomyces is independent of chromosomal architecture. EMBO J 2002;21:2282–91.

[89] Bressan DA, Vazquez J, Haber JE. Mating type–dependent constraints on the mobility of the left arm of yeast chromosome III. J Cell Biol 2004;164:361–71.

[90] Carlton PM, et al. Fast live simultaneous multiwavelength four-dimensional optical microscopy. Proc Natl Acad Sci U S A 2010;107:16016.

[91] Shao S, et al. Long-term dual-color tracking of genomic loci by modified sgRNAs of the CRISPR/Cas9 system. Nucl Acids Res 2016;44:e86.

[92] Heun P, Laroche T, Shimada K, Furrer P, Gasser SM. Chromosome dynamics in the yeast interphase nucleus. Science 2001;294:2181.

[93] Levi V, Gratton E. Bräuchle C, Lamb DC, Michaelis J, editors. Single particle tracking and single molecule energy transfer. Wiley-VCH Verlag GmbH & Co. KGaA; 2009. p. 1–24. ⟨http://onlinelibrary.wiley.com/doi/10.1002/9783527628360.ch1/summary⟩.

[94] Vermot J, Fraser SE, Liebling M. Fast fluorescence microscopy for imaging the dynamics of embryonic development. HFSP J 2008;2:143–55.

[95] Beaudouin J, Mora-Bermudez F, Klee T, Daigle N, Ellenberg J. Dissecting the contribution of diffusion and interactions to the mobility of nuclear proteins. Biophys J 2006;90:1878–94.

[96] Molenaar C, Abdulle A, Gena A, Tanke HJ, Dirks RW. Poly(A)+ RNAs roam the cell nucleus and pass through speckle domains in transcriptionally active and inactive cells. J Cell Biol 2004;165:191–202.

[97] Hajjoul H, et al. High throughput chromatin motion tracking in living yeast reveals the flexibility of the fiber throughout the genome. Genome Res 2013;23:1829–38.

[98] Chenouard N, et al. Objective comparison of particle tracking methods. Nat Meth 2014;11:281–9.

[99] Sbalzarini IF, Koumoutsakos P. Feature point tracking and trajectory analysis for video imaging in cell biology. J Struct Biol 2005;151:182–95.

[100] Savin T, Doyle PS. Static and dynamic errors in particle tracking microrheology. Biophys J 2005;88:623–38.

[101] Hajjoul H, Kocanova S, Lassadi I, Bystricky K, Bancaud A. Lab-on-chip for fast 3D particle tracking in living cells. Lab Chip 2009;9:3054–8.

[102] Backlund MP, Joyner R, Weis K, Moerner WE. Correlations of three-dimensional motion of chromosomal loci in yeast revealed by the double-helix point spread function microscope. Mol Biol Cell 2014;25:3619–29.

[103] Weber SC, Spakowitz AJ, Theriot JA. Nonthermal ATP-dependent fluctuations contribute to the in vivo motion of chromosomal loci. Proc Nat Acad Sci U S A 2012;109:7338–43.

[104] Izeddin I, et al. Single-molecule tracking in live cells reveals distinct target-search strategies of transcription factors in the nucleus. eLife 2014;3:e02230.

[105] Saxton MJ. Single particle tracking. Fundam Concepts Biophys 2009;1:1–33.

[106] Weber SC, Spakowitz AJ, Theriot JA. Bacterial chromosomal loci move subdiffusively through a viscoelastic cytoplasm. Phys Rev Lett 2010;104:238102.

[107] Javer A, et al. Short-time movement of *E. coli* chromosomal loci depends on coordinate and subcellular localization. Nat Commun 2013;4:3003.

[108] Albert B, et al. Systematic characterization of the conformation and dynamics of budding yeast chromosome XII. J Cell Biol 2013;202:201–10.

[109] Wang R, Mozziconacci J, Bancaud A, Gadal O. Principles of chromatin organization in yeast: relevance of polymer models to describe nuclear organization and dynamics. Curr Opin Cell Biol 2015;34:54–60.

[110] Bronshtein I, et al. Loss of lamin A function increases chromatin dynamics in the nuclear interior. Nat Commun 2015;6:8044.

[111] Manzo C, Garcia-Parajo MF. A review of progress in single particle tracking: from methods to biophysical insights. Rep Prog Phys 2015;78:124601.

[112] Metzler R, Klafter J. The random walk's guide to anomalous diffusion: a fractional dynamics approach. Phys Rep 2000;339:1–77.

[113] Doi M, Edwards SF. The theory of polymer dynamics. Oxford: Oxford University Press; 1988.

[114] Burnecki K, Muszkieta M, Sikora G, Weron A. Statistical modelling of subdiffusive dynamics in the cytoplasm of living cells: A FARIMA approach. EPL (Europhys Lett) 2012;98:10004.

[115] Burnecki K, et al. Universal algorithm for identification of fractional Brownian motion. A case of telomere subdiffusion. Biophys J 2012;103:1839–47.

[116] Chen C, et al. Budding yeast chromatin is dispersed in a crowded nucleoplasm in vivo. Mol Biol Cell 2016;27:3357–3368.

[117] Hancock R. Structures and functions in the crowded nucleus: new biophysical insights. Front. Phys 2014;2:53.

[118] Verkman AS. Solute and macromolecule diffusion in cellular aqueous compartments. Trends Biochem Sci 2002;27:27–33.

[119] Seksek O, Biwersi J, Verkman AS. Translational diffusion of macromolecule-sized solutes in cytoplasm and nucleus. J Cell Biol 1997;138:131–42.

[120] Schiessel H. The physics of chromatin. J Phys: Condens Matter 2003;15:R699–774.

[121] Beliveau BJ, et al. Single-molecule super-resolution imaging of chromosomes and in situ haplotype visualization using Oligopaint FISH probes. Nat Commun 2015;6:7147.

[122] Berger AB, et al. High-resolution statistical mapping reveals gene territories in live yeast. Nat Meth 2008;5:1031–7.

[123] Therizols P, Duong T, Dujon B, Zimmer C, Fabre E. Chromosome arm length and nuclear constraints determine the dynamic relationship of yeast subtelomeres. Proc Natl Acad Sci 2010;107:2025.

[124] Wong H, et al. A predictive computational model of the dynamic 3D interphase nucleus. Curr Biol 2012;22:1881–90.

[125] Tjong H, Gong K, Chen L, Alber F. Physical tethering and volume exclusion determine higher-order genome organization in budding yeast. Genome Res 2012;22:1295–305.

[126] Schober H, et al. Controlled exchange of chromosomal arms reveals principles driving telomere interactions in yeast. Genome Res 2008;18:261.

[127] Mirny L. The fractal globule as a model of chromatin architecture in the cell. Chromosome Res 2011;19:37–51.

[128] Bystricky K, Heun P, Gehlen L, Langowski J, Gasser SM. Long range compaction and flexibility of interphase chromatin in budding yeast analysed by high-resolution imaging techniques. Proc Natl Acad Sci U S A 2004;101:16495–500.

[129] Fekete RA, Chattoraj DK. A cis-acting sequence involved in chromosome segregation in Escherichia coli. Mol Microbiol 2005;55:175–83.

[130] Maeshima K, Hihara S, Eltsov M. Chromatin structure: does the 30-nm fibre exist in vivo? Curr Opin Cell Biol 2010;22:291–297.

[131] Miele A, Bystricky K, Dekker J. Yeast silent mating type loci form heterochromatic clusters through silencer protein-dependent long-range interactions. PLoS Genet 2009;5:e1000478.

[132] Roukos V, et al. Spatial dynamics of chromosome translocations in living cells. Science 2013;341:660–4.

[133] King KR, et al. A high-throughput microfluidic real-time gene expression living cell array. Lab Chip 2006;7:77–85.

[134] Normanno D, et al. Probing the target search of DNA-binding proteins in mammalian cells using TetR as model searcher. Nat Commun 2015;6:7357.

[135] von Hippel PH, Berg OG. Facilitated target location in biological systems. J Biol Chem 1989;264:675–8.

[136] Elf J, Li G-W, Xie XS. Probing transcription factor dynamics at the single molecule level in a living cell. Science 2007;316:1191–4.

[137] Dekker J, Mirny L. The 3D genome as moderator of chromosomal communication. Cell 2016;164:1110–21.

[138] Ochiai H, Sugawara T, Yamamoto T. Simultaneous live imaging of the transcription and nuclear position of specific genes. Nucl. Acids Res 2015;43 e127–e127.

GLOSSARY

3C chromosome conformation capture

ANCHOR ANCHOR system is a bipartite system composed of ANCH sequences (derived from bacterial ParS site) that contains nucleation seeds for the binding of OR proteins (derived from ParB protein). Based on Bacteria chromosome partition systems (Burkholderiaecae), three or more ANCHOR systems can be used

EM-CCD electron-multiplying-charge-coupled device. EM-CCD cameras have high sensitivity, low noise, but larger pixel size and are slower than sCMOS detector.

fBM fractional Brownian motion

FISH fluorescent in situ hybridization

FLSM fractional Levy stable motion

FROS fluorescent repressor operator system; FROS combines the expression of a bacterial repressor fused to a fluorescent protein and the integration of operator sequence as tandem arrays at a specific locus

Gene territories volume (μm^3) in which 50% of gene's loci are confined in a cell population analysis

HML and HMR hidden MAT Left and right; correspond to silent mating type cassette—left or right

HTM high-throughput microscopy

MSD mean square displacement; for every time interval(s), the square of the mean displacement observed (μm2) is calculated. Confinement of a locus in areas of the nucleus appears as a plateau at longer time scale, defining a radius of constrains (expressed in μm, see gene territories)

OD obstructed diffusion

OMX microscope "Optical Microscope eXperimental" is a platform for multimodal, multichannel wide-field imaging. OMX can be used for three-dimensional structured illumination microscopy (3D-SIM)

rDNA ribosomal DNA; genes coding ribosomal DNA are organized in loci indirect tandem head-to-tail repeat (100–200-fold).

ROI region of interest

sCMOS or CCD/CMOS cameras detectors consists of complementary metal-oxide-semiconductor (CMOS) readout integrated circuits that are bump bonded to a CCD imaging substrate; sCMOS cameras have small pixel size and high sensitivity

SNR signal-to-noise ratio;

SPT single particle tracking

SunTAg SUperNova tagging system; Sun-Tag is based on antibody to a multimerized protein scaffold

TADs topologically associated domains; TADs represent a chromatin domain within which physical interactions occur more frequently than bulk of the genome

TALE transcription activator like effector proteins

Index

Note: Page numbers followed by "*f*" and "*t*" refer to figures and tables, respectively.

D

E

F

G

H

Q

R

S

T

U

V

W

X

Y

Z

Kolibri 360 GmbH
Ferdinand-Jühlke-Straße 7
99095 Erfurt, Deutschland
produktsicherheit@kolibri360.de